Biosimulation

This practical guide to biosimulation provides the hands-on experience needed to devise, design, and analyze simulations of biophysical processes for applications in biological and biomedical sciences. Through real-world case studies and worked examples, students will develop and apply basic operations through to advanced concepts, covering a wide range of biophysical topics, including chemical kinetics and thermodynamics, transport phenomena, and cellular electrophysiology. Each chapter is built around case studies in a given application area, with simulations of real biological systems developed to analyze and interpret data. Open-ended project-based exercises are provided at the end of each chapter, and with all data and computer codes available online (www.cambridge.org/biosim) students can quickly and easily run, manipulate, explore, and expand on the examples inside. This hands-on guide is ideal for use on senior undergraduate/graduate courses, and also as a self-study guide for anyone who needs to develop computational models of biological systems.

Daniel A. Beard is a Professor in the Biotechnology and Bioengineering Center and the Department of Physiology at the Medical College of Wisconsin. Research in his laboratory is focused on systems engineering approaches to understanding the operation of physiological systems in health and disease. A recent major effort in his group has been on theoretical and experimental characterization of the thermodynamics, kinetics, and electrophysiology of cardiac mitochondria. Additional research interests include nonequilibrium thermodynamics in biochemical networks, mass transport and microvascular exchange in physiological systems, and drug metabolism and physiologically based pharmacokinetics.

CAMBRIDGE TEXTS IN BIOMEDICAL ENGINEERING

Cambridge Texts in Biomedical Engineering provide a forum for high-quality textbooks targeted at undergraduate and graduate courses in biomedical engineering. It covers a broad range of biomedical engineering topics from introductory texts to advanced topics, including biomechanics, physiology, biomedical instrumentation, imaging, signals and systems, cell engineering, and bioinformatics, as well as other relevant subjects, with a blending of theory and practice. While aiming primarily at biomedical engineering students, this series is also suitable for courses in broader disciplines in engineering, the life sciences and medicine.

Biosimulation

Simulation of Living Systems

Daniel A. Beard
Medical College of Wisconsin

CAMBRIDGE
UNIVERSITY PRESS

University Printing House, Cambridge CB2 8BS, United Kingdom

One Liberty Plaza, 20th Floor, New York, NY 10006, USA

477 Williamstown Road, Port Melbourne, VIC 3207, Australia

314-321, 3rd Floor, Plot 3, Splendor Forum, Jasola District Centre, New Delhi - 110025, India

103 Penang Road, #05-06/07, Visioncrest Commercial, Singapore 238467

Cambridge University Press is part of the University of Cambridge.

It furthers the University's mission by disseminating knowledge in the pursuit of
education, learning and research at the highest international levels of excellence.

www.cambridge.org
Information on this title: www.cambridge.org/9780521768238

First published 2012

A catalogue record for this publication is available from the British Library

Library of Congress Cataloging in Publication data
Beard, Daniel A., 1971–
Biosimulation : simulation of living systems / Daniel A. Beard.
    p.   cm. – (Cambridge texts in biomedical engineering)
ISBN 978-0-521-76823-8 (hardback)
1. Biophysics – Computer simulation.   2. Biophysics – Simulation methods.   3. Biomedical
engineering – Computer simulation.   4. Medical sciences – Case studies.   I. Title.
QH505.B35   2012
571.401´13 – dc23   2011046839

ISBN 978-0-521-76823-8 Hardback

Additional resources for this publication at www.cambridge.org/9780521768238

Cambridge University Press has no responsibility for the persistence or
accuracy of URLs for external or third-party internet websites referred to in
this publication, and does not guarantee that any content on such websites is,
or will remain, accurate or appropriate.

Contents

Preface

Research, development, and design in bioengineering, biomedical engineering, biophysics, physiology, and related fields rely increasingly on mathematical modeling and computational simulation of biological systems. Simulation is required to analyze data, design experiments, develop new technology, and simply to attempt to understand the complexity inherent in biological systems.

This book focuses on practical implementation of techniques to study real biological systems. Indeed, whenever possible, specific applications are developed, starting with a study of the basic operation of the underlying biological, biochemical, or physiological system and, critically, the available data. It is hoped that this data-rich exposition will yield a practical text for engineering students and other readers interested primarily in earthy real-world applications such as analyzing data, estimating parameter values, etc. Thus for the examples developed here, important details of underlying biological systems are described along with a complete step-by-step development of model assumptions, the resulting equations, and (when necessary) computer code. As a result, readers have the opportunity, by working through the examples, to become truly proficient in *biosimulation*.

In this spirit of soup-to-nuts practicality, the book is organized around biological and engineering application areas rather than based on mathematical and computational techniques. Where specific mathematical or computational techniques can be conveniently and effective separated from the exposition, they have been and can be found in the Appendices. Computer codes implemented in MATLAB® (The MathWorks, Natick, MA, USA) for all of the examples in the text can be found online at the URL http://www.cambridge.org/biosim.

I am particularly grateful to a number of individuals who provided critical feedback on the text, including Edmund Crampin, Peter Hunter, Muriel Mescam, Gary Raymond, Nic Smith, Matt Thompson, Kalyan Vinnakota, and Fan Wu. Andy Salmon graciously provided the data from his experiments presented in Section 2.2.2. Tom O'Hara provided some guidance on the model analyzed in Section 8.3. Jim Bassingthwaighte's guidance and advice over many years, as well as specific criticism of the text, are gratefully acknowledged.

Finally, I want to give special thanks to my colleagues Henry and Nicholas Beard for helping with the experiments described in Chapter 1.

Extracts

"All is in flux."

Heraclitus (540–480 BCE)

"This application of mathematics to natural phenomena is the aim of all science, because the expression of the laws of phenomena should always be mathematical."

Claude Bernard, Introduction a l'étude de la médecine expérimentale *1865 Flammarion, Paris (English translation from Noble,* Exp. Physiol. *93: 16–26, 2008)*

"Of physiology from top to toe I sing."

Walt Whitman, Leaves of Grass, *1883*

"The human body is a chemical and physical problem, and these sciences must advance before we can conquer disease."

Henry A. Rowland, The Highest Aims of the Physicist, *Address to the American Physical Society, 1899, published in* Science *10: 825–833, 1899*

"We are seeing the cells of plants and animals more and more clearly as chemical factories, where the various products are manufactured in separate workshops. The enzymes act as the overseers. Our acquaintance with these most important agents of living things is constantly increasing. Even though we may still be a long way from our goal, we are approaching it step by step. Everything is justifying our hopes. We must never, therefore, let ourselves fall into the way of thinking 'ignorabimus' ('We shall never know'), but must have every confidence that the day will dawn when even those processes of life which are still a puzzle today will cease to be inaccessible to us natural scientists."

Eduard Buchner, Nobel Lecture, 1907

"To a physician or physiologist at the present day a man's body is a machine, or rather a factory full of machines, all working harmoniously together for the good of the organism."

Ernest N. Starling, The Linacre Lecture on the Law of the Heart, *1915, published by Longmans, Green and Co., London, 1918*

"The mathematical box is a beautiful way of wrapping up a problem, but it will not hold the phenomena unless they have been caught in a logical box to begin with."

John R. Platt, Strong inference. Science, *146: 347–353, 1964*

"People who wish to analyze nature without using mathematics must settle for a reduced understanding."

Richard Feynman

"[This book] is aimed at 'non-believers', that is to say the 90% or so of biochemistry students, and indeed of practicing biochemists, who place enzyme kinetics in the same category as Latin and cold showers, character-building perhaps, but otherwise to be forgotten as quickly as possible."

Paul C. Engel, Enzyme Kinetics: The Steady-State Approach *1977, Chapman & Hall, London*

"Why make models? To think (and calculate) logically about what components and interactions are important in a complex system."

James E. Bailey, Mathematical modeling and analysis in biochemical engineering: past accomplishments and future opportunities. Biotechnol. Prog., *14: p. 8–20, 1998*

"Without data, there is nothing to model; and without models, there is no source of deep predictive understanding."

James B. Bassingthwaighte, The Physiome Project: The macroethics of engineering toward health. The Bridge, *32: 24–29, 2002*

"Over the last half century, we have proceeded by breaking living systems down into their smallest components, the individual genes and molecules. Humpty Dumpty has been smashed into billions of fragments . . . Can we put Humpty Dumpty back together again?"

Denis Noble, The Music of Life: Biology beyond the Genome. *2006, Oxford, New York*

1 Introduction to simulation of biological systems

Overview

This chapter is built around analyzing a real data set obtained from a real biological system to illustrate several complementary approaches to simulation and analysis. The particular system studied (a home aquarium) is a well-mixed chemical reactor. Or, more accurately, the system studied is treated as a well-mixed chemical reactor, a basic modeling paradigm that will appear again and again in this book.

Here, we look at this single physical system from several different perspectives (that is, under different sets of underlying modeling assumptions) with the aim of motivating the reader to undertake the study of the rest of this book. The aim is not to overwhelm the reader with mathematical details that can be found in later chapters. Therefore let us clearly state at the outset: it is not expected or required that the reader follow every detail of the examples illustrated here. Instead, we invite the reader to focus on the basic assumptions underlying the methods applied, and to compare and contrast the results that are obtained based on these different approaches. Proceeding this way, it is hoped that the reader may gain an appreciation of the breath of the field. Furthermore, it is hoped that this appreciation will continue to grow with a study of the rest of this book and beyond.

1.1 Modeling approaches

The number of different approaches to simulating biosystems behavior is perhaps greater than the number of biological systems. The number is at least large enough that a finite and complete list cannot be constructed. Simulation methods may be classified according to the physical systems simulated (for example, cellular metabolism, whole-body drug distribution, or ecological network dynamics), the sets of assumptions used to build a simulation (for example, rapid mixing versus spatial inhomogeneity in chemical reaction systems), or the mathematical/computational formulation of the simulation (for example, systems of ordinary differential equations versus statistical inference networks for describing regulation of gene transcription). A glance at the table of contents reveals that most of this book is organized by biological system or application area. (Modeling assumptions and relevant computational techniques are introduced as necessary.)

These biological systems can (and will!) be studied by applying a variety of sets of assumptions and associated computational methods. Doing this, we will see that the methodology applied to a given system depends strictly on what one thinks one knows about the system in advance, and what one wishes to discover through computational analysis. In the following introductory example we will see that what we can learn (for example, what variables and what parameters we can estimate) depends on the prior knowledge built into a model, including (but not limited to) what data are available for a given system.

1.2 An introductory example: biochemistry of a home aquarium

As our first exemplar modeling study, let us analyze the buildup and reaction of waste materials in a home aquarium, a system that may be familiar to some readers. Ammonia (NH_3), which is toxic to fish, is excreted from fish as a waste product and produced through decomposition of organic matter. In a well-functioning aquarium, nitrifying bacteria in the aquarium filter oxidize ammonia to nitrite (NO_2^-) and oxidize the nitrite to nitrate (NO_3^-). Of these three nitrogen-containing compounds, nitrate is by far the least toxic to fish.

When one sets up a new aquarium, populations of nitrifying bacteria are yet to be established, and concentrations of toxic compounds can temporarily build up. Figure 1.1 plots data collected by the author from his own aquarium following the addition of fish into a previously uninhabited new tank. Here we see that ammonia concentration tends to build up over the first week or more. Once significant populations of bacteria that convert NH_3 to NO_2^- appear, the ammonia declines while the nitrite level increases. Nitrite concentration eventually declines as nitrate begins to appear.[1]

We wish to understand how these three concentrations are related kinetically. To simplify the notation, we introduce the definitions $x_1 = $ [ammonia], $x_2 = $ [nitrite], $x_3 = $ [nitrate] for the concentration variables. As already described, the expected sequence of reaction in this system is $x_1 \rightarrow x_2 \rightarrow x_3$. In fact, that sequence is apparent from the data illustrated in the figure. Ammonia (x_1) peaks around day 10, followed by nitrite (x_2) around day 13. Nitrate concentration (x_3) really picks up following the peak in nitrite, and continues to steadily increase.

---

[1] Ammonia, nitrite, and nitrate exist in aqueous solution in a number of rapidly interconverting forms. For example, at low pH NH_3 is largely protonated to form the ammonium ion NH_4^+. Here, the terms ammonia, nitrite, and nitrate are understood to include all such rapidly converting species of these reactants.

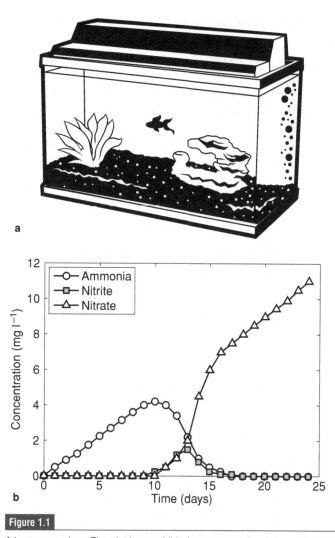

Figure 1.1

A home aquarium. The plot in panel (b) shows ammonia, nitrite, and nitrate concentration versus time in a home aquarium. Concentrations are given in units of mg of nitrogen per liter (mg l^{-1}).

Yet in addition to the reaction sequence, is it possible to obtain additional quantitative information from these data? To do so, let us construct a series of simple models and see what we can find.

1.2.1   First model: a nonmechanistic analysis

In the first model we would like to introduce the minimum number of assumptions that allow us to explain the observed data. The idea is to construct a general set of governing differential equations for x_1, x_2, and x_3 based only on the assumption of mass conservation. Conservation laws are a universal tool for developing models of many systems, including practically all of the examples studied in this book. Here we assume that the volume of the system, V, remains constant. So the rate of change of mass of a substance, $\frac{d(cV)}{dt}$, is equal to $V\frac{dc}{dt}$. In this case, the general differential equation for concentration of a reactant is

$$V\frac{dc}{dt} = \begin{pmatrix} \text{rate of production or} \\ \text{input measured in units} \\ \text{of mass per unit time} \end{pmatrix} - \begin{pmatrix} \text{rate of loss} \\ \text{measured in units of} \\ \text{mass per unit time} \end{pmatrix} \qquad (1.1)$$

or

$$\frac{dc}{dt} = \begin{pmatrix} \text{rate of production or} \\ \text{input measured in units} \\ \text{of mass per unit time} \\ \text{per unit volume} \end{pmatrix} - \begin{pmatrix} \text{rate of loss} \\ \text{measured in units of} \\ \text{mass per unit time} \\ \text{per unit volume} \end{pmatrix}. \qquad (1.2)$$

Applying this general form to the variables x_1, x_2, and x_3 gives

$$\frac{dx_1}{dt} = k(t) - r_{12}(t)$$

$$\frac{dx_2}{dt} = r_{12}(t) - r_{23}(t)$$

$$\frac{dx_3}{dt} = r_{23}(t), \qquad (1.3)$$

where $k(t)$ is the rate of ammonia (x_1) production, and $r_{12}(t)$ and $r_{23}(t)$ are the rates of conversion from ammonia to nitrite and from nitrite to nitrate, respectively. The first equation states that the rate of change of x_1 is equal to the rate of production minus the rate of degradation. Similar statements of mass conservation follow for dx_2/dt and dx_3/dt. Since no processes degrading nitrate are considered, there is no degradation term in the dx_3/dt equation. Because $k(t)$, $r_{12}(t)$, and $r_{23}(t)$ are (so far) assumed to be arbitrary functions, we have (so far) not introduced any assumptions about the rules governing the behavior of these functions. The names and definitions of the model variables are listed in the table below.

Variable	Units	Description
x_1	$mg\,l^{-1}$	concentration of ammonia
x_2	$mg\,l^{-1}$	concentration of nitrite
x_3	$mg\,l^{-1}$	concentration of nitrate
k	$mg\,l^{-1}\,day^{-1}$	rate of ammonia production
r_{12}	$mg\,l^{-1}\,day^{-1}$	rate of nitrite production from ammonia
r_{23}	$mg\,l^{-1}\,day^{-1}$	rate of nitrate production from nitrite

We call this model "nonmechanistic" because it does not invoke any biochemical/biophysical mechanisms to describe the rates of conversion $r_{12}(t)$ and $r_{23}(t)$, or the rate of production $k(t)$. Instead these rates are all allowed to be arbitrary functions.

So what can we do with this simple general model? One useful thing we can do is analyze the data using the model to estimate $r_{12}(t)$, $r_{23}(t)$, and k and test the model assumptions. From Eq. (1.3), we have $r_{23}(t) = dx_3/dt$, which can be numerically approximated using a finite difference approximation

$$\hat{r}_{23}(t) \approx \frac{x_3(t + \Delta t) - x_3(t - \Delta t)}{2\Delta t}. \tag{1.4}$$

Here Δt is the discrete time step over which the data in Figure 1.1 are sampled. (I took one measurement per day, so $\Delta t = 1$ day.) Equation (1.4) is the "central-difference" approximation for the derivative of x_3 with respect to time.[2] Here we use the notation $\hat{r}_{23}$ to denote the approximation (from the data) of r_{23}. Next, given our approximation of $r_{23}(t)$, we can approximate $r_{12}(t)$:

$$\hat{r}_{12}(t) = \frac{dx_2}{dt} + r_{23}(t) \approx \frac{x_2(t + \Delta t) - x_2(t - \Delta t)}{2\Delta t} + \hat{r}_{23}(t). \tag{1.5}$$

Similarly, we can approximate k as a function of time

$$\hat{k}(t) = \frac{dx_1}{dt} + r_{12}(t) \approx \frac{x_1(t + \Delta t) - x_1(t - \Delta t)}{2\Delta t} + \hat{r}_{12}(t). \tag{1.6}$$

Values of $\hat{k}(t)$, $\hat{r}_{12}(t)$, and $\hat{r}_{23}(t)$ computed from the data in Figure 1.1 are plotted in Figure 1.2.[3] From these estimated rates we learn a number of things about this system not immediately apparent from a simple inspection of the raw data. First, we can see that the rate of ammonia production ($\hat{k}(t)$) is estimated to be approximately constant. Moreover, this analysis provides an estimate of the constant k, approximately 0.4 to 0.5 mg l^{-1} day^{-1}. This observation is perhaps not unexpected, because the number and size of the fish remained approximately constant over the course of the experiment, as did the amount of food introduced per day. Therefore we might have expected the rate of ammonia production to be nearly constant. Second, the analysis reveals that nitrite production ($\hat{r}_{12}(t)$) peaks near day 13 while nitrate production peaks shortly after, around day 14. Towards the end of the experiment, all of the reaction rates converge to equal approximately 0.5 mg l^{-1} day^{-1}. Finally, we note that the estimated rates $\hat{k}(t)$, $\hat{r}_{12}(t)$, and $\hat{r}_{23}(t)$ remain positive for the duration of the experiment. This observation makes sense, because under normal conditions neither of the nitrification reactions is expected

[2] Discrete approximations of derivatives are reviewed in Section 9.1 in the Appendices.
[3] Computer codes (implemented in MATLAB) for this and all of the examples in this book can be found online at the URL http://www.cambridge.org/biosim.

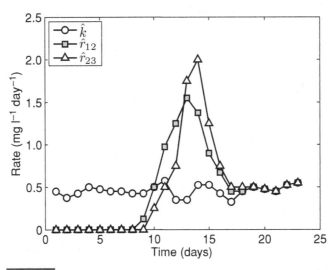

Figure 1.2

Values of $\hat{k}(t)$, $\hat{r}_{12}(t)$, and $\hat{r}_{23}(t)$ estimated from Eqs (1.4)–(1.6) and the data in Figure 1.1. These rates are expressed in units of mass of nitrogen per unit volume per unit time: $mg\,l^{-1}\,day^{-1}$.

to proceed in the reverse direction. Thus the result that $\hat{r}_{12}(t)$ and $\hat{r}_{23}(t)$ remain positive provides a useful check of the physical realism of the model.

To summarize, analyzing the data of Figure 1.1 using the simple model of Eq. (1.3), which invokes no more serious assumption than conservation of mass, provides quantitative estimates of a number of variables that are not directly measured.

1.2.2   Nonmechanistic analysis with noise

The preceding analysis was applied to a relatively noise-free data set, yielding reasonable (and smooth) numerical estimates for the derivatives in Eqs (1.4)–(1.6). However, differentiation has the unfortunate side effect of tending to magnify noise. And since significant measurement noise is often associated with real-world biological signals, analyses that require the estimation of derivatives of biological data are often seriously confounded by noise.[4]

---

[4] The aquarium experiment studied here cannot be regarded as a precisely controlled study. The original data set of Figure 1.1 was collected by the author using a simple consumer kit, with which the concentrations are estimated by visual comparison of the fluid in a test-tube assay with a color chart, possibly introducing bias. Although dilutions and replicates were performed as appropriate, there is a human psychological component to interpreting these assays. Furthermore, because the nitrate assay used is relatively insensitive over the reported concentration range, data were obtained by a combination of interpolation, assuming a constant total nitrate production rate, and colorimetric assay. Given the potential for bias, the reader is encouraged to conduct his or her own experiments in his or her own home laboratory!

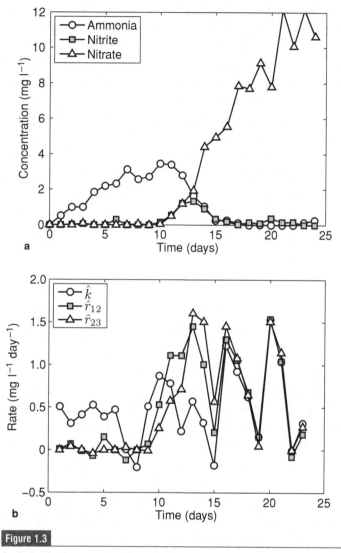

Figure 1.3

Analysis of noisy aquarium data. In panel (a) the data of Figure 1.1 are reproduced with added noise. The analysis of the previous section is reproduced in panel (b), with $\hat{k}(t)$, $\hat{r}_{12}(t)$, and $\hat{r}_{23}(t)$ computed from Eqs (1.4)–(1.6) applied to the noisy data from panel (a).

To illustrate this problem, and to explore some ideas of how to deal with it, we can add some noise to our aquarium data. Figure 1.3(a) shows the same data as those of Figure 1.1, with a relatively small amount of noise added. We can see that the basic trends in the data remain the same, but this data set is less smooth than the previous one.

The values of $\hat{k}(t)$, $\hat{r}_{12}(t)$, and $\hat{r}_{23}(t)$, computed from Eqs (1.4)–(1.6) for these data are plotted in Figure 1.3(b). Here we can see the consequence of differentiating the noisy signals from the upper panel: these estimated rates are terribly noisy, and hardly resemble the estimates illustrated in Figure 1.2. This analysis tells us very little about what we think happened in the experiment. With only a modest amount of random noise added to the signal, we are no longer able to draw any conclusions or confidently estimate any of the unmeasured variables in the system.

To analyze the noisy data effectively requires additional analysis. One place to start is to reexamine the data to look for clues on how to improve our calculations. Is there any trend in the data set of Figure 1.3 that we might be able to take advantage of? After a careful look, the observer's attention might be directed to the fact that over the second half of the experiment the ammonia and nitrite concentrations remain very small compared with nitrate, which continues to grow. If x_1 and x_2 remain constant over some time regime (say in the limit $t \to \infty$ or, more practically, for the last two weeks of the experiment), then in this time window Eq. (1.3) reduces to

$$\frac{dx_1}{dt} = k(t) - r_{12}(t) = 0$$

$$\frac{dx_2}{dt} = r_{12}(t) - r_{23}(t) = 0$$

$$\frac{dx_3}{dt} = r_{23}(t). \tag{1.7}$$

This system of equations tells us that $dx_3/dt = k(t)$ in this time window. Furthermore, observation of the raw data in Figure 1.3 tells us that the rate of growth of nitrate (or dx_3/dt) is approximately constant in this time window.

Therefore, before trying to estimate the rates $k(t)$, $r_{12}(t)$, and $r_{23}(t)$, there is justification for introducing the a priori assumption that $k(t)$ becomes constant *at some point in the experiment*. If we have other reasons (such as those discussed above) to think that $k(t)$ might be constant *throughout the whole experiment*, we might hypothesize that this is the case. Doing so, this hypothesis can be formally built into the analysis as an additional assumption, while being sure to remember that this assumption is a hypothesis that remains to be tested against the data.

To test the hypothesis, we can sum the equations in Eq. (1.3) to obtain

$$\frac{d}{dt}(x_1 + x_2 + x_3) = k(t). \tag{1.8}$$

If indeed $k(t)$ is constant then the sum $x_1 + x_2 + x_3$ should increase at at constant rate throughout the experiment. In Figure 1.4 we plot this sum to test the hypothesis and find that, indeed, $x_1 + x_2 + x_3$ increases at an approximately constant rate. The solid line in the figure represents a line of slope $\hat{k} = 0.45$ mg l^{-1} day^{-1}, which is the estimate of the constant rate of ammonia production obtained from this analysis. With the hypothesis that $k(t)$ is constant not disproved, and an estimate

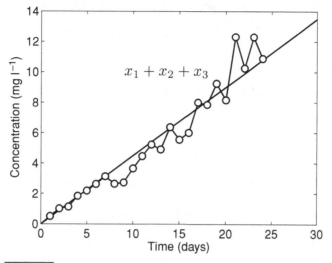

Figure 1.4

Plot of summed data $x_1 + x_2 + x_3$. The solid line has slope 0.45 mg l^{-1} day^{-1}, which provides an estimate of the constant k.

of k in hand, we next continue by using this assumption to help estimate the other rates.

With constant k, Eq. (1.3) provides three equations for the two unknowns $r_{12}(t)$ and $r_{23}(t)$:

$$r_{12}(t) = k - \frac{dx_1}{dt}$$

$$r_{12}(t) - r_{23}(t) = \frac{dx_2}{dt}$$

$$r_{23}(t) = \frac{dx_3}{dt} \tag{1.9}$$

with the equivalent numerical approximation

$$\hat{r}_{12}(t) = \hat{k} - \frac{x_1(t + \Delta t) - x_1(t - \Delta t)}{2\Delta t}$$

$$\hat{r}_{12}(t) - \hat{r}_{23}(t) = \frac{x_2(t + \Delta t) - x_2(t - \Delta t)}{2\Delta t}$$

$$\hat{r}_{23}(t) = \frac{x_3(t + \Delta t) - x_3(t - \Delta t)}{2\Delta t}. \tag{1.10}$$

In general this is an ill-posed problem, and there is no solution (for $\hat{r}_{12}(t)$ and $\hat{r}_{23}(t)$) that satisfies all of the equations. Instead, we can seek a solution that in some way approximately solves Eq. (1.10).

Putting this linear system into matrix-vector form, we have

$$
\begin{bmatrix} 1 & 0 \\ 1 & -1 \\ 0 & 1 \end{bmatrix} \begin{bmatrix} \hat{r}_{12}(t) \\ \hat{r}_{23}(t) \end{bmatrix} = \begin{bmatrix} \hat{k} - \dfrac{x_1(t+\Delta t) - x_1(t - \Delta t)}{2\Delta t} \\ \dfrac{x_2(t+\Delta t) - x_2(t - \Delta t)}{2\Delta t} \\ \dfrac{x_3(t+\Delta t) - x_3(t - \Delta t)}{2\Delta t} \end{bmatrix}. \qquad (1.11)
$$

One approach to computing $\hat{r}_{12}(t)$ and $\hat{r}_{23}(t)$ is to find the solution that minimizes the error between the left-hand and right-hand sides of this equation.

In fact, there exists a handy general solution to problems of this sort when the error is formulated as the sum of squares of differences. These problems are called *least-squares* problems in mathematics, and here we consider the specific problem of minimizing the error in the overdetermined linear system

$$
\mathbb{A}\mathbf{x} = \mathbf{b},
$$

where $\mathbb{A}$ is a matrix in which the number of rows (number of equations) outnumbers the number of columns (number of unknowns). Note that the matrix in Eq. (1.11) is a matrix of this type. The least-squares solution (the vector $\mathbf{x}$ that minimizes $\|\mathbb{A}\mathbf{x} - \mathbf{b}\|^2$) is found as the vector $\mathbf{x}$ that solves the well-posed problem

$$
\mathbb{A}^T\mathbb{A}\mathbf{x} = \mathbb{A}^T\mathbf{b}.
$$

(This least-squares analysis is reviewed in Section 9.2 of the Appendices.)

Applying this formula to Eq. (1.11), we obtain the following estimates for $\hat{r}_{12}(t)$ and $\hat{r}_{23}(t)$.

$$
\hat{r}_{12}(t) = \frac{1}{3}\left(2\hat{k} - 2\frac{x_1(t+\Delta t) - x_1(t - \Delta t)}{2\Delta t} + \frac{x_2(t+\Delta t) - x_2(t - \Delta t)}{2\Delta t} \right.
$$
$$
\left. + \frac{x_3(t+\Delta t) - x_3(t - \Delta t)}{2\Delta t}\right)
$$
$$
\hat{r}_{23}(t) = \left(\hat{r}_{12}(t) + \frac{x_3(t+\Delta t) - x_3(t - \Delta t)}{2\Delta t} - \frac{x_2(t+\Delta t) - x_2(t - \Delta t)}{2\Delta t}\right)\bigg/2.
$$
$$
\qquad (1.12)
$$

(This solution is easy to verify and is the subject of Exercise 1.1.)

Holding $\hat{k}$ constant and computing $\hat{r}_{12}(t)$ and $\hat{r}_{23}(t)$ at each time point from Eq. (1.12), we obtain the estimates plotted in Figure 1.5. This result is a clear improvement over Figure 1.3. Here we are able to capture the peak production times near 13 days for nitrite and nitrate. However, the estimates are still noisier than those obtained for the low-noise case illustrated in Figure 1.2. In addition, in

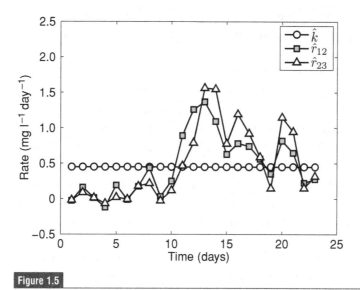

Figure 1.5

Analysis of noisy aquarium data assuming constant $\hat{k}$. The rates $\hat{r}_{12}(t)$ and $\hat{r}_{23}(t)$ are computed from Eq. (1.12), which is the least-squares solution to Eq. (1.11).

Figure 1.5 the rates $\hat{r}_{12}(t)$ and $\hat{r}_{23}(t)$ are estimated to be negative for part of the experiment, a phenomenon that we have argued does not reflect the real behavior of the system.

Further improvements to our analysis of the noisy data would require the introduction of additional (valid) assumptions. These improvements could perhaps include imposing constraints on the signs of the fluxes r_{12} and r_{23} as prior assumptions in the analysis. However, it is worth remembering that a goal of our non-mechanistic anlaysis was to introduce as few assumptions as possible. The results plotted in Figures 1.2 and 1.5 were obtained invoking only the assumption of mass conservation and the observation that k is effectively constant in this system. While it is possible to layer more assumptions onto this analysis, doing so would lead us closer to mechanistic modeling, in which the dynamics of the underlying processes are simulated based on representative models of their behavior.

1.2.3   Mechanistic modeling

Development of a model

Next we take on the challenge of attempting to simulate the kinetics of x_1, x_2, and x_3 based on a mathematical model of the mechanisms at work in transforming ammonia to nitrite and nitrite to nitrate. We call this analysis "mechanistic" modeling because we postulate and test the ability of specific mechanisms governing r_{12}, r_{23}, and other possible variables in the system to explain the observed data.

We note at the outset that it may not always be possible to formally distinguish between "mechanistic" and "nonmechanistic" modeling. Our nonmechanistic analyses presented in the previous section were based on the underlying reaction sequence ammonia $\rightarrow$ nitrite $\rightarrow$ nitrate, which certainly represents mechanistic information about the biochemical operation of the system. In loose terms, what we mean when we say that those analyses are nonmechanistic is that mechanisms of these reactions are not investigated because $r_{12}(t)$ and $r_{23}(t)$ are treated as arbitrary functions. In a mechanistic model we replace these arbitrary functions with explicit (hypothetical) functional dependences on the concentrations x_1, x_2, and x_3, and perhaps other unmeasured (hidden) variables.

A mechanistic model can be built on the basic framework of Eq. (1.3). Adopting our earlier finding that k remains constant, we have to construct functions for r_{12} and r_{23} to compute these variables from the state variables in the model. For example, we might hypothesize that r_{12}, the rate of oxidation of ammonia to nitrite, is proportional to both the level of nitrifying bacteria in the system and the concentration of ammonia. Denoting the nitrifying bacterial level as x_4, the specific model is

$$r_{12} = k_{12}x_4x_1, \tag{1.13}$$

where k_{12} is a constant. This model assumes that the simple *law of mass action* describes the relationship between ammonia concentration and its rate of oxidation. (Models of chemical kinetics are described in some detail in Chapter 5 of this book. For our purposes here, we take Eq. (1.13) as a hypothetical model component that must be tested against the data.)

Similarly, we hypothesize the equation

$$r_{23} = k_{23}x_4x_2 \tag{1.14}$$

for the second reaction. The parameters k_{12} and k_{23} can (we hope) be treated as constants, while x_4 is a variable that changes with time. Specifically, we expect x_4 to increase with time as the population of nitrifying bacteria increases.[5]

More complicated models may be constructed based on more complex kinetic mechanisms. For example, an alternative model could be constructed based on the saturating *Michaelis–Menten* kinetic model described in Chapter 5. Again,

[5] Here we have introduced a further assumption of one homogeneous population of nitrifying bacteria with the ability to oxidize both ammonia and nitrite. This assumption surely represents a simplification of reality. Here we will see that this simplification is reasonable, at least in comparison of model predictions with the particular data set at hand.

whatever model we choose must be treated as a hypothesis that must be tested against the observed data.

We define k_{12} as the maximal possible rate of ammonia oxidation per mass of ammonia, corresponding to the highest possible level of nitrifying bacteria; k_{23} is the maximal possible rate of nitrite oxidation per mass of nitrate. Given this formulation, x_4 is defined to vary between 0 at no oxidative capacity and 1 at the maximal capacity. Therefore the assumption that the nitrifying bacterial population is constrained to not exceed some maximal system capacity is built into the model. The variable x_4 is the bacterial population measured relative to this maximal capacity. To construct a model for how x_4 varies with time, we make the following additional assumptions: (i) the growth rate of x_4 is exponential (proportional to x_4) when x_4 is much smaller than the carrying capacity; (ii) the growth rate goes to zero when bacterial level approaches the carrying capacity, $x_4 \rightarrow 1$; and (iii) the growth rate is proportional to the concentration of "food" $(x_1 + x_2)$ for the bacteria.

A formula that captures these assumptions is

$$r_4 = k_4 x_4 (1 - x_4)(x_1 + x_2). \qquad (1.15)$$

In the absence of other dependences, this equation predicts that the growth rate of x_4 is proportional to $x_4(1 - x_4)$. Dynamic behavior following this equation is called *logistic growth*, which is a canonical model in ecological dynamics.

To complete the model including x_4 as a state variable, an equation for x_4 is appended to the basic governing equations

$$\frac{dx_1}{dt} = k - r_{12}(t)$$

$$\frac{dx_2}{dt} = r_{12}(t) - r_{23}(t)$$

$$\frac{dx_3}{dt} = r_{23}(t)$$

$$\frac{dx_4}{dt} = r_4(t). \qquad (1.16)$$

This model introduces a number of kinetic parameters: k_{12}, k_{23}, and k_4, for which we do not know the values in advance. In our model analysis, we will also treat k, the rate of ammonia production, as an unknown parameter to be estimated. In addition to the values of the kinetic constants, we require initial conditions in order to simulate the system via Eq. (1.16). Concentrations x_1, x_2, and x_3 start at zero at $t = 0$. However, we cannot assume that the initial condition of the new variable x_4 is zero, because $r_4(x_4 = 0) = 0$ according to Eq. (1.15). Thus without some initial "seed" population the predicted level of nitrifying bacterial would remain at zero throughout the experiment. Since we have no measurement of $x_4(t=0) = x_{40}$, we will include its value in our list of unknown parameters and hope to estimate it from the data.

Parameters and variables

Before moving on to the analysis of this model it is worth pausing to review some terminology that requires some definition. Here, it is essential to distinguish between what we mean by a model *variable* and a model *parameter*. Formally, the model at hand has four state variables, x_1 through x_4, with time derivatives listed in Eq. (1.16). Reaction and growth rates r_{12}, r_{23}, and r_4 are variables that depend as simple functions on the state variables. Model parameters $(k, k_{12}, k_{23}, k_4, x_{40})$ are scalar numbers that do not depend on time, and must be estimated by comparing model predictions with measured data.

If a formal mathematical definition for variables versus parameters exists, then this classification probably does not follow it. Instead, we use working definitions for which the distinction is application-specific: variables are the things that the model predicts; parameters are numbers that are fixed to generate a specific realization of a model.

Analysis of the data

The model of Eq. (1.16) is nonlinear, and no simple closed-form solution exists. However, such systems of ordinary differential equations are straightforward to solve computationally. (Implementation of integrating a simple ODE model using a computer is illustrated as an example in Section 9.3 of the Appendices.) For current purposes, it is necessary only to appreciate that, given initial conditions and parameter values, we can easily obtain time courses of the variables $x_1(t)$, $x_2(t)$, $x_3(t)$, and $x_4(t)$ for the model using a desktop computer. Furthermore, adjusting the parameter values to minimize the least-squares difference between the model predictions and the observed data, the optimal parameter estimates are obtained.

Optimal model predications are compared with the noisy data in Figure 1.6, with the optimal parameter values associated with the predictions listed in the figure legend. Given the estimated model parameters, the differential-equation model predictions in Figure 1.6 closely follow the observed data. Yet beyond simply providing a fancy vehicle for making smooth curves go through the data, the mechanistic model provides estimates of the unmeasured variables $r_{12}(t), r_{23}(t)$, and $x_4(t)$. These estimates, which are plotted for the experimental time course in Figure 1.7, agree closely with those obtained from the nonmechanistic analysis. Namely, the rate of ammonia production is estimated to be 0.445 ml l^{-1} day^{-1}, nearly identical to our previous estimate. In addition, the predicted fluxes in

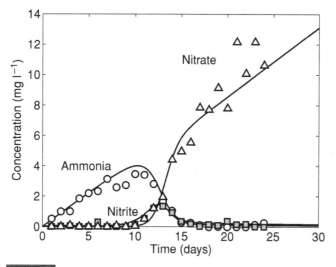

Figure 1.6

Fit of mechanistic model of Eqs (1.13)–(1.16) to data on ammonia, nitrite, and nitrate. Optimal parameter values used for this simulation are $k = 0.445$ mg l^{-1} day^{-1}, $k_{12} = 5.71$ day^{-1}, $k_{23} = 10.0$ day^{-1}, $k_4 = 0.183$ day^{-1}, and $x_{40} = 0.00027$.

Figure 1.7 closely resemble those obtained for the nonmechanistic analysis of the relatively noise-free data.

These findings affirm the ability of this mechanistic model to explain the data. Formally, we treat this agreement between the model and the data as a failure to disprove the model. Indeed, models are rarely, if ever, proved. A failure to disprove means that the model remains a feasible and potentially useful representation of reality.

Figure 1.7(b) shows that with the optimal parameter values, the model predicts that at the end of the 25-day experiment the nitrifying bacteria population reaches neither the carrying capacity of $x_4 = 1$ nor a steady state. The model predicts that as long as k remains constant, the bacterial population will continue to rise asymptotically toward the carrying capacity. Validation of this interesting model *prediction* would require comparison with additional data that are not available. Were such data available, we could use them to challenge the model. That challenge would result in one of two possible outcomes: the model would be either disproved or not disproved by the data.

While a failure to disprove can be a satisfying outcome, in scientific research a successful disproof is an opportunity to postulate new models, or refinements to existing models. While the current model does a nice job of matching our one data set, it is necessary to never lose sight of the fact that all mathematical models are *wrong* in the sense that, at best, they can only imperfectly represent reality. This

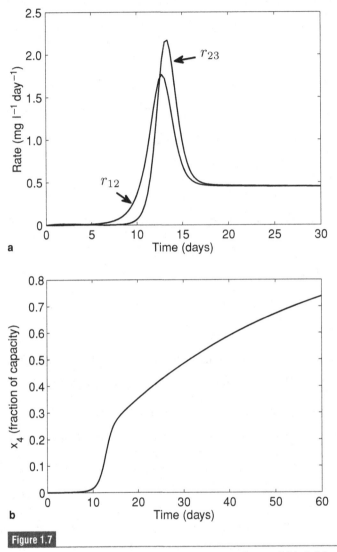

Figure 1.7

Plots of model predictions of (a) reaction rates r_{12} and r_{23} and (b) nitrifying bacterial level x_4.

idea can be briefly explored by comparing the model predictions with the original (much less noisy) data set.

To do this, let us recall that we obtained the data introduced in Figure 1.3 and used in the previous analysis by adding noise to the data of Figure 1.1. The original relatively noise-free data are compared with the mechanistic model predictions in Figure 1.8. (Here we have used the optimal parameter values obtained for the noisy data to obtain this simulation.)

Apparently the noise that confounded our nonmechanistic analysis can obscure deviations between the mechanistic model and reality. Some systematic difference

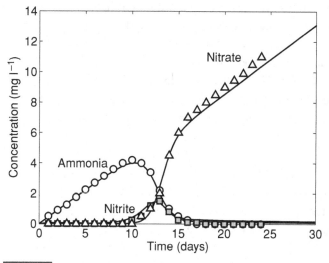

Plot of prediction of Eqs (1.13)–(1.16) and data on ammonia, nitrite, and nitrate. Model and parameters are the same as in Figure 1.6.

between the simulation and the low-noise data can be seen, for example, in the ammonia and nitrite concentrations (x_1 and x_2) exceeding the measured data for later times in the experiment. At day 18 both ammonia and nitrate concentrations reached undetectable levels, while the model predicts $x_1 = 0.25$ mg l^{-1} and $x_2 = 0.15$ mg l^{-1} at $t = 18$ days. (A closer fit to the nitrate data for $t > 15$ days can be obtained by adjusting the parameters to better match these low-noise data.) Whether or not the goodness of the fit of the simulation to the data is *good enough* is a question beyond the scope of the current discussion. For now, let us recall our assertion that all models are wrong. The degree of wrongness tends to vary inversely with the complexity of a model, while its usefulness is a property that emerges as a trade-off in trying to minimize the wrongness without introducing unnecessary complexity.

1.3 Recapitulation and lessons learned

In the preceding sections we have carried out a series of analyses on data on nitrogen reactants obtained from a home aquarium. The first analysis (nonmechanistic analysis of the low-noise data) revealed estimates of the unmeasured reaction fluxes plotted in Figure 1.2. We can have relatively high confidence in these predictions because they do not depend on arbitrary modeling assumptions. Obtaining those predictions required invoking only the principle of conservation of mass, which we have no reason to doubt was obeyed in the author's fish tank.

The nonmechanistic analysis was more difficult when noise was added to the data. Without modifying the computational method to account for the noise, the nonmechanistic prediction of the reaction rates (Figure 1.3(b)) does not tell us much, if anything, about what happened in the real system. Luckily, observations of the data revealed an insight that could be used in our computational analysis. Specifically, we found that ammonia production k is effectively constant. Summing the variables $x_1, x_2,$ and x_3 (Figure 1.4) verified this observation and provided an estimated value for k. By holding k constant, the mass conservation equations (Eq. (1.10) or (1.11)) became overdetermined; the least-squares solution was computed; and the estimated fluxes were plotted in Figure 1.5). In this example we were able to vastly improve our predictions simply by being more clever in our methodology. So the lesson here is very simple: when what you are doing is not working, try to be more clever! (More practically, in the author's case, experience and persistence usually must substitute for cleverness.)

The mechanistic analysis revealed additional information about the biophysical system at the cost of additional uncertainty. The additional information included hypothetical mechanisms and associated parameter values explaining the behavior of the chemical reactions and an estimated time course of the unmeasured variable x_4. Additional uncertainty arises from the fact that the mechanism used in the model (and hence the model itself) remains hypothetical. Agreement between model predictions and data is appropriately interpreted as a failure to disprove the model – not an affirmative proof. Even if these mechanisms do provide a useful, compact, and predictive representation of the system, this representation remains an approximation.

In later chapters the cycle of testing and refining models by comparing simulations to data will be revealed. Remember, all models – even useful ones – are wrong. Yet, as wrong as any model is doomed to be, the best way to understand the behavior of many (if not all) complex systems is to simulate their behaviors, to compare the simulations with experimental observations, and to make progress by constructing models that successively become less wrong.

Problems

1.1   Linear least squares. Show that Eq. (1.12) is the least-squares solution to Eq. (1.10).

1.2   Linear least squares (continued). The least-squares solution

$$\mathbb{A}^T \mathbb{A} \mathbf{x} = \mathbb{A}^T \mathbf{b}$$

to the overdetermined system

$$\mathbb{A}\mathbf{x} = \mathbf{b}$$

can be used to fit smooth functions to noisy data. For example, consider data $\{y_1, y_2, \cdots, y_n\}$ measured at time points $\{t_1, t_2, \cdots t_3\}$. The linear function $y = a_0 + a_1 t$ implies the system of equations

$$\begin{bmatrix} 1 & t_1 \\ 1 & t_2 \\ \vdots & \vdots \\ 1 & t_n \end{bmatrix} \begin{bmatrix} a_0 \\ a_1 \end{bmatrix} = \begin{bmatrix} y_1 \\ y_2 \\ \vdots \\ y_n \end{bmatrix}.$$

Use the least-squares algorithm to find values for a_0 and a_1 for the data set $\{t_i\} = \{1, 2, 3, 4, 5\}$ and $\{y_i\} = \{2.22, 2.67, 2.95, 3.18, 3.57\}$. Plot the data and the linear model fit on the same graph. How would you use this approach to fit a higher-order polynomial to data?

1.3   Open-ended computer exercise. Use a computer to numerically integrate the model in Eq. (1.16) and obtain the time courses plotted in Figure 1.7 (using parameter values listed in the legend). (Recall that tools for numerical integration of ordinary differential equations are described in Section 9.3.)

Next, add noise to the simulated time course data to obtain "pseudo-data" to compare with model predictions. Vary the model parameters (by hand or using a computer algorithm) to try to obtain a "best fit" of the model to the pseudo-data. Here you are free to define "best fit" in any reasonable way (perhaps as the least-square difference between the data and the simulated variables, or perhaps as the most visually satisfying plot of the predictions along with the data). How close are your parameter estimates to the original values used to obtain the pseudo-data? What happens as you change the relative strength of the noise compared with the original signals?

1.4   Open-ended computer exercise. Invent alternative mechanistic models to simulate the aquarium data analyzed in this chapter. For example, invent hypothetical mechanisms for r_{12}, r_{23}, and r_4 different from the ones used in the chapter. Or introduce two separate populations of nitrifying bacteria, one that oxidizes ammonia to nitrite and one that oxidizes nitrite to nitrate.

Explore the behavior of the resulting models. Can you find a model that captures the observed long-time behavior of x_1 and x_2 better than the one presented in the chapter? If so, does your model involve more or fewer adjustable parameters compared with the one in the chapter?

1.5   Experimental project. Set up a new home aquarium. Measure ammonia, nitrite, and nitrate over a one-month period, controlling for constant pH, temperature, and food input. Do the data match the kinetic model developed in Section 1.2.3? Which parameter values have to be adjusted to best fit the data?

2 Transport and reaction of solutes in biological systems

Overview

Transport of mass, into, out of, and within biological systems (including single cells, multicellular organisms, and even ecological systems) is fundamental to their operation. The subject of transport phenomena is treated in great depth in classic texts [10], as well as in books focused on biological systems [62]. Here we explore a number of examples that allow us to see how fundamental transport phenomena are accounted for in a wide range of biological systems. Specifically, we develop and apply basic frameworks for simulating transport in the following sorts of systems:

- Well-mixed systems.[1] The defining characteristic of these systems is that they are fluid systems (often aqueous solutions in biological application) with the solutes of interest distributed homogeneously (i.e., well mixed) over the timescales of interest. An example of a well-mixed system is the aquarium studied in the previous chapter. Other examples are chemical reaction systems inside cells or compartments within cells when spatial gradients of the intracellular reactants do not significantly influence the behaviors that are simulated. Models of well-mixed systems (or models that adopt the well-mixed assumption) do not explicitly account for the spatial distribution of the variables simulated. For biochemical systems this means that, at any given time, concentrations are constant throughout a compartment. The kinetics of such systems are typically described by ordinary differential equations, as in the examples of Section 2.1 of this chapter and in Chapter 3.

  Note that different physical mechanisms may justify the well-mixed assumption in different systems. In cells, molecular diffusion can effectively drive mixing on timescales on the order of seconds or less. In the previous chapter's aquarium, mixing is driven by a mechanical pump that circulates the water. In the human body, the situation is more complex (as we shall see) and the validity of the well-mixed assumption depends on the particular application. Human

---

[1] Sometimes the term *well-stirred systems* is used instead.

beings are most obviously not well-mixed tanks, but for certain applications certain compartments in the body can be treated that way.

- Reaction–diffusion systems. In biochemical systems when chemical transformation and molecular diffusion proceed on similar timescales (and no additional mixing processes occur), important spatial gradients can form. An example of a chemical reactant that tends to be distributed in cells with significant spatial gradients is oxygen, which is soluble only at relatively low concentrations in aqueous media. Low concentration combined with relatively rapid chemical consumption leads to spatial gradients on the typical length scales of cells. In fact, oxygen diffusion is a key factor in constraining the feasible spatial dimensions of cells and driving the evolutionary development of circulatory systems in higher organisms. Spatial gradients are also critically important in molecular signaling in growth and development, where cues driving heterogeneous cell differentiation are encoded in the spatial distribution of specialized signaling molecules.

    Reaction–diffusion systems are described in Section 2.4.2 of this chapter, with several examples.

- Advection–diffusion systems. These are systems where flow influences spatial concentration fields. Important examples arise in organisms with vascular systems–systems of pipes that transport material within plants and animals over distances larger than is possible via molecular diffusion. Sections 2.4.1 and 2.4.3 of this chapter describes the basic mathematical framework for simulating transport in this sort of system.

2.1   Well-mixed systems: compartmental modeling

A simple compartmental system is illustrated by the bucket in Figure 2.1. Here F_{in} and F_{out} denote flow of solvent (for example, water) into and out of the bucket. We assume that the contents of the bucket are mixed instantaneously. This well-mixed assumption implies that while volume ($V(t)$) and concentration of a given solute ($c(t)$) may vary with time, the solution in the bucket remains homogeneously mixed.

The rate of change of solvent (fluid) volume is given by the simple mass balance

$$\frac{dV}{dt} = F_{in} - F_{out}, \qquad (2.1)$$

where flow in and out are given in volume per unit time.

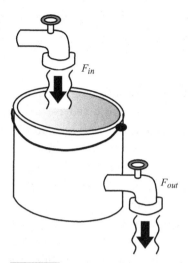

A simple transport system. If the contents of the bucket are well stirred, then solute transport is governed by Eqs (2.1)–(2.3).

The rate of change of mass of solute is given by another mass-balance equation

$$\frac{d(cV)}{dt} = F_{in}c_{in} - F_{out}c, \tag{2.2}$$

where the concentration of solute in the inflow is denoted c_{in}. Therefore $F_{in}c_{in}$ is the rate of mass influx and $F_{out}c$ is the rate of mass efflux. Since the system remains well mixed, the outflow concentration is equal to $c(t)$, the concentration in the bucket. If F_{in}, F_{out}, and c_{in} are prescribed (either as constants or as functions of time), then Eqs (2.1)–(2.2) are a complete set of governing equations for $V(t)$ and $c(t)$.

If we wish we can manipulate these equations to obtain an equation for dc/dt:

$$V\frac{dc}{dt} + c\frac{dV}{dt} = F_{in}c_{in} - F_{out}c$$

$$V\frac{dc}{dt} = F_{in}c_{in} - F_{out}c - F_{in}c + F_{out}c$$

$$V\frac{dc}{dt} = F_{in}(c_{in} - c). \tag{2.3}$$

These mass-balance equations will be used in examples throughout this text. For example, they form the basis of simulating transport of solutes throughout the body in a number of applications, as will be treated in Chapter 3. In the next section we consider transport of water and solutes across porous membranes, also a ubiquitous phenomenon in biological and biotechnological systems.

2.2 Passive flow of water, solutes, and electrical current across membranes

2.2.1 Basic equations

In this section we develop equations for modeling transport through membranes
driven by gradients in pressure, osmolarity, and electrical potential. Specifically, here we are concerned with passive porous membranes that allow material
exchange driven by various thermodynamic driving forces. In treating "passive
porous" membranes we are not considering biological membranes with specialized transport molecules that are associated with complex kinetic mechanisms,
which will be considered in later chapters.

Using Onsager's theory of general relationships between thermodynamic driving forces and fluxes near equilibrium [46, 47], Kedem and Katchalsky [34] developed a broadly relevant and widely used set of formulae for modeling volume
flow, solute flux, and electrical current through membranes in aqueous systems.
The basic idea is that, near equilibrium, fluxes J_i are related to thermodynamic
driving forces X_j via the linear relationship

$$J_i = \sum_i L_{ij} X_j, \tag{2.4}$$

where the L_{ij} are constant. The thermodynamic driving forces (also called *thermodynamic potentials*) are the appropriate free energies associated with the flux
variables, as we will see below. A key component of the theory is that where this
linearity holds, $L_{ij} = L_{ji}$, as a consequence of microscopic reversibility.

Here we are interested in obtaining a relationship between the total volume flux
(J_v), solute flux (J_s), and electrical current (I) and the appropriate driving forces,
illustrated in Figure 2.2. So to apply Eq. (2.4), the challenge is to determine the
proper X's.[2] Kedem and Katchalsky tackled this problem based on expressing
the *dissipation function* in terms of the flux variables of interest. The dissipation
function is the rate of heat dissipation, which is equal to the rate of steady-state
entropy production, which is

$$\Phi = \sum_i X_i J_i,$$

for processes with fluxes and thermodynamic driving forces J_i and X_i. For example, let us assume the solute is a salt with chemical formula $(S_1)_{\nu_1}(S_2)_{\nu_2}$, which
dissociates into ν_1 S_1 ions and ν_2 S_2 ions. For fluxes J_w, J_1, and J_2 (where J_w is
the water flux and J_1 and J_2 are the fluxes of S_1 and ν_2 S_2) the thermodynamic
driving forces and dissipation function are simple to express. The thermodynamic

---

[2] The derivation here may be esoteric compared with the practical aplication of the theory. Readers who wish to are invited
to skip ahead to Eq. (2.13) for the derived flux–force relationship used in the examples in the following section.

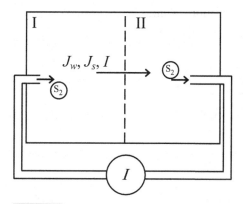

Illustration of a membrane permeable to water and a dissolved salt. Electrical current I is associated with salt S_2 ions (see text) flowing through a circuit in series with the path through the membrane.

driving forces are

$$X_w = \Delta\mu_w = \bar{V}_w(\Delta p - \Delta\Pi)$$

$$X_1 = \Delta\mu_1 = \bar{V}_1\Delta p + RT\,\Delta \ln c_1 + z_1 F\Delta\psi$$

$$X_2 = \Delta\mu_2 = \bar{V}_2\Delta p + RT\,\Delta \ln c_2 + z_2 F\Delta\psi,$$

where the terms on the right-hand side are defined as follows.

Following Figure 2.2, the two sides of the membrane are labeled I and II, and we use the convention that fluxes are defined as positive in the I $\rightarrow$ II direction. The driving force for water transport through the membrane is $\Delta\mu_w = \bar{V}_w(\Delta p - \Delta\Pi)$, where $\Delta p = p^{(I)} - p^{(II)}$ is the mechanical pressure difference and $\Delta\Pi$ is the osmotic pressure difference, which (for dilute solutions) is calculated from the sum of concentrations (c_i) of all dissolved species i on either side of the membrane: $\Delta\Pi = RT(\sum_i c_i^{(I)} - \sum_i c_i^{(II)})$. The factor $\bar{V}_w$ is the partial molar volume of water – essentially the volume per mole of water in the solution.

The driving force for species S_1 is $\Delta\mu_1 = \bar{V}_1\Delta p + RT\,\Delta \ln c_1 + z_1 F\Delta\psi$, where $\Delta \ln c_1 = \ln c_1^{(I)} - \ln c_1^{(II)}$ is the difference in log of concentration of S_1 between the two sides of the membrane. The electrical potential across the membrane is denoted $\Delta\psi = \psi^{(I)} - \psi^{(II)}$, and z_1 is the valence of S_1. Similarly, $\Delta \ln c_2 = \bar{V}_2\Delta p + \ln c_2^{(I)} - \ln c_2^{(II)}$, and z_2 is the valence of S_2. In Eq. (2.5) below, fluxes have units of moles per unit time and thermodynamic driving forces in units of energy per mole. The factors $\bar{V}_1$ and $\bar{V}_2$ are the partial molar volumes of solutes S_1 and S_2 in the solution. The constants R, T, and F are the gas constant, temperature, and Faraday's constant,[3] respectively.

---

[3] Faraday's constant F is a simple unit conversion factor – the number of coulombs of electrical charge per mole of elementary charges – and has the accepted value of $96\,485.339$ C mol^{-1}.

Hence the Onsager relationship may be expressed

$$
\begin{bmatrix} J_w \\ J_1 \\ J_2 \end{bmatrix} = \begin{bmatrix} M_{11} & M_{12} & M_{13} \\ M_{21} & M_{22} & M_{23} \\ M_{31} & M_{32} & M_{33} \end{bmatrix} \begin{bmatrix} \Delta\mu_w \\ \Delta\mu_1 \\ \Delta\mu_2 \end{bmatrix}, \tag{2.5}
$$

where $M_{ij} = M_{ji}$ are constants. Rather than equations for J_w, J_1, and J_2, equations for J_v, J_s, and I are more convenient for many applications. We can find such equations by expressing the dissipation function in terms of these fluxes:

$$
\Phi = J_v \Delta\mu_v + J_s \Delta\mu_s + IE, \tag{2.6}
$$

where we seek the proper expressions for $\Delta\mu_v$, $\Delta\mu_s$, and E. To do this, first we note that the current is expressed in terms of the ion fluxes:

$$
I = F(z_1 J_1 + z_2 J_2). \tag{2.7}
$$

Since the overall electroneutrality of the solution on both sides of the membrane is maintained $(z_1 c_1^{(I)} + z_2 c_2^{(I)} = z_1 c_1^{(II)} + z_2 c_2^{(II)} = 0)$, a nonzero current is maintained only when ions are removed from one side of the membrane and introduced to the other side at some finite rate, as illustrated in Figure 2.2.

Without loss of generality, let us assume that S_2 ions are injected and removed as shown in the figure, while S_1 ions are not. Therefore the total flux of of S_1 ions from I to II is given by

$$
J_1 = \nu_1 J_s, \tag{2.8}
$$

where J_s is the flux of salt through the membrane (in moles per unit time) and ν_1 is the stoichiometric coefficient for ion 1. Combining Eqs (2.7) and (2.8) we obtain

$$
J_2 = \frac{I}{z_2 F} - \frac{z_1 \nu_1}{z_2} J_s = \frac{I}{z_2 F} + \nu_2 J_s. \tag{2.9}
$$

Here we have made use of the fact that electroneutrality requires $z_1 \nu_1 + z_2 \nu_2 = 0$. This equation says that if the current is zero then $J_2 = \nu_2 J_s$ is given by the salt flux through the membrane. If I and z_2 are positive then the term $\frac{I}{z_2 F}$ adds an extra positive term on top of the electroneutral salt flux J_s.

Using these expressions the dissipation function is

$$
\Phi = J_w \Delta\mu_w + \nu_1 J_s \Delta\mu_1 + \nu_2 J_s \Delta\mu_2 + I \frac{\Delta\mu_2}{z_2 F}
$$

$$
= J_w \bar{V}_w (\Delta p - \Delta\Pi) + J_s (\bar{V}_s \Delta p + RT \Delta \ln c_1^{\nu_1} c_2^{\nu_2}) + I \frac{\Delta\mu_2}{z_2 F}, \tag{2.10}
$$

where $\bar{V}_s = v_1 \bar{V}_1 + v_2 \bar{V}_2$. Also noting that the total volume flow is $J_v = \bar{V}_w J_w + \bar{V}_s J_s$, Eq. (2.10) becomes

$$\Phi = J_v(\Delta p - \Delta \Pi) + J_s(RT \Delta \ln c_1^{v_1} c_2^{v_2} + \bar{V}_s \Delta \Pi) + I \frac{\Delta \mu_2}{z_2 F}. \qquad (2.11)$$

Kedem and Katchalsky [34] argue that "in both natural and experimental systems" $\frac{\bar{V}_s \Delta \Pi}{RT \Delta \ln c_1^{v_1} c_2^{v_2}} \ll 1$: hence the dissipation function is

$$\Phi = J_v(\Delta p - \Delta \Pi) + J_s(RT \Delta \ln c_1^{v_1} c_2^{v_2}) + I \frac{\Delta \mu_2}{z_2 F}, \qquad (2.12)$$

and the force–flux relation is

$$\begin{bmatrix} J_v \\ J_s \\ I \end{bmatrix} = \begin{bmatrix} L_{11} & L_{12} & L_{13} \\ L_{21} & L_{22} & L_{23} \\ L_{31} & L_{32} & L_{33} \end{bmatrix} \begin{bmatrix} \Delta p - \Delta \Pi \\ RT \Delta \ln c_1^{v_1} c_2^{v_2} \\ \Delta \mu_2/(z_2 F) \end{bmatrix}, \qquad (2.13)$$

which is the basic formulation developed by Kedem and Katchalsky [34]. Here note that J_v is measured in volume per unit time; J_s in moles per unit time; and I in coulombs per unit time. Hence $\Delta \mu_w = \Delta p - \Delta \Pi$, $\Delta \mu_s = RT \Delta \ln c_1^{v_1} c_2^{v_2}$, and $E = \Delta \mu_2/(z_2 F)$.

In certain applications (including the example below) electrical current is not significant, and we seek a reduced form of Eq. (2.13) with $I = 0$. With $I = 0$, Eq. (2.13) becomes

$$\begin{bmatrix} J_v \\ J_s \end{bmatrix} = \begin{bmatrix} \left(L_{11} - L_{13}^2/L_{33} \right) & (L_{12} - L_{13}L_{23}/L_{33}) \\ (L_{12} - L_{13}L_{23}/L_{33}) & \left(L_{22} - L_{23}^2/L_{33} \right) \end{bmatrix} \begin{bmatrix} \Delta p - \Delta \Pi \\ RT \Delta \ln c_1^{v_1} c_2^{v_2} \end{bmatrix}. \qquad (2.14)$$

Often these equations are expressed using the following definitions:

$$\Delta \Pi_s = RT(c_1^{(I)} + c_2^{(I)} - c_1^{(II)} - c_2^{(II)}),$$

and

$$c_s = \frac{\Delta \Pi_s}{RT \Delta \ln c_1^{v_1} c_2^{v_2}}.$$

Furthermore, we define $(L_{11} - L_{13}^2/L_{33}) = L_p$, which is called the *mechanical permeability* or *filtration coefficient*, $(L_{22} - L_{23}^2/L_{33}) = c_s \omega'$, where ω' is the solute permeability, and $(L_{12} - L_{13}L_{23}/L_{33}) = c_s(1 - \sigma)L_p$, where σ is called the *reflection coefficient*. With these definitions, the flux–force relationship becomes[4]

$$\begin{bmatrix} J_v \\ J_s \end{bmatrix} = \begin{bmatrix} L_p & c_s(1 - \sigma)L_p \\ c_s(1 - \sigma)L_p & c_s \omega' \end{bmatrix} \begin{bmatrix} \Delta p - \Delta \Pi \\ \Delta \Pi_s/c_s \end{bmatrix}. \qquad (2.15)$$

---

[4] In Eq. (2.15) $\Delta \Pi$ represents the total osmotic gradient, while $\Delta \Pi_s$ represents the concentration gradient of the salt that is transported through the membrane.

The filtration coefficient determines the relationship between volume flow and pressure in the absence of a gradient in the transported salt: $L_p = J_v/(\Delta p - \Delta\Pi)$ (i.e., at $\Delta\Pi_s = 0$). The solute permeability determines the solute flux in the absence of a pressure gradient: $\omega' = J_s/(\Delta\Pi_s)$. The prime notation on ω' distinguishes this solute permeability from the permeability measured when current and volume flow equal zero. In other words:

$$\omega = \left(\frac{J_s}{\Delta\Pi_s}\right)_{J_v=0,\,I=0}$$

$$\omega' = \left(\frac{J_s}{\Delta\Pi_s}\right)_{\Delta p-\Delta\Pi=0,\,I=0}.$$

When the reflection coefficient is 1, then the cross terms are zero and the volume and solute fluxes are independent.

With a little bit of algebra[5] it is possible to show that (with $I = 0$)

$$J_v = L_p(\Delta p - \Delta\Pi) + (1-\sigma)L_p\Delta\Pi_s$$

$$J_s = c_s(1-\sigma)J_v + \omega\Delta\Pi_s, \tag{2.16}$$

where $\omega' = \omega + c_s(1-\sigma)^2 L_p$. Equation (2.16) represents one of several specific forms that the Onsanger–Kedem–Katchalsky formulae take for specific applications. This is the formulation that we will use below.

Since these linear phenomenological equations are formulated above to account for total solute flux J_s, the solute permeability and reflection coefficient are measured with respect to total solute flux. In the example below we will see how to generalize these equations to account for more solutes than the single salt considered here.

2.2.2 Example: volume changes in isolated glomeruli

Glomeruli are bundles of relatively leaky capillaries (the smallest blood vessels) that are involved in filtering solutes out of the blood in the kidney. The anatomy of the kidney is illustrated in Figure 2.3. Normally, blood flowing through a glomerulus is at a higher mechanical pressure than the surrounding fluid. Mechanical pressure drives volume and solute flux out of the blood into a structure called the *Bowman's capsule*. Because certain solutes in the blood – notably the protein albumin – have reflection coefficients close to 1 for transport across the glomeruli membrane, plasma in the blood leaving a glomerulus normally has a higher osmotic pressure[6] than either the plasma entering or the fluid in the Bowman's capsule.

[5] See Exercise 2.4.
[6] The contribution to the osmotic pressure from plasma proteins is called the *oncotic* pressure.

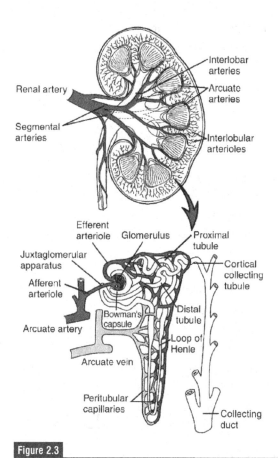

Figure 2.3

Diagram of kidney anatomy. Reprinted with permission from Guyton and Hall *Textbook of Medical Physiology* [25], Chapter 26, page 305, Figure 26-3, Copyright Elsevier.

To measure filtration coefficients, albumin reflection coefficients, and permeabilities associated with glomeruli, glomeruli can be micro-dissected from animal kidneys and studied under controlled conditions under a microscope. The first experiments along these lines were developed and reported by Savin *et al.* [59, 54], in which the buffer medium was rapidly switched to an iso-osmolar solution of dextran, a high-molecular-weight polymer with reflection coefficient equal to 1, in order to estimate the reflection coefficient for albumin. Here we explore data from experiments of Savin *et al.* to illustrate how the developed concepts of water and solute transport apply in this system, as well as from more recent experiments performed by Salmon *et al.* [53].

As we will see, the data obtained from these experiments may be adequately explained using the Kedem–Katchalsky formulae developed above and generalized to account for multiple solutes. To generalize the theory this way, assigning a possibly different reflection coefficient to each dissolved solute, the volume flux across the membrane can be modeled by

$$J_v = L_p \left(\Delta p - \sum_i \Delta \Pi_i \right) + L_p \sum_i (1 - \sigma_i) \Delta \Pi_i$$

$$= L_p \left(\Delta p - \sum_i \sigma_i \Delta \Pi_i \right). \tag{2.17}$$

Recall that in Eq. (2.16) $\Delta \Pi$ includes contributions from all solutes while $\Delta \Pi_s$ is the contribution from the permeant solute. Here we are not distinguishing between permeant and impermeant solutes; impermeant solutes appear in Eq. (2.17) with reflection coefficient $\sigma_i = 1$. The sum $\sum_i$ is over all solutes with osmotic pressure difference across the membrane $\Delta \Pi_i$ associated with solute i. (If $\Delta \Pi_i$ is measured as inside minus outside concentration, then fluxes are defined here as from inside to outside of the glomerulus.) The flux of each solute is modeled[7]

$$J_i = \bar{c}_i (1 - \sigma_i) J_v + \omega_i \Delta \Pi_i, \tag{2.18}$$

where, analogous to the definition of c_s, $\bar{c}_i$ is defined $\bar{c}_i = \Delta c_i / \Delta \ln c_i$.

In the experiments we wish to simulate, the mechanical pressure difference is assumed negligible, and the rate of glomerulus volume change is given by

$$\frac{dV}{dt} = -J_v = +L_p \sum_i \sigma_i \Delta \Pi_i \tag{2.19}$$

and the rate of change of mass of each dissolved solute in the glomerulus is

$$\frac{dQ_i}{dt} = -J_i = -\bar{c}_i (1 - \sigma_i) J_v - \omega_i \Delta \Pi_i. \tag{2.20}$$

The simulation of the experiments described here is further simplified by the fact that $\sigma_i \approx 0$ for all small solutes (not including protein or dextran) in normal glomeruli. Therefore, accounting for the dextran and albumin gradients, the governing equation for volume flux simplifies to

$$\frac{dV}{dt} = -J_v = +L_p \left(\sigma_p \Delta \Pi_p - \sigma_d \Delta \Pi_d \right),$$

where σ_p and $\Delta \Pi_p$ are the reflection coefficient and osmotic difference for albumin; σ_d and $\Delta \Pi_d$ are the reflection coefficient and osmotic difference for dextran.

In the first set of experiments we wish to consider, glomeruli are initially isolated with no dextran present, and the buffer media is rapidly switched to a BSA-free

---

[7] Although this generalization represents standard practice, it is prudent to note that Eq. (2.18) represents an unjustified application of expressions derived in Section 2.2.1 for a binary solute to an arbitrarily more complex solution [57]. The additional, typically unstated, assumption in Eq. (2.18) is that frictional interactions between solute fluxes are transmitted only through the solvent in Eq. (2.17): i.e, there are no cross terms correlating J_i and $\Delta \Pi_j$ ($i \neq j$) in Eq. (2.18).

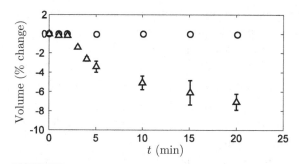

Figure 2.4

Time course of glomerular volume change in response to change to iso-osmolar albumin-free dextran solution. Data from the normal case are plotted as solid circles; data from the case where glomeruli are damaged by treatment with protamine are plotted as triangles. Data are replotted from Savin *et al.* [54].

dextran solution. In this case the equation for volume becomes

$$\frac{dV}{dt} = -J_v = +L_p \left(\sigma_p \Pi_p - \sigma_d \Pi_d \right), \qquad (2.21)$$

assuming that no significant quantities of dextran enter the isolated glomeruli and no significant quantities of albumin build up in the buffer. Here Π_p is the osmotic pressure due to protein in the glomeruli, and Π_d is the osmotic pressure due to dextran in the buffer. From Eq. (2.21), volume flux goes to zero when

$$\frac{\sigma_p}{\sigma_d} = \left(\frac{\Pi_d}{\Pi_p} \right)_{eq}. \qquad (2.22)$$

The notation $(\cdot)_{eq}$ denotes that this is an equilibrium ratio. Since the buffer osmotic pressure remains fixed, the ratio of reflection coefficients is computed

$$\frac{\sigma_p}{\sigma_d} = \frac{\Pi_d}{\left(\Pi_p \right)_{eq}} = \frac{V_f}{V_i}, \qquad (2.23)$$

where V_f / V_i is the ratio of final-to-initial volume of the glomerulus. This expression, which was first applied in this context by Savin *et al.* [54], assumes a linear relationship between albumin (mass per unit volume) and osmotic pressure. In Savin *et al.*, an observation that $V_f < V_i$ was taken as evidence that $\sigma_p < \sigma_d$. This is because, if the albumin reflection coefficient is lower than the dextran reflection coefficient, then (given a relatively large buffer volume) equilibration requires that Π_p increase via loss of water from the glomeruli.

Figure 2.4 shows data obtained by Savin *et al.* from this sort of experiment. In the normal case (solid circles) no volume change is detected, because $\sigma_p \approx \sigma_d \approx 1$. When isolated glomeruli are treated with protamine (a small protein that damages the glomerular capillaries), the glomeruli are found to shrink (data plotted as triangles in the figure) owing to an apparent effect of protamine decreasing

σ_p. However, analysis of the protamine-treated case is not as simple as Eq. (2.23) implies, because this equation invokes the major assumption that neither albumin nor dextran enters or leaves the glomeruli in the experiment. While this may be a reasonable assumption for normal untreated glomeruli, a reduction in albumin reflection coefficient significantly below 1 may be associated with a nonzero permeability that may allow protein to leak out. Thus it is not clear whether this assay provides an accurate estimate of σ_p when σ_p is significantly less than 1. Indeed, we can see from Figure 2.4 that a clear equilibrium is not established over the time course reported here. Furthermore, in the protamine-treatment experiments in Figure 2.4 (triangles), the time course was obtained during protamine incubation. Thus the σ_p starts at close to 1 at the initial time point and gradually decreases over the course of the experiment.

Because of the difficulty in interpreting this experiment, Savin *et al.* [54] used a different strategy to assay σ_p quantitatively. Specifically, they searched (by trial and error) for a dextran concentration (and associated Π_d) that caused no significant volume changes over the first few minutes in glomeruli pre-incubated with protamine. Given the known equilibrium ratio Π_d/Π_p that generates no volume change, Eq. (2.22) gives a direct estimate of σ_p/σ_d.

To include permeation in our analysis, and to estimate filtration and reflection coefficients quantitatively, we make use of a set of experiments reported by Salmon *et al.* [53], where the buffer osmolarity was rapidly switched and the resulting volume changes in the isolated glomeruli were assayed by video microscopy. Figure 2.5 illustrates volume time courses of an individual glomerulus from this study.

During the first 15 s (from time -15 to 0 s) the isolated glomerulus maintained a constant volume, with 1% BSA ($= 1\,\mathrm{mg\,dl^{-1}}$) inside the glomerular capillary. When the buffer is switched to 8% BSA the osmotic gradient drives fluid flow out and the glomerulus shrinks to approximately 90% of its initial value. This modest shrinking reveals that during this protocol BSA permeates into the glomerular capillary, because otherwise the final equilibrated volume at 8% BSA would be 1/8 of the initial volume (assuming the isolated glomerulus does not support a mechanical pressure gradient). Since fluid flow is outward when external BSA is switched from 1% to 8% concentration, BSA influx must be via diffusive permeation. Because the underlying model is nonlinear,[8] the kinetic equations (2.25) are integrated using a computer. Integration of nonlinear ordinary differential equations such as these is a ubiquitous application in simulation, as will be seen in further examples. (Section 9.3 demonstrates how one popular computer package can be used to integrate ordinary differential equations.)

---

[8] Although the flux–force relationship of the Onsager–Kedem–Katchalsky theory is linear, the derived kinetic equations are nonlinear owing to the nonlinearity of the relationship between concentration and chemical potential.

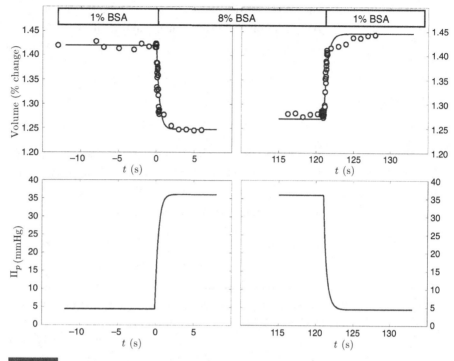

Figure 2.5

Water and solute transport across the glomerular membrane. Data are re-plotted from Salmon *et al.* [53] compared with simulations of the model described by Eqs (2.25). Data are concentration estimated from video microscopy of a single isolated glomeruli, initial equilibrated at 1% BSA ($= 1 \, \mathrm{mg\,dl}^{-1}$) in the external buffer. At $t = 0$ (left panel) the BSA concentration in buffer was rapidly increased to 8%. After re-equilibration for 2 min, the buffer was rapidly switched back to 1% BSA concentration. Model predictions are shown as solid lines. The lower panels show predicted glomerular oncotic pressure $\Pi_p(t) = \alpha c_p(t)$ for the time courses. Parameters used to obtain these simulations are $L_p = 0.013 \, \mathrm{nl\,mmHg}^{-1}\,\mathrm{s}^{-1}$, $\sigma_p = 1$, and $k = 2.9 \, \mathrm{dl\,s}^{-1}$.

Assuming no mechanical pressure gradient, the equations governing this system are

$$\frac{dV}{dt} = -J_v = +L_p \sigma_p \Delta \Pi_p$$

$$\frac{dQ_p}{dt} = -\bar{c}_p(1 - \sigma_p)J_v - \omega_p \Delta \Pi_p, \qquad (2.24)$$

where Q_p is the mass of protein (BSA) in the glomerular capillary. In this example concentrations are conveniently measured in units of $\mathrm{mg\,dl}^{-1}$, or % BSA. Although these units are not standard physical chemical units, in the concentration range studied there exists a linear relationship between BSA concentration and osmotic pressure: $\Delta\Pi_p = \alpha(c_p - c_o)$, where $\alpha = 4.5$ mmHg dl mg^{-1}, and c_p and c_o are the concentrations inside and outside the capillary, measured in units of mg dl^{-1}.

Using this relationship these equations can be cast in the units used here:

$$\frac{dV}{dt} = -J_v = +L_p\sigma_p\alpha(c_p - c_o)$$

$$\frac{dQ_p}{dt} = -\bar{c}_p(1 - \sigma_p)J_v - k(c_p - c_o), \qquad (2.25)$$

where $c_p = Q_p/V$, c_o is the (externally imposed) outside protein concentration, and $k = \alpha\omega_p$ is the permeability in units of volume time^{-1}. The log mean concentration $\bar{c}_p$ is computed $\bar{c}_p = (c_p - c_o)/(\ln c_p - \ln c_o)$.

These equations involve three parameters σ_p, L_p, and k, which may be estimated by matching the simulations predicted by the equations to the data in Figure 2.5. To simulate the first protocol (switching from 1% to 8% BSA), we use the initial conditions $V(0) = 1.42$ nl and $Q_p(0) = (1 \text{ mg dl}^{-1}) \times V(0)$, and parameter values listed in the figure legend.

During the second phase of the experiment (illustrated in the right-hand panels of Figure 2.5) the bathing solution is switched back to 1% BSA. Here the initial volume has increased slightly, from approximately 1.25 nl at 5 s to approximately 1.27 nl at 115 s. Therefore, to simulate this protocol, the initial conditions invoked are $V(0) = 1.27$ nl and $Q_p(0) = (8 \text{ mg dl}^{-1}) \times V(0)$. Parameter values are the same as those used for the first phase of the experiment. The lower panel of the figure illustrates the model-predicted oncotic pressure Π_p over the simulated time courses. In the shrinking protocol, the Π_p increases from approximately 5 to 36 mmHg, as the capillary shrinks and albumin permeates in. Equilibrium is reached when $\Pi_p = \alpha \times (8 \text{ mg dl}) = 36$ mmHg. During the swelling protocol, the process is reversed, with Π_p dropping back to the initial value of $\alpha\times (1 \text{ mg dl}) = 4.5$ mmHg.

Salmon *et al.* used the protocol of Figure 2.5 to estimate L_p in individual glomeruli. Their analysis was based on the initial slope following the switch in external concentration. Specifically, it was assumed that before the concentration changed substantially in the glomerulus the volume flux was initially constant, yielding

$$L_p = \left(\frac{dV/dt}{\alpha\Delta c}\right)_{t=0},$$

where $(c_p - c_o)_{t=0} = \Delta c$, and it is assumed $\sigma_p = 1$. The accuracy of this approximation is explored in Figure 2.6, which shows the data and simulation of the first protocol of Figure 2.5 over a shorter timescale.

Here we can see that data over the first 0.2 or so seconds following the initiation of the osmotic transient are necessary to estimate the initial slope effectively.

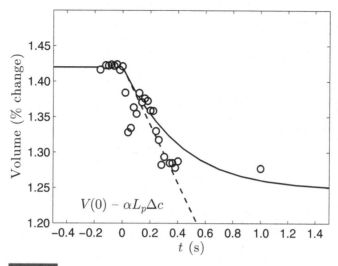

Figure 2.6

Initial glomerular swelling following switch from 1% to 8% BSA at $t = 0$ s. Data and model simulation (solid curve) are same as in Figure 2.6.

(Salmon *et al.* recognized this fact, and went to some trouble to obtain data on the millisecond timescale for these studies.)

2.3 Simulating chemical reactions

So far we have seen examples of how to develop equations to simulate transport of solutes and water between well-mixed aqueous compartments. In this section we use mass-balance relationships to simulate chemical reaction kinetics in well-mixed systems.

2.3.1 Example: synthesis of ethanol from xylose

To illustrate how to simulate chemical reactions, we shall build a computer simulation of the biochemical process of synthesizing ethanol from xylose in a yeast bioreactor. This is an important process, because xylose and other 5-carbon aldose sugars constitute the majority of sugars derived from hydrolysis of hemicellulose (wood fiber), and represent a potentially useful energy source.

Figure 2.7 reproduces data from a study by Petschacher and Nidetzky [50], in which time-course data on xylose, ethanol, and several biochemical intermediates were obtained from a well-stirred oxygen-limited system initially containing approximately 0.11 moles of xylose per gram of cell dry weight of yeast.

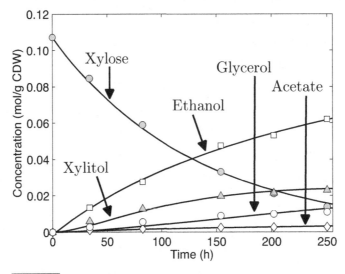

Xylose fermentation. Concentration time courses are plotted for xylose, xylitol, ethanol, glycerol, and acetate in a yeast bioreactor during oxygen-limited operation. Concentrations are plotted in units of moles per gram of yeast cell dry weight (mol/g CDW). Data are obtained from Figure 3A of Petschacher and Nidetzky [50] and apply to the wild-type (BP000) strain. Solid lines are model simulations (see page 39) with parameter values listed in Table 2.2.

Under oxygen-limited conditions, yeast cells metabolize sugars primarily by oxidizing them to form ethanol. Figure 2.7 illustrates that, as xylose is used up, ethanol and several other reactants are formed. We will see below that xylitol is an intermediate in the process of converting xylose to ethanol, while glycerol and acetate are by-products that represent alternative end products for the carbohydrate metabolism.

The solid lines in Figure 2.7 were obtained from a model simulation described below. In our analysis, we first develop a chemical reaction model to simulate the metabolic data obtained from this no-flow bioreactor experiment and a similar experiment using a yeast strain metabolically engineered for optimal fermentation of xylose to ethanol. Later, we apply compartmental transport modeling to explore *optimal* process design of continuous-flow bioreactors.

Biochemical reactions

Yeast cells employ a series of reactions in anaerobically fermenting xylose to produce ethanol. The first reaction is catalyzed by the enzyme xylose reductase:

$$\text{xylose} + \text{NAD(P)H} \rightleftharpoons \text{xylitol} + \text{NAD(P)}, \tag{2.26}$$

where NAD, NADH, NADP, and NADPH are metabolic cofactors.[9] The notation NAD(P)H and NAD(P) indicates that this enzyme can utilize either NADPH or NADH as a reducing cofactor in converting xylose to xylitol.

Xylitol is converted to xylulose via the xylitol dehydrogenase reaction:

$$\text{xylitol} + \text{NAD} \rightleftharpoons \text{xylulose} + \text{NADH}. \tag{2.27}$$

A series of biochemical reactions phosphorylate the sugar xylose and convert between several phosphorylated sugars:

xylose kinase:
2 xylulose + 2 ATP $\rightleftharpoons$ 2 xylulose-5-phosphate + 2 ADP
transketolase:
xylulose-5-phosphate + ribose-5-phosphate
$\rightleftharpoons$ gap + sedoheptulose-7-phosphate
transaldolase:
sedoheptulose-7-phosphate + gap
$\rightleftharpoons$ erythrose-4-phosphate + fructose-6-phosphate
transketolase:
erythrose-4-phosphate + xylulose-5-phosphate

$$\rightleftharpoons \text{gap} + \text{fructose-6-phosphate}. \tag{2.28}$$

Here gap stands for glyceraldehyde-3-phosphate, a 3-carbon phosphorylated sugar. The underlined names in (2.28) are the names of the enzymes that catalyze the reactions listed. The stoichiometry in the first reaction above is expressed so that this set of reactions sums to

$$2\,\text{xylulose} + 2\,ATP + \text{ribose-5-phosphate}$$
$$\rightleftharpoons 2\,\text{fructose-6-phosphate} + \text{gap} + 2\,\text{ADP}. \tag{2.29}$$

Fructose-6-phosphate and glyceraldehyde-3-phosphate can both be converted to pyruvate via glycolysis (see below). The source of ribose-5-phosphate to allow this reaction series to proceed is another series of reactions called the *pentose phosphate pathway*, which utilizes fructose-6-phosphate (a 6-carbon sugar) to generate ribose-5-phosphate (a 5-carbon sugar) and carbon dioxide:

$$\text{fructose-6-phosphate} + 2\,\text{NADP} \rightleftharpoons \text{ribose-5-phosphate} + 2\,\text{NADPH} + \text{CO}_2. \tag{2.30}$$

---

[9] Concentrations of cofactors such as NAD, NADH, ATP, and ADP will not be explicitly simulated here. Also, we note that Eq. (2.26) and all reactions in this section are represented as biochemical reactions that do not necessarily balance charge and mass. Specifically, protons (H^+ ions) and water molecules (H_2O) are formally not listed in biochemical reactions. Also, the reactants are assumed to be sums of rapidly converting related species. Notations and conventions for biochemical reactions are introduced in Chapter 5 of this book and described in detail in Beard and Qian [8].

In a series of reactions called *glycolysis*, fructose-6-phosphate and glyceraldehyde-3-phosphate can both be converted to pyruvate with the following stoichiometries:

$$\text{fructose-6-phosphate} + 3\,ADP + 2\,Pi + 2\,NAD$$
$$\rightleftharpoons 2\,\text{pyruvate} + 2\,\text{NADH} + 3\,\text{ATP} \qquad (2.31)$$

and

$$\text{gap} + \text{Pi} + \text{NAD} + 2\,\text{ADP} \rightleftharpoons \text{pyruvate} + \text{NADH} + 2\,\text{ATP}, \qquad (2.32)$$

where Pi represents inorganic phosphate. Equations (2.29), (2.30), (2.31), and (2.32) sum to

$$2\,\text{xylulose} + 2\,\text{NADP} + 3\,\text{NAD} + 3\,\text{ADP} + 3\,\text{Pi}$$
$$\rightleftharpoons 3\,\text{pyruvate} + 2\,\text{NADPH} + 3\,\text{NADH} + 3\,\text{ATP} + CO_2, \qquad (2.33)$$

giving an overall reaction for xylulose conversion to pyruvate.

In the absence of oxygen, pyruvate may be decarboxylated to acetaldehyde

$$\text{pyruvate} \rightleftharpoons \text{acetaldehyde} + CO_2 \qquad (2.34)$$

and the acetaldehyde converted to ethanol

$$\text{acetaldehyde} + \text{NADH} \rightleftharpoons \text{ethanol} + \text{NAD}. \qquad (2.35)$$

(Incidentally, this reaction is catalyzed by an enzyme called *alcohol dehydrogenase*; running in reverse, this reaction is the first step that our bodies use to detoxify ethanol.) Alternatively, acetaldehyde can undergo a series of reactions to form acetate:

$$\text{acetaldehyde} + \text{NAD} + \text{CoA} \rightleftharpoons \text{acetyl-CoA} + \text{NADH}$$
$$\text{acetyl-CoA} + \text{PPi} + \text{AMP} \rightleftharpoons \text{ATP} + \text{acetate} + \text{CoA}, \qquad (2.36)$$

where acetyl-CoA, CoA (coenzyme-A), AMP, and PPi (pyrophosphate) are more metabolic cofactors.

The final set of reactions that we need to consider is the conversion of xylulose to the by-product glycerol. This occurs through the following reactions:

phosphofructokinase:
fructose-6-phosphate + ATP $\rightleftharpoons$ fructose-1,6-bisphosphate + ADP
aldolase:
fructose-1,6-bisphosphate $\rightleftharpoons$ gap + dhap
triose phosphate isomerase:
$2\,\text{gap} \rightleftharpoons 2\,\text{dhap}$
glycerol-3-phosphate dehydrogenase:
$3\,\text{dhap} + 3\,\text{NADH} \rightleftharpoons 3\,\text{glycerol-3-phosphate} + 3\,\text{NAD}$
glycerol-3-phosphatase:
$3\,\text{glycerol-3-phosphate} \rightleftharpoons 3\,\text{glycerol} + 3\,\text{Pi}.$ $\qquad (2.37)$

Table 2.1: Reduced summary reactions for fermentation model.

Number	Reaction stoichiometry	Derived from
1.	xylose $\rightleftharpoons$ xylitol	Eq. (2.26)
2.	xylitol $\rightleftharpoons$ xylulose	Eq. (2.27)
3.	2 xylulose $\rightleftharpoons$ 3 acetaldehyde	Eqs (2.33)–(2.34)
4.	acetaldehyde $\rightleftharpoons$ ethanol	Eq. (2.35)
5.	acetaldehyde $\rightleftharpoons$ acetate	Eq. (2.36)
6.	2 xylulose $\rightleftharpoons$ 3 glycerol	Eq. (2.38)

Summing all of the reactions in Eqs (2.29), (2.30), and (2.37) yields

$$2\,\text{xylulose} + 3\,\text{ATP} + 2\,\text{NADP} + 3\,\text{NADH}$$
$$\rightleftharpoons 3\,\text{glycerol} + 3\,\text{ADP} + 2\,\text{NADPH} + 3\,\text{NAD} + 3\,\text{Pi} + \text{CO}_2. \quad (2.38)$$

A computational model for xylose fermentation by yeast

In all of the reactions described in the previous section, ATP, ADP, AMP, Pi, PPi, NAD, NADH, NADP, NADPH, CO_2, xylulose-5-phosphate, ribose-5-phosphate, sedoheptulose-7-phosphate, erythrose-4-phosphate, fructose-6-phosphate, glyceraldehyde-3-phosphate, acetyl-CoA, and pyruvate all represent intermediates that will not be explicitly modeled. Ignoring these intermediates, the model will necessarily not account for detailed biochemical processes that depend on changes in these cofactor concentrations. Nevertheless, we wish to construct a model in this framework to simulate the kinetics of the following seven reactants: xylose, xylitol, xylulose, acetaldehyde, ethanol, acetate, and glycerol. The fact is that not enough data are available from the experiments at hand to construct a more detailed model. (Indeed, we show below that it is possible to explain the kinetics of these seven reactants using a model at this reduced level of detail that satisfactorily explains the given data set.)

The lumped reactions generating and consuming these reactants (ignoring the other intermediates) can be constructed from the reactions listed in the previous section. The reduced reaction set is tabulated below in Table 2.1.

Denoting the flux (mass per unit time) through reactions 1 through 6 as J_1 through J_6, we can construct the mass-balance differential equations for the system:

$$\frac{d[\text{xylose}]}{dt} = -J_1$$
$$\frac{d[\text{xylitol}]}{dt} = J_1 - J_2$$
$$\frac{d[\text{xylulose}]}{dt} = J_2 - 2J_3 - 2J_6$$
$$\frac{d[\text{acetaldehyde}]}{dt} = 3J_3 - J_4 - J_5$$

$$\frac{d[\text{ethanol}]}{dt} = J_4$$

$$\frac{d[\text{acetate}]}{dt} = J_5$$

$$\frac{d[\text{glycerol}]}{dt} = 3J_6. \tag{2.39}$$

The numbers multiplying the J's in these equations are the stoichiometric numbers from the reactions in the preceding table. For example, reaction 3 generates 3 acetaldehyde for every 2 xylose consumed. Therefore, the equation for $d[\text{xylulose}]/dt$ contains the term $-2J_3$, while the equation for $d[\text{acetaldehyde}]/dt$ contains the term $+3J_3$.

Note that in the experimental data all concentrations are given in total moles in the reaction system per cell dry weight, ignoring the distribution of reactants in solutions versus in cells. This model also ignores this distribution: the concentrations in Eq. (2.39) are total (intra- plus extracellular) concentrations.

The next step is to determine functions that effectively model the chemical reaction fluxes (J_1 through J_6) as functions of the concentrations in the system (just like we did in Section 1.2.3 of the previous chapter). Given that the model ignores a large number of intermediate reactants and does not distinguish between intracellular and extracellular concentrations, it is by no means guaranteed that an effective model (that can simulate the observed data) exists in this framework. Bearing that in mind, let us at least give it a try.

After a certain amount of trial and error,[10] we may postulate simple mass-action kinetics governing reactions 1, 4, 5, and 6:

$$J_1 = k_1[\text{xylose}]$$
$$J_4 = k_4[\text{acetaldehyde}]$$
$$J_5 = k_5[\text{acetaldehyde}]$$
$$J_6 = k_6[\text{xylulose}], \tag{2.40}$$

where k_1, k_4, k_5, and k_6 are adjustable parameters. Reactions 2 and 3 in this model take slightly more complicated forms:

$$J_2 = k_2[\text{xylitol}] - k_{-2}[\text{xylulose}][\text{ethanol}]$$
$$J_3 = k_3[\text{xylulose}] - k_{-3}[\text{acetaldehyde}][\text{ethanol}] \tag{2.41}$$

where k_2, k_3, k_{-2}, and k_{-3} are adjustable parameters. These equations assume that the fluxes through reactions 2 and 3 are slowed through buildup of xylose

---

[10] By "trial and error" I am referring to the process of trying different model formulations in an attempt to find a model that can fit the data. The model presented here is the simplest one (in terms of the complexity of the kinetics and number of adjustable parameters) that I was able to find to match the data. That does not mean the model is unique.

Table 2.2: Parameter values for the model of Section 2.3.1.

Parameter	Value (BP000)	Value (BP10001)	Units
k_1	7.67×10^{-3}	8.87×10^{-3}	h^{-1}
k_2	3.60	13.18	h^{-1}
k_3	0.065	0.129	h^{-1}
k_4	0.867	0.497	h^{-1}
k_5	0.045	0.027	h^{-1}
k_6	1.15×10^{-3}	0.545×10^{-3}	h^{-1}
k_{-2}	88.0	87.7	$g\,h^{-1}\,mol^{-1}$
k_{-3}	99.0	99.9	$g\,h^{-1}\,mol^{-1}$

and acetaldehyde under the given experimental conditions. Since xylulose and acetaldehyde are the products of these two reactions, they appear in the negative terms in these flux expressions.

Although ethanol does not participate directly in either reaction 2 or 3, buildup of ethanol does impair the ability of yeast to continue to utilize carbohydrates. One way this may occur is through a buildup of cellular NADH and/or NADPH. The conversion of xylose to ethanol involves a net increase in one or both of these reduced factors, and a net decrease in the oxidized factors NAD and/or NADP. This buildup makes Eq. (2.27), which is reaction 2, less favorable to proceed in the forward direction. (It could also make other reactions less favorable, but incorporating such additional effects in the model does not lead to any significant improvement in its predictions in terms of comparisons with the data sets studied here.) A buildup of NADH will also make Eq. (2.33) less favorable to proceed in the forward direction, justifying the form of J_3 in Eq. (2.41).

This model can be simulated using a computer to integrate the differential equations. To do so, values for the adjustable parameters must be determined. Here, we determine values for the parameters by minimizing the differences between the model predictions and the experimental measurements in Figure 2.7. Computer implementation of the simulation and parameter estimation for this specific example is detailed in Sections 9.3 and 9.4 of the Appendices, which illustrate these techniques using the MATLAB computing platform.

Figure 2.7 plots model simulations obtained using the optimal parameter estimates along with the experimental data. The parameter values used for these simulations are listed in Table 2.2; data and parameter values for Figure 2.7 correspond to the wild-type BP000 strain.

As is apparent from Figure 2.7, this simple model does a good job of matching the observed experimental data. Note that model simulations of five of the seven state variables closely match the experimental measurements at the five sampled

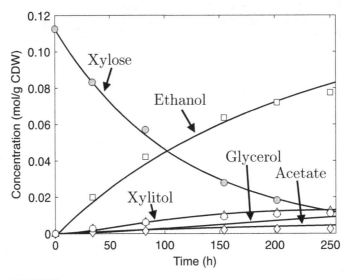

Figure 2.8

Xylose fermentation with metabolically engineered yeast. Concentration time courses are plotted for xylose, xylitol, ethanol, glycerol, and acetate in a yeast bioreactor during oxygen-limited operation. Concentrations are plotted in units of moles per gram of yeast cell dry weight (mol/g CDW). Data are obtained from Figure 3B of Petschacher and Nidetzky [50] and apply to the engineered (BP10001) strain. Solid lines are model simulations (see text) with parameter values listed in Table 2.2.

time points. Since no measures of xylulose or acetaldehyde were provided, there is no basis for comparison of the model predictions of those variables.

In addition to the experiments on the wild-type yeast strain, Petschacher and Nidetzky developed a yeast strain (BP10001) engineered to increase the preference of xylose reductase – Eq. (2.26) – for NADH over NADPH. The idea is to prevent the buildup of NADH and reduce the related inhibition of xylitol dehydrogenase: Eq. (2.27).

Fits of the model to data obtained from the engineered strain are shown in Figure 2.8. The estimated parameters for this fit (Table 2.2) indicate that k_2 and k_3, the effective rate constants associated with reactions 2 and 3, are higher in the engineered strain than in the wild type. This finding is consistent with reaction 2 and 3 being less inhibited in the engineered strain compared with the control, and is consistent with the expected behavior of the engineered strain.

In addition, reactions 4, 5, and 6 are predicted to proceed with effective forward rate constants significantly lower in the engineered strain than in the wild-type control. The observed effect on reaction 6 could perhaps be interpreted as a consequence of NADPH building up in the engineered strain more than in the control. (Remember it is expected that in the control strain reaction 1 will consume NADPH more favorably than in the engineered strain.) However, reaction 4 consumes NADH while reaction 5 generates NADH. Based on our simple analysis, we would not necessarily expect these reactions to shift in the same direction.[11] Furthermore, we note that our data fits result in an increase in the forward rate constants (k_2 and k_3) rather than a decrease in the reverse rate constants (k_{-2} and k_{-3}).

Therefore, while the expected phenomenon of slowing production of ethanol in the wild type (inhibition of reaction 2 due to buildup of NADH) is apparent in our model analysis, the expected mechanism (through the effective k_{-2} rate constant) is not captured. Yet the model does predict the expected increase in the favorability of this reaction in the engineered strain compared with the control. Important changes in other reactions are predicted as well. For example, reaction 3 (which is the net oxidation of xylulose to acetaldehyde) is predicted to be significantly enhanced in the engineered strain.

Given the phenomenological nature of the current model (ignoring many important intermediates and reactions, not accounting for intracellular versus extracellular reactants, and not accounting for realistic chemical kinetics), we should not expect all of the details of the biochemical mechanisms acting in these experiments to emerge from the analysis. That said, it is clear that the model, which makes predictions at the phenomenological level at which it was built, successfully predicts the time course data from these experiments. Thus with some confidence we can apply the model to simulate the kinetics of the state variables under experimental conditions and in concentration ranges that are similar[12] to those obtained in these experiments.

Process design

Rather than the *batch process* bioreactor modeled in the preceding section, chemical manufacturing is often done using continuous-flow reactors, where there are continuous influx of substrate materials and outflux of products. Industrial bioreactors may have multiple distinct components for filtration, extraction, mixing, reaction, and other chemical processes. Reactions may occur in a series of pipes, porus media, or other complex geometrical arrangements, and may involve specialized materials. Thus, in addition to operating under continuous flow, the assumption of rapid mixing may not apply.

Here, we study the operation of a continuous-flow bioreactor that takes the form of a rapidly mixed reaction vessel with a steady (constant) flow in equal to the

---

[11] In reality there are many reactions in yeast consuming and generating NADH and NADPH not considered in this simple model. The reducing capacity of NADPH may be transferred to NAD (forming NADH) or vice versa.

[12] The word "similar" is not precisely defined here. Remember, the conditions under which any model works (is predictive) are subject to experimental verification.

flow out. (In Section 2.4.1 we investigate a spatially distributed continuous-flow bioreactor.) Assuming that xylose is input at the concentration used in the study analyzed above, the governing equations (2.39) are modified to account for flow in and out of the reactor:

$$\frac{d[\text{xylose}]}{dt} = -J_1 + (x_o - [\text{xylose}]) \cdot F/V$$

$$\frac{d[\text{xylitol}]}{dt} = J_1 - J_2 - [\text{xylitol}] \cdot F/V$$

$$\frac{d[\text{xylulose}]}{dt} = J_2 - 2J_3 - 2J_6 - [\text{xylulose}] \cdot F/V$$

$$\frac{d[\text{acetaldehyde}]}{dt} = 3J_3 - J_4 - J_5 - [\text{acetaldehyde}] \cdot F/V$$

$$\frac{d[\text{ethanol}]}{dt} = J_4 - [\text{ethanol}] \cdot F/V$$

$$\frac{d[\text{acetate}]}{dt} = J_5 - [\text{acetate}] \cdot F/V$$

$$\frac{d[\text{glycerol}]}{dt} = 3J_6 - [\text{glycerol}] \cdot F/V, \tag{2.42}$$

where the input xylose concentration is x_o and the inflow concentrations of all other reactants are assumed zero. The flow and volume of the reactor are denoted F and V, respectively.

If we take the reactor conditions (cell density and input substrate conditions) as given design constraints, we might be interested in determining the design that maximizes the yield of ethanol under steady-state conditions. We know that if the flow is zero (and thus the rate of substrate delivery is zero), then the steady-state rate of ethanol production will be zero. At high flows the rate of ethanol production may be finite, but the ethanol concentration in the outflow may be too dilute for practical extraction. Thus we expect there to be a trade-off between production rate and concentration.

Since we are primarily interested here in illustrating basic principles rather than engineering a real chemical manufacturing process, let us assume that cell density of yeast is fixed at whatever value was used in the Petschacher and Nidetzky experiment [50]. Therefore we can use the kinetic models and associated rate constants developed above. As above, concentrations are measured as mass of reactant per unit biomass of yeast. The dynamics of the system depends on F/V, the flow per unit volume, and not otherwise on either variable F or V. Hence we define the ethanol production rate as $[\text{ethanol}] \cdot F/V$, the flux of ethanol out of the system, measured in mass of product per unit biomass of yeast per unit time.

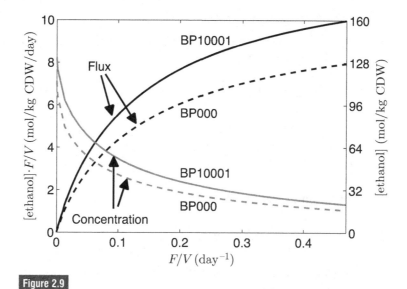

Figure 2.9

Steady-state production of ethanol using the BP000 and BP10001 strains of yeast in a continuous-flow well-mixed bioreactor. Variables are predicted using the model of Eq. (2.42). Steady-state flux ([ethanol]·F/V) and concentration ([ethanol]) are plotted as functions of F/V using the left-hand and right-hand ordinates, respectively. Model simulations (see text) with parameter values listed in Table 2.2.

Figure 2.9 shows the model-predicted steady-state ethanol production per unit volume as a function of F/V, for both yeast strains studied above. (Note that here we are using units of mass of ethanol per kilogram of biomass, and we are reporting time in units of days. These are more convenient units than those reported for the original data sets in Figures 2.7 and 2.8.) The predictions reported here are made simply by integrating the model equations until they reach a steady-state long-time limit. The different curves (solid lines for BP10001 and dashed lines for BP000) are obtained by using the two different parameter sets in Table 2.2 to simulate the two different strains.

Several interesting phenomena are illustrated in Figure 2.9. First, both ethanol production rate and concentration are predicted to be higher for the engineered strain (BP10001) at any given flow per unit volume. Second, the maximal production rate occurs in the limit of high F/V, where the output concentration is relatively low. Thus the trade-off between production rate and concentration is apparent. This is the nature of this well-mixed system. At fixed biomass and input substrate concentration, reaction progress is time-limited. Reactions go near completion only for low flows ($F/V < 0.1$ day^{-1}). Production rate is increased with flow, but at the cost of lower concentration and increased waste of substrate. In Section 2.4.1 we explore how this design limitation may be partially overcome with a spatially distributed continuous flow bioreactor.

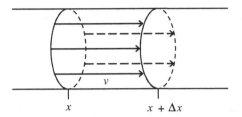

Figure 2.10

Illustration of plug flow in a pipe.

2.4   Distributed transport modeling

2.4.1   Flowing systems: the advection equation

Net fluid motion – or *flow* – is a ubiquitous phenomenon in biological systems.
Material dissolved or suspended in flowing fluid moves along with the fluid carrying it. Examples include oxygen bound to hemoglobin in red blood cells in
blood, transport of solutes in microfluidic systems, and the movement of nutrients
throughout oceans and lakes. The process of suspended or dissolved material being
carried along with bulk fluid motion is called *advection*, and the equation governing advective transport is called the *advection equation*. In this section we derive
the one-dimensional advection equation and the more general multidimensional
equation. We also show how to use the advection equation to simulate chemical
transport in a flowing chemical system.

Consider a pipe with fluid moving with plug flow with velocity v, as illustrated in
Figure 2.10. The term *plug flow* means that the velocity is the same at all positions
in the pipe. To derive the one-dimensional advection equation, we consider two
positions along the length of the pipe, x and $x + \Delta x$. We further assume that there
are no radial gradients in solution concentration (the concentration of a solute at
any position along the length of the pipe is constant across the cross section) and
denote concentration as a function of time and position as $c(x, t)$.[13] The mass
flux through the cross section at position x is $v \cdot c(x, t) \cdot A$, where A is the cross-
sectional area. The rate of change of mass of solute between the two positions x
and $x + \Delta x$ is equal to the mass flux in minus the mass flux out:

$$\dot{Q} = Av\left[c(x, t) - c(x + \Delta x, t)\right], \qquad (2.43)$$

where Q is the mass of solute between x and $x + \Delta x$.

---

[13] The assumptions of plug flow and no radial concentration gradients are not necessary for the derivation of the one-
dimensional advection equation, but do simplify it.

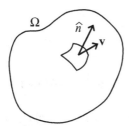

Figure 2.11

Enclosed region Ω with flow field **v** and unit normal surface vector $\hat{n}$.

If we define $\bar{c}(x, \Delta x; t)$ as the average concentration between x and $x + \Delta x$, then

$$\frac{d\bar{c}}{dt} = \frac{\dot{Q}}{A \cdot \Delta x} = v\frac{[c(x, t) - c(x + \Delta x, t)]}{\Delta x}.$$ (2.44)

In the limit $\Delta x \to 0$, $\bar{c}(x, \Delta x; t) = c(x, t)$, and Eq. (2.44) gives

$$\frac{\partial c}{\partial t} = -v\frac{\partial c}{\partial x},$$ (2.45)

which is the one-dimensional advection equation.

More generally, we can consider a surface Ω in a multidimensional flow field, as illustrated in Figure 2.11. The mass flux out of the volume $\mathcal{V}$ enclosed by the surface Ω is equal to

$$\dot{Q} = \oint_\Omega \hat{n} \cdot \mathbf{v}c \, d\Omega$$ (2.46)

where $\dot{Q}$ is the rate of change of mass in the volume and is equal to $-\frac{\partial}{\partial t}\int_\mathcal{V} c(\mathbf{x}, t)\,d\mathcal{V}$ if the velocity is measured relative to a frame in which the surface is fixed. Hence we have

$$\int_\mathcal{V} \frac{\partial}{\partial t}c\,d\mathcal{V} = -\oint_\Omega \hat{n} \cdot \mathbf{v}c\,d\Omega.$$ (2.47)

Applying Gauss's divergence theorem, we have

$$\int_\mathcal{V} \frac{\partial}{\partial t}c\,d\mathcal{V} = -\int_\mathcal{V} \nabla \cdot (c\mathbf{v})\,d\mathcal{V}.$$ (2.48)

Since the bounding surface is arbitrary,

$$\frac{\partial}{\partial t}c = -\nabla \cdot (c\mathbf{v}) = -\mathbf{v} \cdot \nabla c - c\nabla \cdot \mathbf{v}.$$ (2.49)

For incompressible fluid flow $\nabla \cdot \mathbf{v} = 0$ and Eq. (2.49) reduces to[14]

$$\frac{\partial}{\partial t} c = -\mathbf{v} \cdot \nabla c, \tag{2.50}$$

which is the conventional formulation of the advection equation. For one space dimension, Eq. (2.50) reduces to Eq. (2.45).

This derivation is easily extended to include additional processes such as chemical reactions transforming mass in the control volume. Combining advection and chemical reactions yields the general equation

$$\frac{\partial}{\partial t} c_i = -\mathbf{v} \cdot \nabla c_i + r_i(\mathbf{c}(\mathbf{x}, t), \mathbf{x}, t), \tag{2.51}$$

where c_i is the ith entry in the vector of concentration variables $\mathbf{c}(\mathbf{x}, t)$, and $r_i(\mathbf{c}(\mathbf{x}, t), \mathbf{x}, t)$ is the rate of generation of the ith reactant by chemical reactions. For example, imagine the chemical reaction system of Section 2.3.1 occurring in a flow field with velocity $\mathbf{v}(\mathbf{x}, t)$. The governing equations for this system become:

$$\frac{\partial [\text{xylose}]}{\partial t} = -\left(v_x \frac{\partial [\text{xylose}]}{\partial x} + v_y \frac{\partial [\text{xylose}]}{\partial y} + v_z \frac{\partial [\text{xylose}]}{\partial z} \right) - J_1$$

$$\frac{\partial [\text{xylitol}]}{\partial t} = -\left(v_x \frac{\partial [\text{xylitol}]}{\partial x} + v_y \frac{\partial [\text{xylitol}]}{\partial y} + v_z \frac{\partial [\text{xylitol}]}{\partial z} \right) + J_1 - J_2$$

$$\frac{\partial [\text{xylulose}]}{\partial t} = -\left(v_x \frac{\partial [\text{xylulose}]}{\partial x} + v_y \frac{\partial [\text{xylulose}]}{\partial y} + v_z \frac{\partial [\text{xylulose}]}{\partial z} \right)$$
$$+ J_2 - 2J_3 - 2J_6$$

$$\frac{\partial [\text{acetaldehyde}]}{\partial t} = -\left(v_x \frac{\partial [\text{acetaldehyde}]}{\partial x} + v_y \frac{\partial [\text{acetaldehyde}]}{\partial y} \right.$$
$$\left. + v_z \frac{\partial [\text{acetaldehyde}]}{\partial z} \right) + 3J_3 - J_4 - J_5$$

$$\frac{\partial [\text{ethanol}]}{\partial t} = -\left(v_x \frac{\partial [\text{ethanol}]}{\partial x} + v_y \frac{\partial [\text{ethanol}]}{\partial y} + v_z \frac{\partial [\text{ethanol}]}{\partial z} \right) + J_4$$

$$\frac{\partial [\text{acetate}]}{\partial t} = -\left(v_x \frac{\partial [\text{acetate}]}{\partial x} + v_y \frac{\partial [\text{acetate}]}{\partial y} + v_z \frac{\partial [\text{acetate}]}{\partial z} \right) + J_5$$

$$\frac{\partial [\text{glycerol}]}{\partial t} = -\left(v_x \frac{\partial [\text{glycerol}]}{\partial x} + v_y \frac{\partial [\text{glycerol}]}{\partial y} + v_z \frac{\partial [\text{glycerol}]}{\partial z} \right) + 3J_6, \tag{2.52}$$

where v_x, v_y, and v_z are the components of $\mathbf{v}$.

---

[14] See Exercise 2.5 at the end of this chapter.

The method of lines

The method of lines (also called the *method of characteristics*) is a scheme for solving one-dimensional (that is, with one space dimension) hyperbolic equations, such as the one-dimensional advection equation invoked in the next section. In one dimension Eq. (2.51) becomes

$$\frac{\partial}{\partial t} c_i = -v \frac{\partial c_i}{\partial x} + r_i(\mathbf{c}(x,t), x, t),$$

which can be shown to be equivalent to

$$v \frac{d\mathbf{c}}{d\zeta} = \mathbf{r}(\mathbf{c}(\zeta), \zeta, t(\zeta)), \quad \zeta_0 \leq \zeta \leq \zeta_1, \tag{2.53}$$

where $\zeta = x + vt + a$ defines trajectories in the x–t plane.[15] (Different values of the constant a define different trajectories.)

Equation (2.53) represents a convenient formulation of the problem, because it effectively reduces a partial differential equation to an ordinary differential equation. To obtain a solution at a particular point in the x–t space, the equation is integrated over the range of $\zeta = \zeta_0$ to ζ_1 defined by

$$\zeta_0 = \begin{cases} x_1 - vt_1, & x_1 \geq vt_1 \\ 0, & x_1 < vt_1 \end{cases}$$

$$\zeta_1 = \begin{cases} x_1, & x_1 \geq vt_1 \\ vt_1, & x_1 < vt_1 \end{cases}, \tag{2.54}$$

where (x_1, t_1) is the point in the x–t space associated with the value $c(\zeta_1)$. The function $t(\zeta)$ in Eq. (2.53) is defined

$$t(\zeta) = \begin{cases} (\zeta - \zeta_0)/v, & x_1 \geq vt_1 \\ t_1 - (x_1 - \zeta)/v, & x_1 < vt_1 \end{cases}.$$

(See Section 9.5.) Solutions $\mathbf{c}(\zeta_1)$ $(= \mathbf{c}(x_1, t_1))$ are obtained by integrating Eq. (2.53) from $\zeta = \zeta_0$ to ζ_1. For $x_1 \geq vt_1$ the initial condition is specified by the initial condition $\mathbf{c}(x, 0)$:

$$\mathbf{c}(\zeta_0) = \mathbf{c}(x_1 - vt_1, 0).$$

For $x_1 < vt_1$, the initial condition is specified by the boundary condition $\mathbf{c}(0, t)$:

$$\mathbf{c}(\zeta_0) = \mathbf{c}(0, t_1 - x_1/v).$$

For example, if there are no reactions, then $d\mathbf{c}/d\zeta = 0$ and we have

$$\mathbf{c}(x, t) = \mathbf{c}(x - vt, 0), \quad x \geq vt$$

$$\mathbf{c}(x, t) = \mathbf{c}(0, t - x/v), \quad x < vt \tag{2.55}$$

---

[15] For a detailed derivation see Section 9.5 of the Appendices.

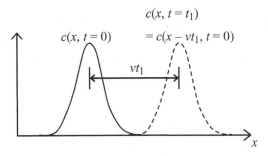

Figure 2.12

Behavior governed by $\frac{\partial c_i}{\partial t} = -v\frac{\partial c_i}{\partial x}$ (one-dimensional advection with no reaction). The solution is given by Eq. (2.55).

and solutions simply translate in space at constant velocity v, as illustrated in Figure 2.12.

Application of this methodology (with nonzero $\mathbf{r}$) is clarified with an example in the following section.

Process design in a distributed system

The efficiency of a chemical manufacturing process can be significantly enhanced using a distributed flowing bioreactor. As an idealized example, consider the chemical process described in Section 2.3.1 occurring in a long pipe with plug flow. Either the yeast (at a given biomass density) is injected with the inflowing substrates, or the cells are somehow fixed to a medium that is embedded in the pipe.

Here we assume that the biomass density and the input substrate concentrations are the same as in the previous analyses of ethanol production in Section 2.3.1. We denote the pipe length and velocity as L and v, respectively. For comparison with the results illustrated in Figure 2.9, the flow per unit volume is therefore v/L.

The governing equations for this system are

$$\frac{\partial[\text{xylose}]}{\partial t} = -v\frac{\partial[\text{xylose}]}{\partial x} - J_1$$

$$\frac{\partial[\text{xylitol}]}{\partial t} = -v\frac{\partial[\text{xylitol}]}{\partial x} + J_1 - J_2$$

$$\frac{\partial[\text{xylulose}]}{\partial t} = -v\frac{\partial[\text{xylulose}]}{\partial x} + J_2 - 2J_3 - 2J_6$$

$$\frac{\partial[\text{acetaldehyde}]}{\partial t} = -v\frac{\partial[\text{acetaldehyde}]}{\partial x} + 3J_3 - J_4 - J_5$$

$$\frac{\partial[\text{ethanol}]}{\partial t} = -v\frac{\partial[\text{ethanol}]}{\partial x} + J_4$$

$$\frac{\partial [\text{acetate}]}{\partial t} = -v \frac{\partial [\text{acetate}]}{\partial x} + J_5$$

$$\frac{\partial [\text{glycerol}]}{\partial t} = -v \frac{\partial [\text{glycerol}]}{\partial x} + 3J_6, \tag{2.56}$$

from reducing Eq. (2.52) to one-dimensional flow. Defining the characteristics $\zeta = x + vt + a$, we can transform this system to the ODE system of Eq. (9.9) and integrate using the boundary condition at the inflow that $c(0, t)$ is equal to x_o for xylose and equal to zero for all reactants. In terms of the transformed equation for $c(\zeta)$ (for $x < vt$) the boundary condition becomes the initial condition

$$c(\zeta = 0) = c(0, t - x/v)$$

and Eq. (2.53) is integrated over $\zeta = 0$ to $\zeta = vt$ to find $c(x, t)$. For $x \geq vt$ we assume that the initial concentrations in the pipe are all zero[16] and

$$c(\zeta = x - vt) = 0$$

and Eq. (9.9) is integrated over $\zeta = x - vt$ to $\zeta = x$ to obtain $c(x, t)$.

The model-predicted concentration profiles for [ethanol], given $v/L = F/V = 1/240 \text{ h}^{-1} = 0.1 \text{ day}^{-1}$, are plotted in Figure 2.13. In the steady state, the x-axis is equivalent to the time axis for the closed (no-flow) bioreactor (Figure 2.7). Here the flow per unit volume is set so that the transport time is 240 h, and therefore the reactions proceed to approximately the same level of completion as for the simulation and data illustrated in Figure 2.8. Note that this flow per unit volume is substantially higher than that necessary to obtain the same efficiency of conversion of xylose to ethanol in the well-mixed reactor. In the well-mixed reactor (see Figure 2.9) the output ethanol concentration at $F/V = 0.1 \text{ day}^{-1}$ is approximately 54 mol/kg CDW, while here the steady-state output concentration is approximately 80 mol/kg CDW. This is because in the distributed system the flowing mixture is separated according to the extent to which the reactions have proceeded. In the well-mixed reactor substrates are continuously mixed with products, undermining efficiency.

While it is clear that efficiency is improved in the distributed system compared with the well-mixed reactor, it is important to remember that the model studied here is an idealization. In any real reactor both realistic flow profiles (non-plug flow) and molecular diffusion contribute to mixing that is not considered here. The one-dimensional advection equation is a simplification that applies with accuracy that is problem-specific.

---

[16] In other words, $c(x, 0) = 0$ on $0 \leq x \leq L$.

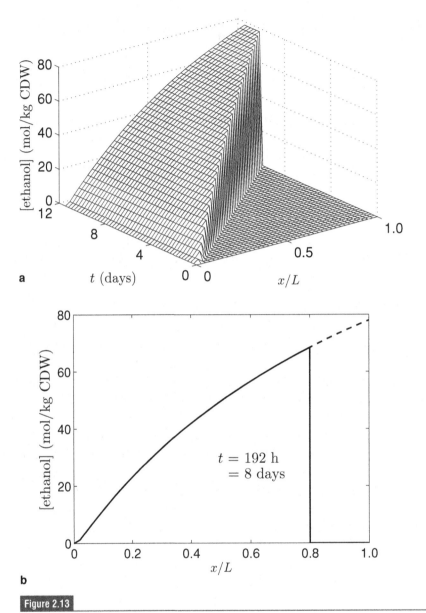

a

b

Figure 2.13

Solution to Eq. (2.56) given the initial condition of $\mathbf{c} = 0$ on $0 \leq x \leq L$, and input concentrations ($\mathbf{c}(0, t)$) equal to those used in the well-mixed bioreactor (Figure 2.9). The flow per unit volume is take to be $v/L = 1/240\ \text{h}^{-1}$, or 0.1 day^{-1}. Part (a) shows ethanol concentration profiles evolving in time. The steady state is reached at $t = L/v = 10$ days. Part (b) shows how the steady-state solution builds up as a wave traveling through the pipe at velocity v. At time $t = 0.8\,L/v$, the concentration wave has traveled 80% of the length of the pipe.

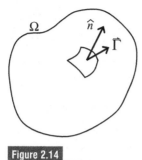

Figure 2.14

Enclosed region Ω with mass flux density $\vec{\Gamma}$ and unit normal surface vector $\hat{n}$.

2.4.2   Reaction–diffusion systems

Chemical processes that appear random at the microscopic molecular scale can result in coherent deterministic behavior when viewed at a larger scale. An example is molecular diffusion, which arises from the random thermal motion of individual molecules in a system that results in the average motion of an ensemble of molecules away from regions of relatively high density (or concentration) to neighboring regions of relatively low density. The tendency for molecules to move from regions of high concentration to neighboring regions of lower concentration is captured by Fick's law of diffusion, which states that the mass flux density $\vec{\Gamma}$ due to molecular diffusion is proportional to the negative of the concentration gradient:

$$\vec{\Gamma} = -D\nabla c, \tag{2.57}$$

where D is the molecular diffusion coefficient. Fick's law of diffusion can be used to derive a partial differential equation governing concentration fields called the *diffusion equation*.[17] In Eq. (2.57) diffusion is assumed to be isotropic (the same in all directions). More generally the molecular diffusivity takes the form of a diffusion tensor $\mathbb{D}$, and Fick's law of diffusion becomes $\vec{\Gamma} = -\mathbb{D}\nabla c$. Here we consider isotropic diffusion, where formally $\mathbb{D}$ is a diagonal matrix with equal entries D along the diagonal, resulting in Eq. (2.57).

Analogous to the derivation of the three-dimensional advection equation, let us consider a volume $\mathcal{V}$ bound in space by the surface Ω, as illustrated in Figure 2.14. If $\vec{\Gamma}$ is the mass flux density (in mass per unit time per unit area) then mass

---

[17] Both Fick's law of diffusion and the diffusion equation can be derived from a statistical thermodynamic treatment of molecular motion. The key assumptions are that the solution is dilute, and that there are no internal forces that tend to make the molecules either stick to or repel one another, or external forces such as electromagnetic or gravitational fields that influence molecular motion. In many cases it is straightforward to modify the diffusion equation to account for additional forces.

conservation implies

$$\int_V \frac{\partial}{\partial t} c \, dV = - \oint_\Omega \vec{\Gamma} \cdot \hat{n} \, d\Omega.$$

Again applying Gauss's divergence theorem:

$$\int_V \frac{\partial}{\partial t} c \, dV = - \int_V \nabla \cdot \vec{\Gamma} \, dV.$$

And again noting that the bounding surface Ω is arbitrary, we have $\partial c / \partial t = -\nabla \vec{\Gamma}$. Substituting Eq. (2.57), we have

$$\frac{\partial c}{\partial t} = \nabla \cdot D\nabla c, \tag{2.58}$$

which is the diffusion equation.

Equation (2.58) can be used to simulate how concentration fields evolve due to diffusion. In the example below we will also see how additional terms representing chemical reactions may be added to Eq. (2.58), to simulate reaction–diffusion systems.

Example: fluorescence recovery after photobleaching

Fluorescence recovery after photobleaching (FRAP for short) is a technique used to probe the the diffusion and binding interactions of molecules inside cells.[18] Essentially what happens in a FRAP experiment is that a molecule of interest inside a cell is tagged with a fluorescent marker, such as a green fluorescent protein (GFP), a commonly used probe that can be introduced into cells in a variety of ways.[19] The spatial concentration field in a cell or within a cell membrane of GFP (and whatever molecule is attached to GFP) can be measured using fluorescent confocal microscopy. In a FRAP experiment, a laser pulse is used to bleach (permanently wipe out the fluorescence of) molecules within a specified region. That is the "photobleaching" part of the experiment. After bleaching, the observed fluorescence recovers as unbleached molecules move into the bleached region. That is the "recovery" part of the experiment.

The shape of the FRAP recovery curve is influenced by the diffusivity of the tagged molecules, as well as any binding reactions that occur in the cell. To analyze data from a FRAP experiment effectively requires the capability to simulate the experiment and to test hypotheses and estimate biophysical parameters by comparing simulations with the measured data.

---

[18] FRAP is also called FPR: fluorescence photobleaching recovery [3].

[19] Isolation of a naturally occurring GFP from a jellyfish was first reported in 1962 [55]. Many related proteins, including blue, cyan, and yellow fluorescent proteins, have been engineered in the years since.

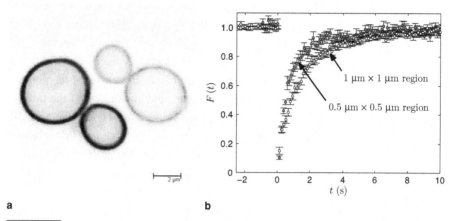

a **b**

Figure 2.15

Illustration of EGFP:Ras2 localization on yeast cell membrane and FRAP experimental data. Image in (a) shows a confocal image of EGFP:Ras2 fluorescence, with the fluorophore localized to the cells' membranes. Panel (b) shows FRAP curves (normalized fluorescence, $F(t)$) obtained for two different bleach areas, as denoted. Image (a) reproduced with permission from Vinnakota *et al.* [64].

Vinnakota *et al.* [64] performed FRAP experiments on a strain of yeast with the Ras2 protein modified with an enhanced GFP attached. (Ras 2 is an important protein involved in cellular signaling.) Like Ras2, the particular EGFP:Ras2 construct tends to localize in the cell membrane, as illustrated in Figure 2.15, which shows images obtained from a confocal plane slicing through several cells. (Note that in the image in Figure 2.15(a) the optical point-spread function exaggerates the apparent width of the cell membrane.)

The experiments of Vinnakota involved bleaching and imaging membrane-bound EGFP:Ras2. Time-course data on recovery after photobleaching from this study are plotted in Figure 2.15(b) for two different sizes of bleached area: 1 μm × 1 μm and 0.5 μm × 0.5 μm. Fluorescence intensity is plotted with background subtracted and normalized with respect to imaged whole-cell fluorescence. Notice that fluorescence recovers more rapidly where a smaller area is bleached. Data from these two different bleach areas were used by Vinnakota *et al.* to analyze the dynamics of EGFP:Ras2 trafficking and to help distinguish between two different models describing the dynamics of EGFP:Ras2 in these cells.

In the absence of any transport/reaction processes other than diffusion, the equation governing fluorophore concentration is

$$\frac{\partial c_m}{\partial t} = D_m \left(\frac{\partial^2}{\partial x^2} + \frac{\partial^2}{\partial y^2} \right) c_m, \ t > T_b, \tag{2.59}$$

which describes the two-dimensional diffusion within the membrane, with concentration in the membrane denoted c_m. Here we assume this equation applies over $t > T_b$, where photobleaching occurs over the interval $0 \leq t \leq T_b$. (An equation governing the photobleaching process is given below.) The spatial directions x and y are a coordinate set within the two-dimensional membrane. (In this analysis the membrane is treated as a plane, and any effects of the three-dimensional shape of the cell are assumed to be not important.) Here molecular diffusion is assumed to be homogeneous within the membrane: that is, D does not depend on x or y.

An alternative model accounts for EGFP:Ras2 insertion in and removal from the membrane as a passive kinetic process:

$$\frac{\partial c_m}{\partial t} = D_m \left(\frac{\partial^2}{\partial x^2} + \frac{\partial^2}{\partial y^2} \right) c_m - k_{off} c_m + k_{on} c_c, \quad t > T_b, \qquad (2.60)$$

where c_c is the concentration in the cytosol, and k_{off} and k_{on} are mass-action rate constants. Equation (2.60) assumes linear kinetics governing transport between the membrane and the cell cytosol. Note that in this two-dimensional equation c_m is measured in units of mass per unit area; c_c in units of mass per unit volume; k_{off} in units of inverse time; and k_{on} in units of length per unit time. We further assume that diffusion within the cytosol is rapid compared with timescales of diffusion within the membrane and insertion in and removal of EGFP:Ras2. Thus concentration gradients inside the cell are rapidly abolished on the timescale of our analysis and $c_c(t)$ does not depend on spatial location. With this assumption, conservation of mass yields

$$V_c c_c + A_m \langle c_m \rangle = M = \text{constant}, \qquad (2.61)$$

where V_c and A_m are the volume and surface area of the cell, respectively. Brackets $\langle \cdot \rangle$ indicate averaging over the whole cell surface. Substituting Eq. (2.61) into Eq. (2.60) yields

$$\frac{\partial c_m}{\partial t} = D_m \left(\frac{\partial^2}{\partial x^2} + \frac{\partial^2}{\partial x^2} \right) c_m - k_{off} c_m + k_{on} \left(\frac{M}{V_c} - \frac{A_m}{V_c} \langle c_m \rangle \right), \quad t > T_b, \quad (2.62)$$

an equation in which the cytoplasm concentration does not explicitly appear.

For both models the rapid bleaching process is simulated as a first-order kinetic process [64]:

$$\frac{\partial c_m}{\partial t} = D_m \left(\frac{\partial^2}{\partial x^2} + \frac{\partial^2}{\partial y^2} \right) c_m - \alpha K(x, y) c_m, \quad 0 \leq t \leq T_b, \qquad (2.63)$$

where α is the bleach rate constant, $K(x, y)$ is the normalized bleaching light distribution, and T_b is the duration of bleaching. (Since the bleaching is relatively rapid, we do not consider exchange during the bleaching interval.)

The function $K(x, y)$ is given by the convolution of the laser beam point-spread function and the bleach region window. For a square domain and a Gaussian point spread function, $K(x, y)$ is [64]:[20]

$$K(x, y) = \frac{1}{4} \left[\text{erf}\left(\frac{W + 2x}{\sqrt{2}\omega_0}\right) \text{erf}\left(\frac{W + 2y}{\sqrt{2}\omega_0}\right) \right.$$
$$+ \text{erf}\left(\frac{-W + 2x}{\sqrt{2}\omega_0}\right) \text{erf}\left(\frac{-W + 2y}{\sqrt{2}\omega_0}\right)$$
$$- \text{erf}\left(\frac{-W + 2x}{\sqrt{2}\omega_0}\right) \text{erf}\left(\frac{W + 2y}{\sqrt{2}\omega_0}\right)$$
$$\left. - \text{erf}\left(\frac{W + 2x}{\sqrt{2}\omega_0}\right) \text{erf}\left(\frac{-W + 2y}{\sqrt{2}\omega_0}\right) \right], \qquad (2.64)$$

where ω_0 is the beam waist, which is 0.248 μm in the experiments simulated here, and W is the width of the square bleached region, which is assumed to be centered at $x = 0$, $y = 0$. The function erf($\cdot$) is a special function called the *error function*, which is defined:

$$\text{erf}(x) = \frac{2}{\sqrt{\pi}} \int_0^x e^{-s^2} ds.$$

In the remainder of this section we illustrate how models assuming transport via Eqs (2.59) and (2.62) behave. Our simulations are obtained using a finite-difference approximation over a defined simulation domain. Section 9.6 of the Appendices shows how to implement this numerical approximation and obtain the results presented here.

Simulation without exchange

In our first set of simulations we consider transport without exchange, governed by Eqs (2.59) and (2.63). With the initial condition $c(x, y, t = 0) = C_0$ over the whole domain, the predicted concentration profiles immediately following bleaching ($t = T_b$) and at several other times are shown in Figure 2.16. Here the bleached window size is set at $W = 1$ μm, corresponding to the data labeled "1 μm × 1 μm region" in Figure 2.15. The bleaching time (from the Vinnakota *et al.* experiments) is $T_b = 133$ ms. At time $t = T_b$ the concentration of fluorophore in the bleached window is nearly wiped out with the simulation parameters used (see below). At $t = 1$ s (after 0.867 s of recovery), unbleached fluorophore has diffused into the bleached window, increasing the concentration. By 3 s, the bleached spot has largely recovered to the background concentration. These

---

[20] See Exercise 2.6.

$t = T_b$:

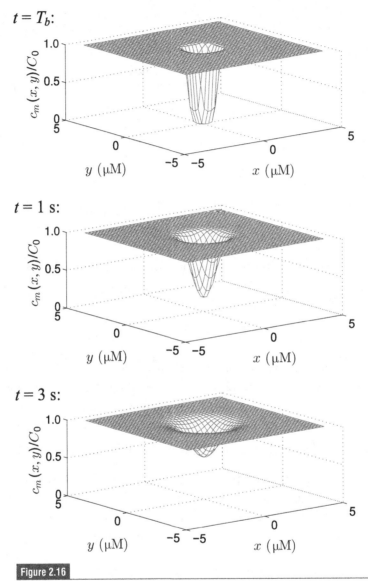

$t = 1$ s:

$t = 3$ s:

Figure 2.16

Evolution of two-dimensional concentration profiles predicted by simulating Eqs (2.59) and (2.63) on a periodic domain. Parameters are as defined in the text for the 1 μm × 1 μm bleach window.

simulations were obtained with the first-order bleaching rate constant $\alpha = 40\ \text{s}^{-1}$, and EGFP:Ras2 diffusion coefficient in the membrane $D = 0.075\ \mu\text{m}^2\,\text{s}^{-1}$. The total membrane area is set to $A_m = 78.5\ \mu\text{m}^2$ and the total simulated domain is a square region of dimensions $\sqrt{A_m} \times \sqrt{A_m}$, following Vinnakota *et al.* [64]. (The numerical algorithm and associated code given in Section 9.6 can be used to produce the concentration profiles shown here.)

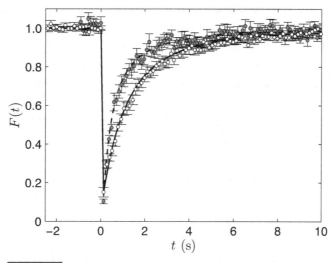

Figure 2.17

Comparison of experimental and simulated FRAP data. The data for the 1 μm × 1 μm and 0.5 μm × 0.5 μm bleach windows are the same as shown in Figure 2.15. Model simulations are shown as solid and dashed lines for the 1 μm × 1 μm and 0.5 μm × 0.5 μm windows, respectively. Simulation parameter values are given in the text.

Although it is not easy to see in the concentration profiles plotted in Figure 2.16, the mean fluorophore concentration is less than the initial concentration C_0 because of loss due to bleaching. (In a real experiment, total mass of fluorophore continuously diminishes owing to bleaching from excitation for confocal imaging.) The FRAP curves $F(t)$ are defined as mean concentration in the bleached window, normalized to mean concentration in the wider field of view. In the experiments of Vinnakota *et al.* $F(t)$ is computed

$$F(t) = \frac{\langle c_m(x, y, t) \rangle_{W \times W}}{\langle c_m(x, y, t) \rangle_{2 \times 2}}, \tag{2.65}$$

where $\langle c_m(x, y, t) \rangle_{W \times W}$ is the mean concentration in the $W \times W$ bleaching window and $\langle c_m(x, y, t) \rangle_{2 \times 2}$ is the mean concentration in a 2 μm × 2 μm imaging window.[21] As the fluorophore concentration on the membrane recovers to a uniform distribution, $F(t)$ returns to its original value of 1, as in Figure 2.15.

Simulated FRAP curves for both the 1 μm × 1 μm and 0.5 μm × 0.5 μm bleach windows are compared with experimental data in Figure 2.17. For both cases all simulation parameters are the same, with the exception of α and W (which sets the bleaching window size). For the 0.5 μm × 0.5 μm window, α is set to 50 s^{-1}, which is slightly higher than the value of 40 s^{-1} used to simulate the 1 μm × 1 μm

---

[21] Experimentally estimating normalized concentration also require the background fluorescence to be substracted from the obtained images. For details see [64].

window data. Theoretically, if the bleaching process is first-order as modeled in
Eq. (2.63), the value of α should not vary between the two different simulations.
However, there is no particular reason to believe that the kinetics of bleaching are
as simple as in Eq. (2.63). Regardless, the purpose here is to simulate the recovery
curve; Eq. (2.63) provides a reasonable means to simulate the initial bleaching for
this purpose.

Since both curves are matched by the same underlying model, with two-
dimensional diffusion as the only transport process, and with the same value of
D, we can conclude that the data are consistent with the model without exchange.
Next, we simulate transport with exchange modeled by Eq. (2.62) to determine
whether any additional information may be obtained from this experiment.

Simulation with exchange

To model combined diffusion and exchange, Eq. (2.59) is replaced by Eq. (2.62)
as the governing equation for the recovery process. Since we have shown that
the experimental data are effectively matched by the model without exchange
($k_{on} = k_{off} = 0$), we now wish to determine whether a model with significant finite
exchange might also be consistent with the data.

From independent data (see analysis in [64]) we know that the ratio of total
mass of EGFP:Ras2 in the membrane to total mass in the cytosol is approximately
1000. This equilibrium ratio determines a relationship between k_{off} and k_{on}:

$$\frac{k_{on}}{k_{off}} = 1000 \cdot \frac{V_c}{A_m}.$$

For exchange to have a significant influence over the observed timescale of $F(t)$
recovery, we choose $k_{off} = 1/10$ s^{-1}. This means that the off rate time constant
is 10 s, somewhat longer than the timescales of $F(t)$ recovery observed in the
FRAP experiments. Given the cell geometry parameter used by Vinnakota et al.
[64] ($A_m = 78.5$ μm^2, $V_c = 65.4$ μm^3), and given this value of k_{off}, we have
$k_{on} = 83.3$ μm s^{-1}.

Using these values, and simulating bleaching and recovery according to
Eqs (2.62) and (2.63) with periodic boundary conditions over the domain defined
above, we find that recovery is significantly faster (for a given value of D) than for
the model that does not account for exchange. This makes sense, because exchange
of fluorophore with a rapidly mixing compartment (the cytosol) is expected to speed
up the recovery to uniform fluorophore distribution in the membrane. By reducing
D from 0.075 to 0.060 μm^2 s^{-1} we obtain the simulated FRAP curves shown in
Figure 2.18.

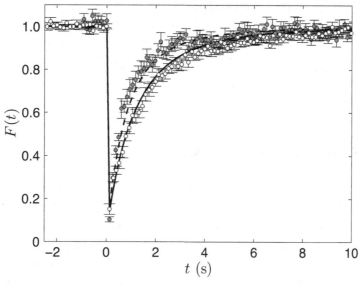

Figure 2.18

Simulated FRAP curves for model with diffusion and exchange. The data for the 1 μm × 1 μm and 0.5 μm × 0.5 μm bleach windows are the same as shown in Figures 2.15 and 2.17. Model simulations are shown as solid and dashed lines for the 1 μm × 1 μm and 0.5 μm × 0.5 μm windows, respectively. Simulation parameter values are given in the text.

These simulations capture the approximate timescale of the FRAP data. However, the exchange process tends to make the two simulated $F(t)$ curves closer together, and hence less dependent on bleach window size. In fact, the only way to match the data with this model is to set k_{off} and k_{on} to values so small that the model behavior is indistinguishable from the model without exchange. Therefore it may be concluded that the kinetics of EGFP:Ras2 exchange between the membrane and the cytosol do not detectably influence the experimental data analyzed here.

2.4.3 Advection–reaction–diffusion systems

In this chapter we have developed theories and methods for simulating systems with advection and reaction and systems with diffusion and reaction. Here, we say a few words about advection–diffusion–reaction systems.

As we have seen, Eq. (2.50) simulates transport by advection and Eq. (2.58) simulates transport by diffusion. Putting advection and diffusion together with reaction processes, we have the general advection–diffusion–reaction equation:

$$\frac{\partial c}{\partial t} = \mathbf{v} \cdot \nabla c + \nabla \cdot D \nabla c + \mathbf{r}. \tag{2.66}$$

Here the function $\mathbf{r}(c(\mathbf{x}, t), \mathbf{x}, t)$ captures any chemical reaction processes. A treatment of systems governed by this equation is beyond the scope of this chapter.

2.5 Recapitulation and lessons learned

In this chapter we have studied a number of crucial phenomena in biological systems: (1) transport by flow in well-mixed and distributed systems; (2) water and solute transport across passive porous membranes; (3) transport by diffusion; and (4) chemical reaction kinetics. Most importantly, we have put these processes together to build models of several real biological systems. We have demonstrated how to simulate at least some of these examples using computers. We have seen examples of how to estimate model parameters by matching simulations to experimental data. Hence, at this stage we already boast a nontrivial set of tools for simulating and analyzing biological systems.

We have applied these tools to analyze data from real systems, including water and solute transport across the glomeruli membrane, fermentation reactors for production of ethanol from xylose by yeast, and to probe intracellular protein transport using FRAP. We have explored engineering trade-offs in bioreactor design by developing the mathematical techniques to simulate transport and reaction phenomena. In one example – simulating volume changes in isolated glomeruli – we applied the basic equations of water and solute movement across membranes introduced by Kedem and Katchalsky to analyze experimental data. Another example – continuous flow bioreactor systems – was studied in some detail by developing the basic governing equations and methods for solving them, estimating kinetic parameters from fitting experimental data, and simulating and analyzing performance of basic bioreactor designs. In yet another example – analyzing the dynamics of GFP-tagged proteins in yeast – a mathematical model of a complex experiment was constructed, computer code to simulate the model was developed, and model simulations were used to analyze experimentally obtained data. Looking back, it is apparent that these applications drew on techniques and knowledge from biochemistry, mathematics, physics, and engineering, in a demonstration of the multidisciplinary nature of applications in biosimulation. It pays to have some competence in all of these areas, or at least not to be afraid to dive into any and all of them.

Problems

2.1 Washout experiment. Consider the special case of Eqs (2.1) and (2.3) where $F_{in} = F_{out} = $ constant. In this case V remains constant. Given some initial mass of solute M_o (such that $c(0) = M_o/V$) and an inflow concentration $c_{in} = 0$, how quickly does the solute wash out of the system? (Find the solution to Eq. (2.3) given these conditions.) How does the timescale of washout depend on the flow and volume? Does it depend on M_o?

2.2 Mixed tanks in series. Imagine two constant-volume mixed tanks connected so that the outflow of the first tank is the inflow of the second:

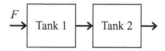

Applying Eq. (2.3) we have

$$\frac{dc_1}{dt} = \frac{F}{V_1}(c_{in} - c_1)$$

and

$$\frac{dc_2}{dt} = \frac{F}{V_2}(c_1 - c_2),$$

where V_1 and V_2 are the volumes and c_1 nd c_2 are the concentrations of a solute in tank 1 and tank 2, respectively. Assuming $c_{in}(t) = 0$ and $c_1(0) = M_o/V_1$, what is $c_1(t)$? (This is the solution to the previous exercise.) Given this solution, what is $c_2(t)$? What is the mean transit time for washout of solute from the composite system? How does it depend on V_1, V_2, and F? (The mean transit time is defined $\bar{t} = \int_0^\infty tc_2(t)dt / \int_0^\infty c_2(t)dt$.)

2.3 What is the form of $c_2(t)$ in the preceding problem for the special case where $V_1 = V_2$?

2.4 Kedem and Katchalsky formulae. Show that with $I = 0$, Eq. (2.13) becomes Eq. (2.14). Derive Eq. (2.16) and show that $\omega = \omega' - c_s(1 - \sigma)^2 L_p$. Show that in the limit $c_p = Q_p/V \to c_o$,

$$\frac{dQ_p}{dt} = -\left[\frac{Q_p/V - c_o}{\ln\left(Q_p/Vc_o\right)}\right](1 - \sigma_p)J_v - k(Q_p/V - c_o)$$

from Eq. (2.25) becomes

$$\frac{dQ_p}{dt} = -c_o(1 - \sigma_p)J_v.$$

2.5 Incompressible fluid. Following along lines similar to that in Section 2.4.1, derive an equation for the rate of change of fluid density, $\rho(\mathbf{x}, t)$. Show that for incompressible flow ($\frac{\partial \rho}{\partial t} = 0$), the divergence of the velocity is zero ($\nabla \cdot \mathbf{v}(\mathbf{x}, t) = 0$).

2.6 Optical sampling. Assume a Gaussian point spread function

$$f(x, y) = \exp\left[-2\frac{(x^2 + y^2)}{\omega_0}\right]$$

and a region-of-interest (ROI) shape function

$$B(x, y) = \begin{cases} 1, & -W/2 \leq x, y \leq +W/2 \\ 0, & \text{otherwise.} \end{cases}$$

Show that the two-dimensional convolution of $f(x, y)$ and $B(x, y)$ is $K(x, y)$ given in Eq. (2.64).

2.7 Numerical methods for diffusion. Show that Eq. (9.11),

$$(\nabla^2 c_m)_{i,j} \approx \frac{1}{h_x^2}\left[C_{i-1,j} - 2C_{i,j} + C_{i+1,j}\right] + \frac{1}{h_y^2}\left[C_{i,j-1} - 2C_{i,j} + C_{i,j+1}\right],$$

found in the Appendices is indeed a numerical approximation to

$$\nabla^2 c_m = \left(\frac{\partial^2}{\partial x^2} + \frac{\partial^2}{\partial x^2}\right) c_m$$

on the two-dimensional Cartesian grid defined in Section 9.6. (This is the numerical approximation used for the Laplace operator in Eqs (2.59), (2.62), and (2.63) to simulate the FRAP experiments in Section 2.4.2.)

2.8 Computer exercise. Show that Eq. (9.21) solves Eq. (9.20). Use this solution and the algorithm outlined in Section 9.6 to modify the code given in Section 9.6 to simulate the exchange process by Eq. (2.62). Reproduce the simulations plotted in Figure 2.18.

2.9 Open-ended computer exercise. Modify the system of Eq. (2.56) to include dispersion terms of the form $D\partial^2 c/\partial x^2$, where D is the dispersion coefficient. The resulting system may be simulated over a small time step Δt by using the method of lines to simulate the advection and reaction processes (ignoring dispersion) then applying an algorithm to simulate dispersion.

(This sort of strategy is called *Strang operator splitting* [37].) Implement an algorithm along these lines to simulate a distributed continuous-flow reactor with dispersion for the fermentation process studied in this chapter. For what values of D (given values of velocity and length, v and L) does dispersion substantially influence the behavior of the reactor?

3 Physiologically based pharmacokinetic modeling

Overview

Pharmacokinetics is the study of how substances such as drugs and other xeno-biotic compounds are transported within living organisms, particularly human beings and animals used as research models. A pharmacokinetic model for a particular drug provides the ability to simulate introduction of the drug into the body (through injection into soft tissue, intravenous administration, absorption from the gut), transport via the circulatory system and delivery to various tissues, excretion (e.g., by the kidney), and perhaps transformation via chemical reactions. Pharmacokinetic models can be useful in predicting and understanding the timescales and concentrations at which drugs appear and disappear from various regions of the body, for setting dosing guidelines for pharmaceuticals and exposure guidelines for toxic compounds, and as a basic research tool for guiding experimental design and analyzing data, as we will see in the examples in this chapter.

Physiologically based pharmacokinetic (PBPK) models are distinguished by their basis in the anatomical and physiological characteristics of the simulated organism. The structure of a PBPK model is derived directly from anatomy and physiology: transport and reactions are simulated in virtual organs that represent the individual organs of the body, and are connected according to the plumbing of the circulatory system; simulated flows mimic physiological flows of blood, gases, and possibly other fluids. Traditional pharmacokinetic models, in contrast, tend to lump flows, organs, and physiological processes together in models that invoke fewer variables and parameters in an effort to minimize the complexity necessary to simulate a given data set or type of data set. Thus PBPK models require more information (in the form of physiological function and data) to build and parameterize than simpler pharmacokinetic models. One advantage of PBPK modeling is that interpretation of simulation results is more straightforward when model components are clearly tied to real physiological entities than when they are not. An additional advantage is that physiological and pathophysiological differences that result from growth, aging, disease, etc. are implemented explicitly,

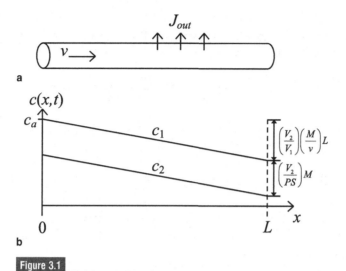

Figure 3.1

Solute flux out of a one-dimensional capillary. The model is diagrammed in panel (a). Panel (b) illustrates the solution to Eq. (3.2) for constant consumption M, given by Eq. (3.3).

allowing *predictions* from these effects to be made, as is illustrated in current PBPK applications.

3.1   Introduction to lumped compartmental PBPK modeling

3.1.1   Basic equations

In this chapter we develop and apply methods for lumped *compartmental* PBPK modeling. Models of this sort are distinguished from *spatially distributed* models that explicitly account for spatial concentration gradients at the cell, tissue, and/or organ levels. To gain an appreciation of when and where the lumped compartmental approximation is reasonable (not to mention what exactly is the approximation), let us consider the spatially dependent loss of a consumed substrate from blood flowing through a capillary. In a capillary, transport is governed by the advection equation (neglecting molecular diffusion):

$$\frac{\partial c_1(x, t)}{\partial t} = -v \frac{\partial c_1(x, t)}{\partial x} - \frac{J_{out}(x, t)}{V_1}, \tag{3.1}$$

where v is the blood velocity, J_{out} is the mass flux of solute out of the capillary, and V_1 is the volume of the capillary. Here we are treating the capillary as a one-dimensional system where the space variable x represents position along the major axis of the vessel, as illustrated in Figure 3.1.

Denoting concentration outside the capillary as $c_2(x, t)$, and assuming that solute flux is proportional to concentration difference (i.e., assuming simple linear permeability), we have $J_{out} = PS(c_1 - c_2)$ and

$$\frac{\partial c_1}{\partial t} = -v \frac{\partial c_1}{\partial x} - \frac{PS}{V_1}(c_1 - c_2)$$

$$\frac{\partial c_2}{\partial t} = +\frac{PS}{V_2}(c_1 - c_2) - M. \qquad (3.2)$$

Here we have assumed no advection (no flow-driven transport) in the extra-capillary space. The parameter PS represents the product of the effective permeability and surface area of the barrier separating the two regions, capillary and extra-capillary. The ratio V_1/V_2 is the ratio of the volumes of the two regions, and M is the rate of consumption of the solute. If M is constant, Eq. (3.2) has the steady-state solution

$$c_1(x) = c_A - \frac{V_2 M}{V_1 v} x$$

$$c_2(x) = c_A - \frac{V_2 M}{V_1 v} x - \frac{V_2}{PS} M, \qquad (3.3)$$

where $c_A = c_1(0)$ is the concentration of solute at the input (arterial) end of the capillary. These steady-state concentration profiles assuming a capillary length of L ($0 \leq x \leq L$) are plotted in Figure 3.1(b). (This model and its predictions make sense only when $\left(\frac{V_2 M}{V_1 v}\right) L \leq c_A$ and thus the predicted concentrations remain positive.)

A lumped compartmental model can be obtained from the distributed model by seeking equations for mean concentration in each region. Defining $\bar{c}(t) = \frac{1}{L} \int_0^L c(x, t) dx$, we have from Eq. (3.2)

$$\frac{d\bar{c}_1}{dt} = -\frac{v}{L} [c_1(L, t) - c_1(0, t)] - \frac{PS}{V_1}(\bar{c}_1 - \bar{c}_2)$$

$$= \frac{F}{V_1}(c_A - c_V) - \frac{PS}{V_1}(\bar{c}_1 - \bar{c}_2)$$

$$\frac{d\bar{c}_2}{dt} = +\frac{PS}{V_2}(\bar{c}_1 - \bar{c}_2) - M. \qquad (3.4)$$

Here we have defined $F/V_1 = v/L$, which is the flow per unit volume into the capillary space, and $c_V = c_1(L, t)$, which is the outflow concentration at the venous end of the capillary. Equation (3.4) governs the simple two-compartment model illustrated in Figure 3.2.

Equation (3.4) is a system of ordinary differential equations that does not include space as an independent variable. Thus, given the assumptions invoked in Eq. (3.2),

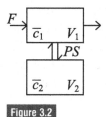

Two-compartment model derived from the distributed model of Figure 3.1 and Eq. (3.2). This model is governed by the ODE system of Eq. (3.4).

the average regional concentrations in a spatially distributed model may be represented by a system of ordinary differential equations of the form of Eq. (3.2). The key assumption necessary in obtaining (3.4) from (3.2) is that the equations are linear in the concentration variables. The consumption term may be zero order (does not depend on concentration), as is assumed here, or first order (depends linearly on concentration). With these assumptions this derivation can easily be generalized to more complex geometries, and to account for more distinct regions in the tissue.

We will see below that practical application of Eq. (3.4) typically requires the introduction of additional assumptions that may technically contradict the assumptions used to derive the equations. For example, while in some cases it may make sense to specify $c_A(t)$ in Eq. (3.4) as a model input, the output venous concentration $c_V(t)$ is governed by the regional transport processes meant to be captured by the model. Therefore in the lumped compartmental model $c_V(t)$ needs to be expressed as some function of the other concentration variables in the model. The output venous concentration is typically taken as the mean capillary concentration $\bar{c}_1(t)$: an approximation that is not consistent with the spatial averaging used to obtain Eq. (3.4), and therefore not expected to be generally accurate.

Furthermore, while linear distributed systems may be effectively reduced to lumped compartmental systems in many circumstances, it is an unavoidable fact that few biological process are accurately modeled as linear phenomena. For the lumped compartmental approach to apply reasonably and generally to nonlinear transport phenomena, a different criterion must be met. Namely, if the transport processes at play do not result in important spatial concentration gradients – if the solutes of interest obtain essentially constant concentrations in the regions of interest – then it is not necessary to account explicitly for the continuous spatial distribution of the solutes. Returning to the linear model of Figure 3.1, if the *fractional extraction* (the fraction of input solute that is extracted from the blood as it passes from the input to the output of the capillary) is much less than 1, then the spatial concentration profile will be approximately constant. In this case equations such as (3.4) arise directly from applying mass conservation principles in the context of well-mixed compartments. For example, the term $\frac{F}{V_1}(c_A - c_V)$ is analogous to the right-hand side of the last equation in (2.3), which accounts for flow of solute in and out of a constant-volume container. The other terms simulate passive permeation between the two compartments in this model.

3.1.2   Comparison of distributed and lumped transport models

To gain an appreciation for when and how a lumped compartmental model approximates a spatially distributed system, let us compare a system governed by Eq. (3.2) with one governed by Eq. (3.4). To make the analysis concrete, let us set model volume and flow parameters as follows: $V_1 = 0.05\,\mathrm{l}$, $V_2 = 0.95\,\mathrm{l}$, and $F = 1\,\mathrm{l\,min^{-1}}$.

To simulate the lumped model we use $c_V = \bar{c}_1$ in Eq. (3.4), yielding

$$\frac{d\bar{c}_1}{dt} = \frac{F}{V_1}\left[c_A(t) - \bar{c}_1\right] - \frac{PS}{V_1}(\bar{c}_1 - \bar{c}_2)$$

$$\frac{d\bar{c}_2}{dt} = +\frac{PS}{V_2}(\bar{c}_1 - \bar{c}_2) - M. \tag{3.5}$$

Thus we have introduced the approximation introduced above that $c_V(t) = c_1(L, t) \approx \bar{c}_1(t)$, a simplification that allows us to express the lumped compartmental model as a self-contained system of equations. This approximation may become reasonable in a number of limiting cases. First, if the flux of solute across the capillary wall is low enough that the solute concentration remains constant in an element of blood as it flows from one end of the capillary to the other, and the input concentration remains approximately constant, then an approximately constant concentration in the capillary will be established within the time it takes for blood to flow from one end of the capillary to the other. This capillary transit time is equal to V_1/F, which is equal to 3 s for the parameter values introduced above. Thus if consumption is insignificant ($M = 0$), the approximation $c_V(t) = \bar{c}_1$ will be valid when $c_A(t)$ remains approximately constant over timescales of V_1/F and permeation is low ($PS/F \ll 1$). For higher permeability values, it is necessary that the timescale of arterial concentration changes be compared with the tissue transit time, $(V_1 + V_2)/F$, which is the mean time it takes a permeant solute molecule to travel from the arterial input to venous output of the capillary and is equal to 60 s for the parameter values used here. Thus, again if $M = 0$, we expect the lumped approximation to be valid when $c_A(t)$ remains constant on timescales greater than $(V_1 + V_2)/F$, regardless of the value of PS.

To explore the validity of the lumped approximation and compare the predictions of the lumped versus compartmental models, let us assume that the initial solute concentrations are zero ($c_1(x, 0) = c_2(x, 0) = 0$) and that the solute concentration builds up in the blood with a time constant k: $c_A(t) = c_o(1 - e^{-kt})$, where c_o is set to the arbitrary value of 1 mM. To highlight a limiting case where differences between the predictions of the distributed and lumped models are apparent, we first consider model predictions where the arterial concentration changes on the timescale $1/k = 60$ s. Thus $1/k$ is significantly greater than the capillary flow transit time, but equal to the tissue mean transit time.

Model predictions for this case, with no consumption ($M = 0$), are plotted in Figure 3.3 for two different values of PS. Solid lines correspond to predictions of the distributed model and dashed lines to those of the compartmental model. We can see that the compartmental model predicts lower concentrations than predicted by the distributed model in both compartments over this simulation time window. This difference is explained by the fact that in the distributed model there is a delay associated with solute transport from the arterial to the venous end of the capillary. The lumped model assumes that the compartments are instantaneously mixed and, as a result, the solute outflux in the early part of the simulation is enhanced compared with the distributed model. In other words, for this linear case, the difference between the distributed- and lumped-based predictions is due to the approximation $c_V = \bar{c}_1$ in the lumped model. As expected, the lumped approximation is better for the lower-permeability case (Figure 3.3(b)), for which the approximation $c_V = \bar{c}_1$ is better than for the higher-permeability case. The approximation becomes increasingly better as PS/F is set to lower values. In fact it is somewhat surprising that the lumped approximation is really not bad with $PS/F = 1/2$, which is a far cry from $PS/F \ll 1$. Although here we are not conducting a detailed error analysis, it is safe to conclude that, with arterial concentrations changing on the timescale of a minute or longer (and with the volume and flow parameters set as defined above), the lumped compartmental model reasonably approximates the mean concentrations predicted by the distributed model (error less than 10% or so) when $PS/F < 1/2$.

In general, the lumped approximation tends to become progressively better as the timescales of the underlying transport phenomena are such that the regions in the distributed model become more rapidly mixed compared with the rate at which the input arterial solute concentration is changing. In typical PBPK applications, solute concentrations in the blood change on the order of minutes to hours while transport times in individual organs and tissues are on the order of minutes or less. Therefore the lumped compartmental approximation may be valid as long as the effective consumption or productions rates are relatively low (compared with tissue transport rates).

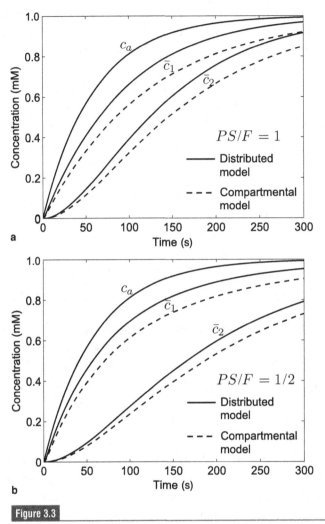

Figure 3.3

Comparison of distributed and lumped compartmental model predictions for blood-tissue solute transport. Models are simulated according to Eq. (3.2) – distributed model – and Eq. (3.5) – compartmental model. Plotted are the predicted mean concentrations in the capillary space, $\bar{c}_1(t)$, and in the extra-capillary space, $\bar{c}_2(t)$. The initial conditions are set $c_1(x, 0) = c_2(x, 0) = \bar{c}_1(0) = \bar{c}_2(0) = 0$. Volumes and flow are set to $V_1 = 0.05$ l, $V_2 = 0.95$ l, and $F = 1$ l min^{-1}, and the PS is set to (a) F and (b) $F/2$. The input arterial concentration is assumed to increase according to $c_A(t) = (1 - e^{-kt})$, where $1/k = 60$ s.

To summarize, the questions of whether and when lumped compartmental approximations may be applied to simulate distributed transport processes are questions of relative timescales. There exist two major requirements for lumped compartmental modeling in physiological transport modeling. First, the concentrations in blood change slowly compared with organ/tissue transport timescales.

Second, mass production/consumption of solutes of interest is slow compared with the rate of advective mass flux into and out of an organ or tissue. A classic example of a nonlinear phenomenon that does not fit into the lumped framework is oxygen transport, because advective oxygen transport to most tissues is nearly matched by metabolic consumption rates, violating the second requirement. Transport of many drugs (which are consumed and cleared from the blood relatively slowly), on the other hand, satisfies both of these requirements quite well.

3.1.3 Quasi-steady model reduction

Often in PBPK modeling applications, multi-compartment models are further simplified to single-compartment models by application of additional approximations. When appropriate, such approximations conveniently reduce the number of state variables in a model, reducing computational complexity and expense. To understand when and how multi-compartmental models may be simplified to an effective single compartment, let us consider the linear two-compartmental system described by the following system of equations:

$$\frac{dc_1}{dt} = \frac{F}{V_1}(c_A - c_1) - \frac{PS}{V_1}(c_1 - c_2/\lambda)$$
$$\frac{dc_2}{dt} = +\frac{PS}{V_2}(c_1 - c_2/\lambda). \tag{3.6}$$

Here we have dropped the overbar notation on c_1 and c_2 and we have introduced the parameter λ, which is called the *partition coefficient* and is a measure of the different apparent solubilities of the solute in the two compartments.

To analyze this system, we may recast the equations in terms of scaled unitless time and concentration variables: $\tau = tF/(\lambda V_2)$, $s_1 = c_1/c_o$, $s_2 = c_2/c_o$, $s_a = c_A/c_o$. The constant c_o is a reference concentration value that turns out to be arbitrary in this analysis. In terms of unitless variables, Eq. (3.6) becomes

$$\epsilon \frac{ds_1}{d\tau} = s_a - s_1 - \frac{PS}{F}(s_1 - s_2/\lambda)$$
$$\frac{ds_2}{d\tau} = +\frac{PS}{F}(\lambda s_1 - s_2), \tag{3.7}$$

where $\epsilon = V_1/(\lambda V_2)$, which is a small parameter when $\lambda V_2 \gg V_1$. In the limit $\epsilon \to 0$, the first equation above becomes *singular*, which means that the leading-order derivative becomes unbounded when the right-hand side is finite. Taking

$\epsilon = 0$, we can approximate the behavior of Eq. (3.7) by solving for s_1

$$s_1 = \frac{s_a + \frac{PS}{F}\frac{s_2}{\lambda}}{1 + PS/F}$$

and substituting into the equation for s_2

$$\frac{ds_2}{d\tau} = \frac{PS/F}{1 + PS/F}\left(\lambda s_a - s_2\right),$$

or in terms of the original variables

$$c_1 = \frac{c_A + \frac{PS}{F}\frac{c_2}{\lambda}}{1 + PS/F}$$

$$\frac{dc_2}{dt} = \frac{F}{V_2}\left(\frac{PS/F}{1 + PS/F}\right)(c_A - c_2/\lambda). \tag{3.8}$$

Essentially, these equations are consistent with the assumption that the equation for dc_1/dt in (3.6) attains a quasi-steady state ($dc_1/dt = 0$) on some timescale. On this timescale c_1 is computed as an explicit function of c_2. Thus solutions to Eq. (3.8) cannot satisfy arbitrary initial conditions for both c_1 and c_2. Mathematically this is because, in taking $\epsilon = 0$, we have reduced the governing equations from a second-order system to a first-order system. Because ϵ is finite, the quasi-steady approximation for the dc_1/dt equation must break down on some relatively rapid timescale, which is the time it takes for a quasi-steady state to be achieved. We can see this by defining a rapid timescale $\theta = Ft/V_1$, which yields

$$\frac{ds_1}{d\theta} = s_a - s_1 - \frac{PS}{F}\left(s_1 - s_2/\lambda\right)$$

$$\frac{ds_2}{d\theta} = +\epsilon\frac{PS}{F}\left(\lambda s_1 - s_2\right). \tag{3.9}$$

From this equation, we see that when $\epsilon PS/F \ll 1$, s_2 remains approximately constant over the timescale defined by $t \lesssim V_1/F$ as long as $\lambda = \mathcal{O}(1)$. With $s_2 =$ constant, the equation for s_1 has the solution

$$s_1(\theta) = \frac{s_2}{\lambda}\left(\frac{PS/F}{1 + PS/F}\right)\left(1 - e^{-\beta\theta}\right) + e^{-\beta\theta}\int_0^\theta e^{+\beta u}s_a(u)du + s_1(0)e^{-\beta\theta},$$

$$\tag{3.10}$$

where $\beta = (1 + PS/F)$. In terms of the original variables,

$$c_1(t) = \frac{c_2}{\lambda}\left(\frac{PS/F}{1 + PS/F}\right)\left(1 - e^{-\alpha t}\right) + e^{-\alpha t}\int_0^\theta e^{+\alpha t}c_A(u)du + c_1(0)e^{-\alpha t},$$

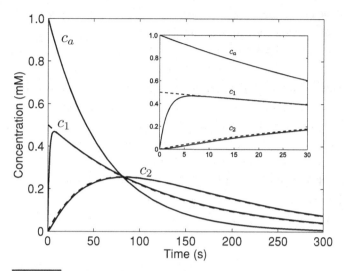

Figure 3.4

Comparison of two-compartment and reduced one-compartment models for blood/tissue solute transport. Solid lines correspond to solutions of the full two-compartment system of Eq. (3.6); dashed lines correspond to the reduced one-compartment model of Eq. (3.8). See text for parameter values.

where $\alpha = F(1 + PS/F)/V_1$. If $c_2(t)$ is constant, then for $t \gg 1/\alpha$, c_1 obtains the value

$$c_1(t \gg 1/\alpha) = \frac{c_A + \frac{PS}{F\lambda}s_2}{1 + PS/F},$$

which is the quasi-steady state solution for the long-timescale solution of Eq. (3.8)!

The behaviors of the full second-order model and the reduced first-order model of Eq. (3.8) are compared over these short and long timescales in Figure 3.4 to understand how the reduced model performs. As in the previous section, we choose physiologically reasonable parameter values $V_1 = 0.05$ l, $V_2 = 0.95$ l, $F = 1$ l min^{-1}, $PS = F$, and $\lambda = 1$, consider the problem with initial conditions $c_1(0) = c_2(0) = 0$, and assume $c_A(t)$ varies over some prescribed timescale. For these parameter values $\epsilon = 0.05$, and the timescale of relaxation to the quasi-steady state is $1/\alpha = 1.5$ s.

Note that this figure compares the full solution to the two-state variable model (Eq. (3.6)) with the long-timescale one-state variable model of Eq. (3.8). This slow-timescale model is the basis for quasi-steady-state model reduction used in PBPK modeling described in Section 3.3. We will see that a simplified version of Eq. (3.8) in the limit $PS/F \to \infty$ is sometimes used for what is termed *flow-limited* transport. In Section 3.4.3 we will see how the flow-limited transport equations arise and how they may be applied.

In Figure 3.4, we set $c_A(t) = c_o e^{-kt}$ with $1/k = 60$ s, which allows us to observe the two different timescales. Initially there is an instantaneous perturbation to the system, and we can see (in the inset) that the full model solution (solid lines) to c_1 approaches the quasi-steady solution (dashed lines) on the timescale of $1/\alpha = 1.5$ s. Once the quasi-steady state is achieved after a few seconds, the full model and the reduced model follow nearly identical kinetics, because the time constant of the $c_A(t)$ forcing function $(1/k)$ is much bigger than the reduction-to-quasi-steady time constant $(1/\alpha)$. In many PBPK applications we are not concerned with timescales as rapid as seconds, and the reduced quasi-steady approximations are applied with reasonable confidence.

3.1.4   Lumped versus distributed transport modeling: a final word

As we have seen, not all physiological transport phenomena can be cast in terms of lumped compartmental modeling. We have demonstrated that this is true even when considering only linear phenomena. To reemphasize the example of oxygen transport (an irreducibly distributed process), oxygen is highly extracted from the blood in the systemic circulation, and thus there are significant concentration gradients in most tissues. Owing to nonlinear binding to proteins in the blood and tissues, and to potentially nonlinear kinetics of consumption, the transport of oxygen in tissue is a nonlinear phenomenon. Therefore the lumped compartmental approach generally cannot be applied to simulate physiological oxygen transport; but it may be applied to transport of many endogenous substances, including many metabolic substrates and drugs.

However, whether justified by homogenization of linear equations or by the fact that spatial concentration gradients are relatively small, lumped compartmental models that take the form of ordinary differential equations with time as the independent variable are the basis of the majority of applications in pharmacokinetic modeling. In the examples of PBPK modeling in this chapter, individual organs are modeled as multiple-compartment systems, analogous to Eq. (3.5). We will consider transport phenomena that operate on timescales of many minutes to hours, and the lumped approximations will be applied effectively.

Yet before getting into the details of whole-body PBPK simulation, a review of the mammalian circulatory system – the basic plumbing that lies at the heart of PBPK modeling – is in order.

3.2   Overview of the mammalian circulatory system

In the introduction to Chapter 6 of his classic text *Physiology and Biophysics of the Circulation* [14] Burton explains:

> A sensible man, if he were appointed to be a member of a river conservation authority, would make it his first task to become thoroughly acquainted with the whole of the river bed, the sources of the rivulets, their successive confluences, the depth and width of the contributing streams, the volume of water, the volume flow and the fall in height in all the branches and in the final river. So it should be with the student of the circulation. The first requirement is thorough familiarity with the vascular bed.

The circulatory system, in which blood travels through the various tissues of the body, is shown schematically in Figure 3.5. The major pump is the heart, which in mammals is a four-chambered organ. Blood is pumped out of the heart into arteries and returns in vessels called veins. The arteries form a treelike structure that branches successively into smaller vessels leading into the *microcirculatory vessels* – blood vessels as small as a few micrometers in diameter that are the primary sites of exchange of solutes between the blood and surrounding tissues. Blood traveling through microcirculatory vessels flows into *venules* (small veins), which feed into progressively larger veins that connect the circulation back to the heart.[1]

The two chambers on the right side of the heart collect blood returning from the *systemic circulation* – the circulation to all tissues of the body excluding the lungs – and pump it through the pulmonary arteries into the lungs. The blood returning to the right side of the heart from the systemic circulation is largely deoxygenated, having supplied the tissues of the body with oxygen. In the lungs it is reoxygenated before returning to the heart in the pulmonary vein. The pulmonary vein feeds into the left side of the heart, which pumps the oxygenated blood out through the aorta, the major vessel that feeds the arteries of the systemic circulation. Oxygen and other solutes are delivered to the tissues, and waste products (including carbon dioxide) are delivered to the blood in the microcirculation, which feeds into the systemic veins, leading back to the right side of the heart.

3.3   Whole-body PBPK simulation

Figure 3.6 shows a block diagram of a whole-body PBPK model that is constructed based on the circulatory plumbing outlined above. Each block in

---

[1] In the early 1600s William Harvey, invoking the principle of conservation of mass, demonstrated that arterial blood and venous blood are connected through a continuous circulation. A century later Marcello Malpighi saw the smallest microvessels under a microscope, verifying Harvey's claim that there existed a continuous circulation connecting arteries and veins.

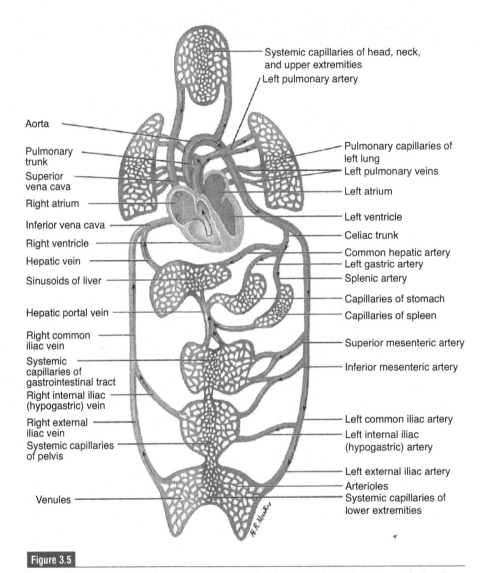

Aorta

Pulmonary trunk

Superior vena cava

Right atrium

Inferior vena cava

Right ventricle

Hepatic vein

Sinusoids of liver

Hepatic portal vein

Right common iliac vein

Systemic capillaries of gastrointestinal tract

Right internal iliac (hypogastric) vein

Right external iliac vein

Systemic capillaries of pelvis

Venules

Systemic capillaries of head, neck, and upper extremities

Left pulmonary artery

Pulmonary capillaries of left lung

Left pulmonary veins

Left atrium

Left ventricle

Celiac trunk

Common hepatic artery

Left gastric artery

Splenic artery

Capillaries of stomach

Capillaries of spleen

Superior mesenteric artery

Inferior mesenteric artery

Left common iliac artery

Left internal iliac (hypogastric) artery

Left external iliac artery

Arterioles

Systemic capillaries of lower extremities

Figure 3.5

Diagram of the mammalian circulatory system. Reprinted with permission from HarperCollins. from Tortora and Grabowski [61].

this diagram represents an organ that we will treat as a lumped compartmental system. A lumped compartmental whole-body model takes the form of a system of ordinary differential equations governing solute transport in the compartments in this diagram.

Here we apply the reduced-compartment formalism of Section 3.1.3, where in a given compartment the tissue concentration is simulated according to the equation for dc_2/dt in Eq. (3.8) and the solute concentration in the blood is given by the

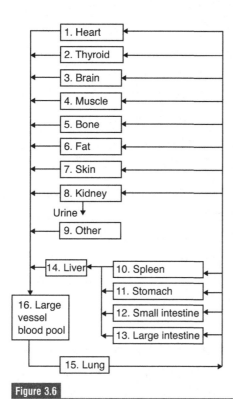

Figure 3.6

Block diagram of the mammalian circulatory system.

expression for $c_1(t)$. The variable $c_1(t)$ for a given compartment will represent the solute concentration in the microcirculatory blood in the organ/tissue that the compartment represents. The overall model of Figure 3.6 lumps all large-vessel blood into a single compartment that includes arteries, veins, and the four chambers of the heart. (The "heart" compartment represents the myocardium: the muscle tissue of the heart.)

To connect the individual compartments into a whole-body model, the output concentrations of certain compartments give the input concentrations of others. For example (as will be detailed below), the flow-weighted sum of the outputs of the heart, thyroid, brain, and other compartments of the systemic circulation provide the input into the large vessel blood pool. Therefore, to apply Eq. (3.8) to simulate the individual compartments, we need to determine an appropriate expression for the output blood concentration in the reduced compartment formalism invoked. Assuming that the output concentration is given by $c_1(t)$ would result in a model that does not conserve mass. To build a mass-conserving model, let us consider a generic reduced compartment with input and output concentration denoted c_{in} and c_{out}, respectively, as illustrated in Figure 3.7.

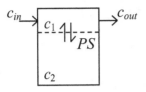

Figure 3.7

Mass balance in a reduced compartmental model with permeability limitation. The variables c_1 and c_2 are governed by Eq. (3.8). In order to balance mass, c_{out} is computed according to Eq. (3.11).

If F is the flow into the compartment, then the dynamic mass balance in the reduced compartment is given by

$$F(c_{in} - c_{out}) = V_1 \frac{dc_1}{dt} + V_2 \frac{dc_2}{dt}. \tag{3.11}$$

Taking c_1 from Eq. (3.8):

$$c_1 = \frac{c_{in} + \frac{PS}{F} \frac{c_2}{\lambda}}{1 + PS/F}$$

and solving for c_{out}, we have

$$c_{out} = c_{in} - \frac{1}{F} \left[\frac{V_1}{1 + PS/F} \frac{dc_{in}}{dt} + \left(\frac{V_1 PS/F}{(1 + PS/F)\lambda} + V_2 \right) \frac{dc_2}{dt} \right]. \tag{3.12}$$

This approach fails when Eq. (3.12) predicts that c_{out} is negative. (In practice this is an issue when the rate of change of the input concentration is rapid, such as during an injection of solute into the arterial blood.) To ensure that negative values for c_{out} are not generated, the governing equations for the compartment illustrated in Figure 3.7 are modified for cases when the dynamic mass balance of Eq. (3.12) generates negative values for c_2:

$$c_1(t) = \frac{c_{in} + \frac{PS}{F} \frac{c_2}{\lambda}}{1 + PS/F}$$

$$c_{out}(t) = \begin{cases} g, & g \geq 0 \\ 0, & g < 0 \end{cases}$$

$$g = c_{in} - \frac{1}{F} \left[\frac{V_1}{1 + PS/F} \frac{dc_{in}}{dt} + \left(\frac{V_1 PS/F}{(1 + PS/F)\lambda} + V_2 \right) \right.$$
$$\left. \times \frac{F}{V_2} \left(\frac{PS/F}{1 + PS/F} \right) (c_{in} - c_2/\lambda) \right]$$

$$\frac{dc_2(t)}{dt} = \begin{cases} \frac{F}{V_2} \left(\frac{PS/F}{1+PS/F} \right) (c_{in} - c_2/\lambda), & c_{out} > 0 \\ \frac{(1+PS/F)Fc_{in} - V_1 dc_{in}/dt}{V_1 PS/(\lambda F) + (1+PS/F)V_2}, & c_{out} = 0 \end{cases} . \tag{3.13}$$

Table 3.1: Basic equations for PBPK modeling.

Permeability-limited	Flow-limited
$g = c_{in} - \frac{1}{F}\left\{ \frac{V_1}{1+PS/F}\frac{dc_{in}}{dt} + \left[\frac{V_1 PS/F}{(1+PS/F)\lambda} + V_2 \right] \right.$ $\left. \times \frac{F}{V_2}\left(\frac{PS/F}{1+PS/F}\right)(c_{in} - c_2/\lambda) \right\}$	$g = \left(1 + \frac{V_1^{(x)}}{\lambda V_2^{(x)}}\right)\frac{c_2}{\lambda} - c_{in}^{(x)}\frac{V_1^{(x)}}{\lambda^{(x)}V_2^{(x)}}$
$g < 0:$	
$c_1(t) = \frac{c_{in} + \frac{PS}{F}\frac{c_2}{\lambda}}{1+PS/F}$	$c_1^{(x)}(t) = c_2/\lambda$
$c_{out}(t) = 0$	$c_{out}(t) = 0$
$\frac{dc_2(t)}{dt} = \frac{(1+PS/F)Fc_{in} - V_1 dc_{in}/dt}{V_1 PS/(\lambda F) + (1+PS/F)V_2}$	$\frac{dc_2(t)}{dt} = \frac{Fc_{in}}{V_1/\lambda + V_2}$
$g \geq 0:$	
$c_1(t) = \frac{c_A + \frac{PS}{F}\frac{c_2}{\lambda}}{1+PS/F}$	$c_1(t) = c_2^{(x)}/\lambda$
$c_{out}(t) = g$	$c_{out}(t) = 0$
$\frac{dc_2(t)}{dt} = \frac{F}{V_2}\left(\frac{PS/F}{1+PS/F}\right)(c_A - c_2/\lambda)$	$\frac{dc_2(t)}{dt} = \frac{F}{V_2}(c_{in} - c_2/\lambda)$

Equation (3.13) is an algorithm for computing the kinetics for the two-compartmental model of Figure 3.7. The variable g is the output concentration c_{out} that would be obtained via Eq. (3.12). As long as this value is positive then the rate of change of the state variable (dc_2/dt) is computed according to Eq. (3.8). If Eq. (3.12) predicts a negative value for output concentration, c_{out} is set to zero and the mass balance of Eq. (3.11) is used[2] to obtain an equation for dc_2/dt that is consistent with $c_{out} = 0$.

These equations are summarized in Table 3.1. (The heading "Permeability-limited" refers to the current case, where the permeability is assumed to be finite.) These are the basic equations for PBPK transport modeling when two-compartment systems may be reduced to one state variable using the quasi-steady model reduction of Section 3.1.3. When additional compartments and/or additional processes, such as chemical reactions, are considered, then the equations must be further modified. Since this quasi-steady approach to PBPK modeling was introduced only recently [60], its potential is yet to be fully investigated. In the remainder of this chapter we explore how the approach may be applied.

Denoting the solute concentration in the aorta as $c_A(t)$, we apply these equations to 13 compartments of the systemic circulation (representing 1. heart, 2. thyroid, 3. brain, 4. muscle, 5. bone, 6. fat, 7. skin, 8. kidney, 9. other, 10. spleen, 11. stomach,

---

[2] See Exercise 3.4.

12. small intestine, 13. large intestine) as follows:

$$c_1^{(x)}(t) = \frac{c_A + \frac{PS^{(x)}}{F^{(x)}} \frac{c_2^{(x)}}{\lambda^{(x)}}}{1 + PS^{(x)}/F^{(x)}}$$

$$c_{out}^{(x)}(t) = \begin{cases} g^{(x)}, & g^{(x)} \geq 0 \\ 0, & g^{(x)} < 0 \end{cases}$$

$$g^{(x)} = c_A - \frac{1}{F^{(x)}} \left[\frac{V_1^{(x)}}{1 + PS^{(x)}/F^{(x)}} \frac{dc_A}{dt} + \left(\frac{V_1^{(x)} PS^{(x)}/F^{(x)}}{(1 + PS^{(x)}/F^{(x)})\lambda^{(x)}} + V_2^{(x)} \right) \right.$$

$$\left. \times \frac{F^{(x)}}{V_2^{(x)}} \left(\frac{PS^{(x)}/F^{(x)}}{1 + PS^{(x)}/F^{(x)}} \right) \left(c_A - c_2^{(x)}/\lambda^{(x)} \right) \right]$$

$$\frac{dc_2^{(x)}(t)}{dt} = \begin{cases} \frac{F^{(x)}}{V_2^{(x)}} \left(\frac{PS^{(x)}/F^{(x)}}{1 + PS^{(x)}/F^{(x)}} \right) \left(c_A - c_2^{(x)}/\lambda \right), & c_{out}^{(x)} > 0 \\ \frac{(1 + PS^{(x)}/F^{(x)})F^{(x)}c_A - V_1^{(x)} dc_A/dt}{V_1^{(x)} PS^{(x)}/(\lambda F^{(x)}) + (1 + PS^{(x)}/F^{(x)})V_2^{(x)}}, & c_{out}^{(x)} = 0 \end{cases} . \tag{3.14}$$

Here we have introduced the superscript notation to specify the index number of a compartment. Thus in this set of equations "(x)" indexes the 13 compartments named above, according to the numbering defined in Figure 3.6. (This set of 13 compartments does not include the liver (#14 in Figure 3.6). Equations for the liver are developed below.)

The solute concentration in the aorta, which is simulated as the output concentration of the lung, is simulated from an explicit (nonreduced) two-compartment model of the lung:

$$\frac{dc_1^{(15)}}{dt} = \frac{dc_A}{dt} = \frac{F_o}{V_1^{(15)}} (c_{PA} - c_A) - \frac{PS^{(15)}}{V_1^{(15)}} \left(c_A - c_2^{(15)}/\lambda^{(15)} \right)$$

$$\frac{dc_2^{(15)}}{dt} = +\frac{PS^{(15)}}{V_2^{(15)}} \left(c_A - c_2^{(15)}/\lambda^{(15)} \right), \tag{3.15}$$

where $c_{PA}(t)$, the pulmonary arterial solute concentration, is the output of the large vessel pool, which is simulated as a single well-mixed tank:

$$\frac{dc_{PA}}{dt} = F_o (c_V - c_{PA}) / V^{(16)}, \tag{3.16}$$

where F_o is the cardiac output, which is equal to the total flow through the pulmonary and systemic circulations; $V^{(16)}$ is the volume of the large vessel pool; and c_V is the venous concentration exiting the systemic compartments and entering the large vessel pool. (We incorporate an explicit two-compartment model for the lung – Eq. (3.15) – in this generic PBPK model so that this generic form could be applied to simulate transport of a gas that is exchanged with the air in the lung. Gas exchange is driven by concentration changes occurring on timescales of seconds, faster than would typically justify the application of the quasi-steady approximation used for other compartments.)

The hepatic circulation is somewhat more complex than that of the 13 other compartments simulated by Eq. (3.14), because the outputs of the spleen and gut feed into the portal vein, which feeds as an input into the liver, as illustrated in Figures 3.5 and 3.6. There is also an input to the liver from the hepatic artery, which branches directly off a major artery coming off the aorta. Thus the input to the liver is the flow-weighted sum of the feeding concentrations:

$$c_{in}^{(14)} = \left(c_A F^{(hepatic)} + c_{out}^{(10)} F^{(10)} + c_{out}^{(11)} F^{(11)} + c_{out}^{(12)} F^{(12)} + c_{out}^{(13)} F^{(13)} \right) / F^{(14)},$$

(3.17)

where $F^{(hepatic)}$ is the hepatic artery flow and $F^{(14)} = F^{(hepatic)} + F^{(10)} + F^{(11)} + F^{(12)} + F^{(13)}$. The transport equations for the liver are

$$c_1^{(14)}(t) = \frac{c_{in}^{(14)} + \frac{P S^{(14)}}{F^{(14)}} \frac{c_2^{(14)}}{\lambda^{(14)}}}{1 + P S^{(14)}/F^{(14)}}$$

$$c_{out}^{(14)}(t) = \begin{cases} g^{(14)}, & g^{(14)} \geq 0 \\ 0, & g^{(14)} < 0 \end{cases}$$

$$g^{(14)} = c_{in}^{(14)} - \frac{1}{F^{(14)}} \left\{ \frac{V_1^{(14)}}{1 + P S^{(14)}/F^{(14)}} \frac{dc_{in}^{(14)}}{dt} \right.$$

$$+ \left[\frac{V_1^{(14)} P S^{(14)}/F^{(14)}}{(1 + P S^{(14)}/F^{(14)})\lambda^{(14)}} + V_2^{(14)} \right]$$

$$\left. \times \frac{F^{(14)}}{V_2^{(14)}} \left(\frac{P S^{(14)}/F^{(14)}}{1 + P S^{(14)}/F^{(14)}} \right) \left(c_{in}^{(14)} - c_2^{(14)}/\lambda^{(14)} \right) \right\}$$

$$\frac{dc_2^{(14)}(t)}{dt} = \begin{cases} \frac{F^{(14)}}{V_2^{(14)}} \left(\frac{P S^{(14)}/F^{(14)}}{1+P S^{(14)}/F^{(14)}} \right) \left(c_{in}^{14} - c_2^{(14)}/\lambda \right), & c_{out}^{(14)} > 0 \\ \frac{(1+P S^{(14)}/F^{(14)}) F^{(14)} c_{in}^{14} - V_1^{(14)} dc_{in}^{14}/dt}{V_1^{(14)} P S^{(14)}/(\lambda F^{(14)}) + (1+P S^{(14)}/F^{(14)}) V_2^{(14)}}, & c_{out}^{(14)} = 0, \end{cases}$$

(3.18)

where $dc_{in}^{(14)}/dt$ is computed

$$\frac{dc_{in}^{(14)}}{dt} = \left(F^{(hepatic)} \frac{c_A}{dt} + F^{(10)} \frac{c_{out}^{(10)}}{dt} + F^{(11)} \frac{c_{out}^{(11)}}{dt} + F^{(12)} \frac{c_{out}^{(12)}}{dt} + F^{(13)} \frac{c_{out}^{(13)}}{dt} \right) / F^{(14)}.$$

The systemic venous concentration is computed from the flow-weighted sum of the output of all the systemic compartments draining into the vena cava:

$$c_V = \left(c_{out}^{(1)} F^{(1)} + c_{out}^{(2)} F^{(2)} + c_{out}^{(3)} F^{(3)} + c_{out}^{(4)} F^{(4)} + c_{out}^{(5)} F^{(5)} \right.$$

$$\left. + c_{out}^{(6)} F^{(6)} + c_{out}^{(7)} F^{(7)} + c_{out}^{(8)} F^{(8)} + c_{out}^{(14)} F^{(14)} + c_{out}^{(9)} F^{(9)} \right) / F_o,$$

which is used as an input to Eq. (3.16).

The set of equations presented in this section may be used to simulate the movement of intravascular and extravascular solutes through the multi-compartment representation of the body illustrated in Figure 3.6. The total number of differential equations is 17, one for each of the 15 compartments other than the lung, and two equations for the lung. In the following examples we apply these general equations (with some additions and modifications) to analyze pharmacokinetic data from rats.

3.4 Example: uptake and delivery of matrine in rat

3.4.1 A PBPK model for rat

Gao and Law [19] measured the concentration kinetics in rat of matrine – an alkaloid component of certain botanical drugs under investigation for their pharmaceutical properties. Here we use the data reported in their study to validate our PBPK modeling framework, and to demonstrate its application to a real data set.

Figure 3.8 shows data on concentration of matrine in plasma, kidney, spleen, brain, heart, muscle, lung, liver, and fat following an oral dose of 15 mg matrine per kg of body weight [19].

To simulate the experiment of Gao and Law, a few customizations of the generic model are necessary. First, to simulate oral delivery of the drug, we need to introduce an additional compartment (region #17) to simulate the contents of the gut, which is assumed to exchange via passive diffusion with the small intestine (organ #12):

$$\frac{dc^{(17)}}{dt} = k_a \left(c_2^{(12)} / \lambda^{(12)} - c_2^{(17)} + r_{in}(t) \right) / V^{(17)}. \tag{3.19}$$

The function $r_{in}(t)$ represents input of the drug into the gut; k_a is the absorption rate.[3]

---

[3] We model r_{in} as a delayed Gaussian function: $r_{in}(t) = \frac{D}{\sqrt{\pi}\tau} \exp\left\{ -[(t - t_d)/\tau]^2 \right\}$, where the delay t_d is set to 5 min, following Gao and Law [19], and D is the dose measured in mass of drug ingested. The time constant τ is arbitrary, and does not influence model predictions as long as it is significantly smaller than t_d. We set $\tau = 30$ s.

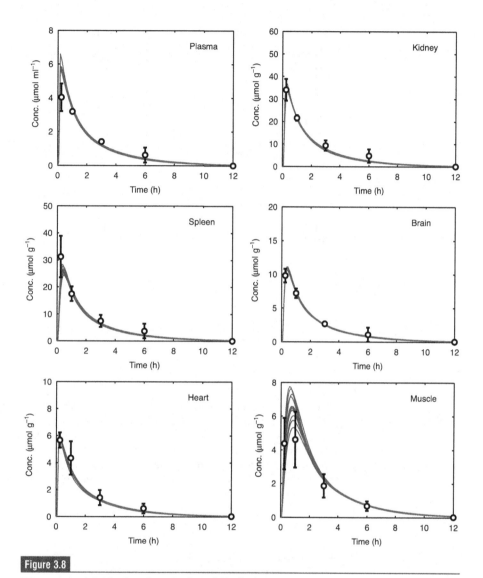

Figure 3.8

Time course of distribution of matrine in rat following an oral dose for Parameter Class 1. (Figure and legend continued on next page.)

With a mass flux of $-k_a(c_2^{(12)}/\lambda^{(12)} - c_2^{(17)})$ out of the gut contents into the small-intestine wall, the equations for organ #12 are modified:[4]

$$c_1^{(12)} = \frac{c_A + \frac{PS^{(12)}}{F^{(12)}} \frac{c_2^{(12)}}{\lambda^{(12)}}}{1 + PS^{(12)}/F^{(12)}}$$

---

[4] See Exercise 3.5.

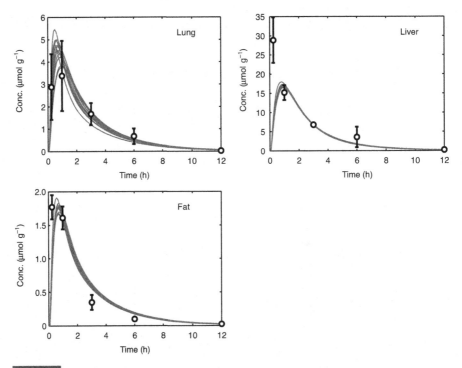

(continued) Time course of distribution of matrine in rat following oral dose for Parameter Class 1. Rats were given an oral dose of 15 mg per kg body weight at $t = 0$. Measured data are plotted as mean value with error bars corresponding to standard error. The simulated concentration time courses are plotted as solid lines. Twelve different simulated time courses are plotted, corresponding to 12 different sets of estimated parameters. Each of the 12 parameter sets results in simulations that match the data equally well. The means plus/minus standard deviations of the estimated parameter values are reported in Table 3.2. Data are from Gao and Law [19] and are plotted as mean value plus/minus standard error.

$$\frac{dc_2^{(12)}}{dt} = \frac{F^{(12)}}{V_2^{(12)}} \left(\frac{PS^{(12)}/F^{(12)}}{1 + PS^{(12)}/F^{(12)}} \right) \left(c_A - c_2^{(12)}/\lambda^{(12)} \right)$$

$$- \frac{k_a}{V_2^{(12)}} \left(c_2^{(12)}/\lambda^{(12)} - c_2^{(17)} \right)$$

$$c_{out}^{(12)} = c_A - \frac{1}{F^{(12)}} \left\{ \frac{V_1^{(12)}}{1 + PS^{(12)}/F^{(12)}} \frac{dc_A}{dt} \right.$$

$$\left. + \left[\frac{V_1 PS^{(12)}/F^{(12)}}{(1 + PS^{(12)}/F^{(12)})\lambda^{(12)}} + V_2^{(12)} \right] \frac{dc_2^{(12)}}{dt} + k_a \left(c_2^{(12)}/\lambda^{(12)} - c_2^{(17)} \right) \right\}.$$

$$(3.20)$$

Similarly the equations for the kidney compartment are modified to account for a mass flux out through incorporation into urine:

$$c_1^{(8)} = \frac{c_A + \frac{PS^{(8)}}{F^{(8)}} \frac{c_2^{(8)}}{\lambda^{(8)}}}{1 + PS^{(8)}/F^{(8)}}$$

$$\frac{dc_2^{(8)}}{dt} = \frac{F^{(8)}}{V_2^{(8)}} \left(\frac{PS^{(8)}/F^{(8)}}{1 + PS^{(8)}/F^{(8)}} \right) \left(c_A - c_2^{(8)}/\lambda^{(8)} \right) - \frac{k_f c_1^{(8)}}{V_2^{(8)}}$$

$$c_{out}^{(8)} = c_A - \frac{1}{F^{(8)}} \left\{ \frac{V_1^{(8)}}{1 + PS^{(8)}/F^{(8)}} \frac{dc_A}{dt} \right.$$

$$\left. + \left[\frac{V_1 PS^{(8)}/F^{(8)}}{(1 + PS^{(8)}/F^{(8)})\lambda^{(8)}} + V_2^{(8)} \right] \frac{dc_2^{(8)}}{dt} + k_f c_1^{(8)} \right\}, \qquad (3.21)$$

where k_f is the filtration rate.[5]

3.4.2 Model parameters

The volumes and flows used to simulate the 18-compartment PBPK model for matrine distribution in rat are listed in Table 3.2. For this exercise, the small vessel volume (V_1) of all organs is set equal to 4% of the tissue volume, although real vascular volumes can deviate significantly from this assumption. The total blood volume ($= \sum_{i=1}^{16} V_1^{(i)}$) is equal to 29 ml, and the total volume of all tissues plus blood ($= \sum_{i=1}^{16} V_1^{(i)} + \sum_{i=1}^{15} V_2^{(i)}$) is 363.2 ml. Assuming an average density of 1.05 g ml^{-1}, the total mass of the animal simulated is 381.4 g. Total cardiac output is assumed $F_o = 1$ ml s^{-1}. The flows and volumes listed in this table are assigned to correspond to a representative standard laboratory rat, and are approximately consistent with values reported in a number of sources, including Brown et al. [12], Peng et al. [49], and Shin et al. [56].

Values for the PS and λ parameters, k_a, and k_f were estimated based on fitting model-predicted concentration time courses to the data on concentration in plasma, kidney, spleen, brain, heart, muscle, lung, liver, and fat measured at 15 min, 1 h, 3 h, 6 h, and 12 h following oral dosing, as illustrated in Figure 3.8. It turns out (perhaps not surprisingly) that the nine data curves do not provide enough information to uniquely identify the 32 adjustable parameters that are invoked in this model. As a result, it is possible to find many different sets of parameter estimates that can fit the data equally well.

---

[5] To keep the presentation compact, these equations for transport in the gut and kidney do not invoke the check to ensure that the output concentrations remain positive. Here the inequality can be ignored, because in executing this model for the conditions used the concentrations do not become negative.

Table 3.2: Parameter values associated with Parameter Class 1 used in the PBPK model for matrine distribution in rat. Mean plus/minus standard deviation are reported for PS, λ, k_a, and k_f (see text). (The flow listed for the liver is the hepatic flow; the total flow to the liver is the sum of the hepatic, spleen, stomach, small intestine, and large intestine flows.)

Organ	V_2 (ml)	V_1 (ml)	Flow (% of F_o)	PS (ml s^{-1})	λ
1. Heart	1.13	0.0451	5	0.0776 ± 0.0359	1.07 ± 0.043
2. Thyroid	0.66	0.0264	0.1	0.0025 ± 0.0021	1.11 ± 0.965
3. Brain	1.39	0.0558	1.5	0.0289 ± 0.0070	2.17 ± 0.058
4. Muscle	144.1	5.7656	35	0.560 ± 0.337	1.71 ± 0.086
5. Bone	24.9	0.9946	6	0.139 ± 0.095	5.89 ± 1.10
6. Fat	40.4	1.6144	2	0.128 ± 0.155	0.443 ± 0.016
7. Skin	57.7	2.3064	6	0.172 ± 0.061	3.12 ± 0.403
8. Kidney	2.77	0.1106	19	1.43 ± 0.523	37.1 ± 4.14
9. Other	35.67	1.4260	7.7	0.267 ± 0.123	7.03 ± 0.831
10. Spleen	0.838	0.0335	1.7	0.0479 ± 0.029	5.34 ± 0.206
11. Stomach	1.822	0.0729	1.4	0.0267 ± 0.0168	0.945 ± 0.437
12. Sm. int.	6.78	0.2710	10	0.203 ± 0.059	0.308 ± 0.108
13. La. int.	1.66	0.0665	3	0.0416 ± 0.0203	0.977 ± 0.803
14. Liver	13.9	0.5564	1.6	0.0101 ± 0.0003	2.58 ± 0.031
15. Lung	1.432	0.0573	100	0.0017 ± 0.0008	1.16 ± 0.073
16. Large vessels	—	15.60	100	—	—

$$k_f = 0.490 \pm 0.041 \quad \text{ml s}^{-1}$$
$$k_a = 0.00063 \pm 8 \times 10^{-5} \quad \text{ml s}^{-1}$$

This issue may be investigated by conducting a Monte Carlo (random based) search of the parameter space. Doing so, a number of different sets of parameter values can be found that can match the measured data reasonably well. In fact, several distinct classes of parameter value sets can be found that result in qualitatively different predicted model behavior. Values and model behaviors associated with one such class of parameter estimates (here referred to as "Parameter Class 1") are summarized in Figure 3.8 and Table 3.2. The parameter statistics reported in Table 3.2 are obtained from 12 different sets of parameter values from Parameter Class 1; model predictions for each of the 12 parameter sets are plotted in along with the data in Figure 3.8. As is apparent from Table 3.2, the range of values for the estimates is relatively large for many parameters. For the eight organs for which concentration data are provided, the partition coefficients ($\lambda^{(x)}$) are estimated with reasonable precision. For other tissues/organs, the range of estimated values is relatively high, and the partition coefficient values are not estimated precisely.

Furthermore, as is indicated in the table, the majority of the PS values (even for organs for which concentration data are provided) are not well estimated. This is because the comparison between model predictions and data is not particularly sensitive to the values of PS used for most of the organs simulated in the model. The PS value that shows the most sensitivity to the data is that of liver, for which the standard deviation is only approximately 3% of the mean estimate. However, the liver is the organ for which the model and the data are in poorest agreement. Thus we should not put terribly much confidence in the estimated PS values for any of the organs/tissues listed in Table 3.2. In fact, some experimentation with the model reveals that we may set many of these poorly estimated PS values to arbitrarily high values without significantly influencing the computed difference between the model and the data. (We will explore this phenomenon in more detail in the following section on *flow-limited transport*.)

Model predictions and associated parameter value estimates for an additional parameter class (distinct from Parameter Class 1 and here referred to as "Parameter Class 2") are reported in Figure 3.9 and Table 3.3.

One major difference between the model predictions in Figure 3.8 (Parameter Class 1) and Figure 3.9 (Parameter Class 2) is that the data from the liver (particularly the first data point) are better matched by Parameter Class 2. Two major differences in parameter estimates are seen in the estimated filtration and absorption rates, k_f and k_a. Parameter Class 1 predicts relatively fast filtration and slow absorption compared with Parameter Class 2. These differences in a much higher plasma concentration for Parameter Class 2 at times precede the initial 15-minute sampling point. Thus the physiological parameters and associated model predictions are significantly different between the two classes of estimated model predictions. Parameter Class 2 provides a slightly better match to the data than Parameter Class 1. However, given the relatively large variability in the data, we should be cautious about discarding the predictions of Parameter Class 1 in favor of Parameter Class 2.

The issue of model discrimination is addressed in the final section of this chapter. For now, let us postpone the question of which parameter class is more reasonable.

3.4.3 Flow-limited transport

Before attempting to discriminate between the two distinct classes of model parameters, it is worthwhile to follow up on the observation that the model behavior, in comparison with the data, is not particularly sensitive to the PS parameter values. As mentioned above, computer experimentation reveals that many of the estimated PS values may be replaced with arbitrarily large values without

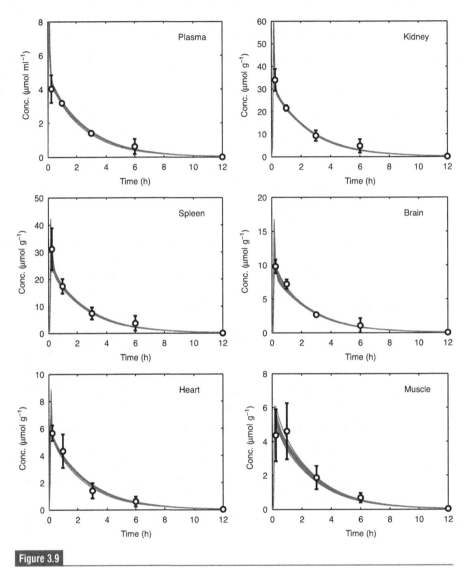

Figure 3.9

Time course of distribution of matrine in rat following an oral dose for Parameter Class 2. (Figure
and legend continued on next page.)

significantly impacting on the goodness of the fit to the data. This phenomenon is
illustrated in Figure 3.10, where *all* of the PS values are set to the arbitrarily high
value of $10 \ \mathrm{ml \, s^{-1}}$. All other parameter values are taken as the mean estimates
from Parameter Class 1 (the mean values given in Table 3.2).

The model predictions plotted in Figure 3.10 are really not significantly worse
than those of Figure 3.8, which uses the finite estimates of PS of Parameter Class
1. In fact, repeating the experiment of setting all the PS's to high values for

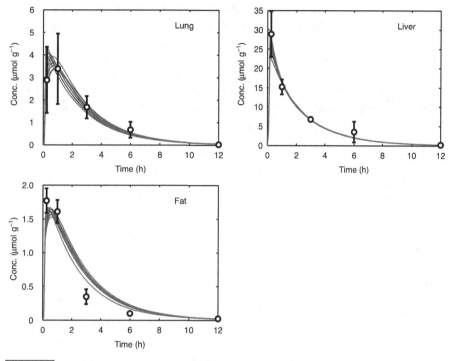

(continued) Time course of distribution of matrine in rat following oral dose for Parameter Class 2. Rats were given an oral dose of 15 mg per kg body weight at $t = 0$. Measured data are plotted as mean value with error bars corresponding to standard error. The simulated concentration time courses are plotted as solid lines. Twelve different simulated time courses are plotted, corresponding to 12 different sets of estimated parameters. Each of the 12 parameter sets results in simulations that match the data equally well. The means plus/minus standard deviations of the estimated parameter values are reported in Table 3.3. Data are from Gao and Law [19] and are plotted as mean value plus/minus standard error.

Parameter Class 2 gives much the same results (not shown). We can conclude from these computational experiments that a *flow-limited* model for several (perhaps all) organs is consistent with the data. (This does not mean that the permeabilities necessarily are effectively infinite, only that these model predictions are not sensitive to whether or not these PS's are finite or effectively infinite.)

In an effort to simplify the model by reducing the number of parameters necessary to describe the observed concentration kinetics, it is possible to develop a formalized simplification to treat flow-limited transport, and apply it to simulate the experiments of Gao and Law on drug transport in rat.

In the limit of large PS/F the governing equations for a given organ or tissue approach a limiting behavior where the exact value of PS is not important, as long as $PS \gg F$. This is formally what we mean by the *flow-limited* regime, where the

Table 3.3: Parameter values associated with Parameter Class 2 used in the PBPK model for matrine distribution in rat. Mean plus/minus standard deviation are reported for PS, λ, k_a, and k_f (see text). (The flow listed for the liver is the hepatic flow; the total flow to the liver is the sum of the hepatic, spleen, stomach, small intestine, and large intestine flows.)

Organ	V_2 (ml)	V_1 (ml)	Flow (% of F_o)	PS (ml s^{-1})	λ
1. Heart	1.13	0.0451	5	0.173 ± 0.120	1.27 ± 0.038
2. Thyroid	0.66	0.0264	0.1	0.0026 ± 0.0016	1.58 ± 0.963
3. Brain	1.39	0.0558	1.5	0.0432 ± 0.0065	1.96 ± 0.052
4. Muscle	144.1	5.7656	35	0.946 ± 0.547	1.14 ± 0.066
5. Bone	24.9	0.9946	6	0.151 ± 0.086	1.74 ± 0.42
6. Fat	40.4	1.6144	2	0.073 ± 0.037	0.429 ± 0.017
7. Skin	57.7	2.3064	6	0.211 ± 0.107	2.61 ± 6.76
8. Kidney	2.77	0.1106	19	1.49 ± 0.992	7.12 ± 0.31
9. Other	35.67	1.4260	7.7	0.242 ± 0.160	1.34 ± 0.275
10. Spleen	0.838	0.0335	1.7	0.043 ± 0.018	5.37 ± 0.184
11. Stomach	1.822	0.0729	1.4	0.035 ± 0.019	1.01 ± 0.785
12. Sm. int.	6.78	0.2710	10	0.403 ± 0.137	0.415 ± 0.181
13. La. int.	1.66	0.0665	3	0.068 ± 0.057	1.07 ± 0.345
14. Liver	13.9	0.5564	1.6	0.242 ± 0.160	4.66 ± 0.132
15. Lung	1.432	0.0573	100	0.013 ± 0.0004	0.997 ± 0.056
16. Large vessels	—	15.60	100	—	—

$$k_f = 0.0053 \pm 0.0044 \quad \text{ml s}^{-1}$$
$$k_a = 0.207 \pm 0.603 \quad \text{ml s}^{-1}$$

PS value is high enough that (other than being significantly greater than F) the exact value of PS does not influence model behavior. In this limit, the governing equations can be reduced to[6]

$$c_1^{(x)}(t) = c_2^{(x)}/\lambda^{(x)}$$

$$c_{out}^{(x)}(t) = \begin{cases} g^{(x)}, & g^{(x)} \geq 0 \\ 0, & g^{(x)} < 0 \end{cases}$$

$$g^{(x)} = \left(1 + \frac{V_1^{(x)}}{\lambda V_2^{(x)}}\right) - c_{in}^{(x)} \frac{V_1^{(x)}}{\lambda^{(x)} V_2^{(x)}}$$

$$\frac{dc_2^{(x)}(t)}{dt} = \begin{cases} \frac{F^{(x)}}{V_2^{(x)}}\left(c_{in}^{(x)} - c_2^{(x)}/\lambda^{(x)}\right), & c_{out}^{(x)} > 0 \\ \frac{F^{(x)}c_{in}^{(x)}}{V_1^{(x)}/\lambda^{(x)}+V_2^{(x)}}, & c_{out}^{(x)} = 0 \end{cases} . \tag{3.22}$$

[6] See Exercises 3.4 and 3.6.

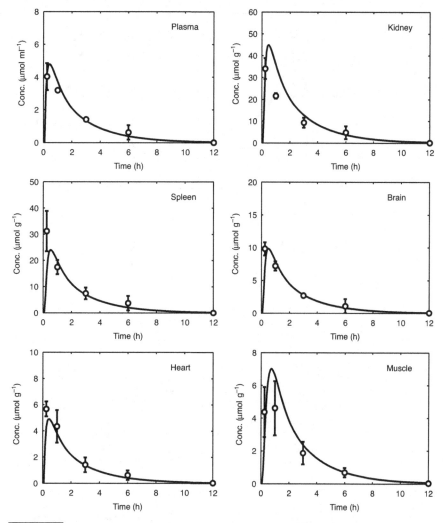

Figure 3.10

Matrine transport with PS values set to arbitrarily high values. (Figure and legend continued on next page.)

When valid, this simplification significantly reduces the complexity of the governing equations and effectively reduces the number of adjustable parameters.

The flow-limited equations for kidney and small intestine are (from Eqs (3.21) and (3.20)):

$$\frac{dc_2^{(8)}}{dt} = \frac{F^{(8)}}{V_2^{(8)}}\left(c_A - c_2^{(8)}/\lambda^{(8)}\right) - \frac{k_f c_2^{(8)}}{\lambda^{(8)} V_2^{(8)}}$$

$$c_{out}^{(8)} = c_A - \frac{1}{F^{(8)}}\left[\left(\frac{V_1^{(8)}}{\lambda^{(8)}} + V_2^{(8)}\right)\frac{dc_2^{(8)}}{dt} + k_f c_1^{(8)}\right] \qquad (3.23)$$

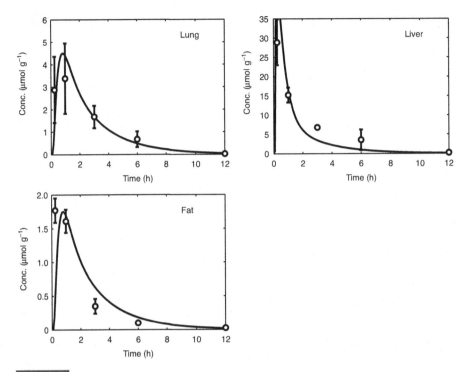

(continued) Matrine transport with PS values set to arbitrarily high values. The time course of distribution of matrine in rat following an oral dose (15 mg matrine per kg of body weight) is plotted along with simulations from the PBPK model, where all PS values are set to arbitrarily high values. Other parameter values are taken as the mean values from Table 3.2. Data are from Gao and Law [19] and are plotted as mean value plus/minus standard error.

and

$$\frac{dc_2^{(12)}}{dt} = \frac{F^{(12)}}{V_2^{(12)}} \left(c_A - c_2^{(12)}/\lambda^{(12)} \right) - \frac{k_a}{V_2^{(12)}} \left(c_2^{(12)}/\lambda^{(12)} - c_2^{(17)} \right)$$

$$c_{out}^{(12)} = c_A - \frac{1}{F^{(12)}} \left[\left(\frac{V_1^{(12)}}{\lambda^{(12)}} + V_2^{(12)} \right) \frac{dc_2^{(12)}}{dt} + k_a \left(c_2^{(12)}/\lambda^{(12)} - c_2^{(17)} \right) \right].$$

$$(3.24)$$

To determine how well an entirely flow-limited model (a model where all permeation other than absorption from the gut is assumed to be flow limited) can match the measured data, we construct a flow-limited model by combining Eq. (3.22) for heart (organ #1), thyroid (#2), brain (#3), muscle (#4), bone (#5), fat (#6), skin (#7), other (#9), spleen (#10), stomach (#11), small intestine (#12), large intestine (#13), liver (#14), and lung (#15), Eqs (3.23) for kidney (#8), Eqs (3.24) for small intestine (#12), Eq. (3.19) for the gut contents (#17), and Eq. (3.16) for the large-vessel pool. The resulting model has 15 fewer parameters than the more general model presented above, because the PS products do not appear in the equations. The model also has one fewer state variable, because the equation for dc_A/dt in Eq. (3.15) is replaced here by the output concentration of the flow-limited lung compartment, which is given by the explicit expression for $c_{out}^{(15)}$, from Eq. (3.22):

$$\frac{dc_2^{(15)}}{dt} = \frac{F_o}{V_2^{(15)}} \left(c_{PA} - c_2^{(15)}/\lambda^{(15)} \right)$$

$$c_{out}^{(15)} = c_A = \left(1 + \frac{V_1^{(15)}}{\lambda^{(15)} V_2^{(15)}} \right) \frac{c_2^{(15)}}{\lambda^{(15)}} - c_{PA} \frac{V_1^{(15)}}{\lambda^{(x)} V_2^{(15)}}. \qquad (3.25)$$

Once again, computational experimentation reveals two parameter classes, analogous to Parameter Classes 1 and 2 introduced above. Model predictions and associated parameter estimates for this flow-limited model are given in Figure 3.11 and Table 3.4 for Parameter Class 1 and Figure 3.12 and Table 3.5 for Parameter Class 2. Generally, the flow-limited model agrees well with the measured data, although here the model predictions do not come as close to the mean concentration measurements as in Figures 3.8 and 3.9. This should not be surprising, because the model fits in Figures 3.8 and 3.9 are associated with 32 adjustable parameters, while the flow-limited model has only 17 parameters. Even though the mean squared error is slightly higher here, given the relatively large error bars on the data, it is not possible to rule out the flow-limited model as a reasonable explanation for the observed transport kinetics.[7]

As for the non-flow-limited case (Tables 3.2 and 3.3), the major difference between the two parameter classes is in the estimated values of k_f and k_a. Also, as for the more general non-flow-limited case, Parameter Class 2 matches the mean liver data slightly better.

Again, for both parameter classes the values of the λ's are estimated within a relatively narrow range for organs for which the time-course data are available. These values are largely consistent across both flow-limited and non-flow-limited versions of the model and both parameter classes. The only significant difference is in the estimates for the λ values for kidney. Parameter Class 1 (with higher k_f estimates) is associated with higher $\lambda^{(8)}$ estimates. The kidney partition coefficient is a component of a very simplified model representation of kidney

---

[7] It may have become a bit difficult to keep track of the different versions of the model at this point. As a reminder, we have two general parameter value classes: Class 1 and Class 2. These classes are associated with significantly different estimates of k_a and k_f, as reported in Tables 3.2 and 3.3. We also have non-flow-limited (permeability limited) and flow-limited versions of the model. Figures 3.8 and 3.9 illustrate non-flow-limited model predictions for the two parameter classes; Figures 3.11 and 3.12 illustrate flow-limited model predictions for the two parameter classes.

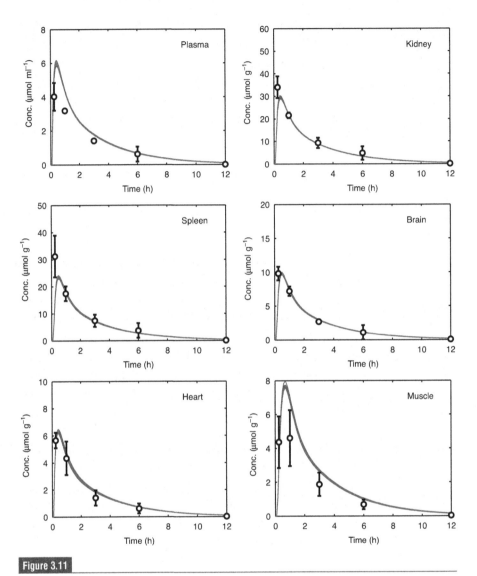

Figure 3.11

Optimized-fit flow-limited time course of distribution of matrine in rat following an oral dose for Parameter Class 1. (Figure and legend continued on next page.)

filtration, and thus we expect estimated values of k_f and $\lambda^{(8)}$ to be statistically related.

Although the underlying models are the same, the two parameter classes (whether applied to the flow-limited or the non-flow-limited case) predict that fundamentally different physiological phenomena are responsible for the observed kinetics of matrine concentration in rat. Parameter Class 1 predicts that the absorption from the gut is relatively slow ($1/k_a$ is approximately 30 min), while filtration

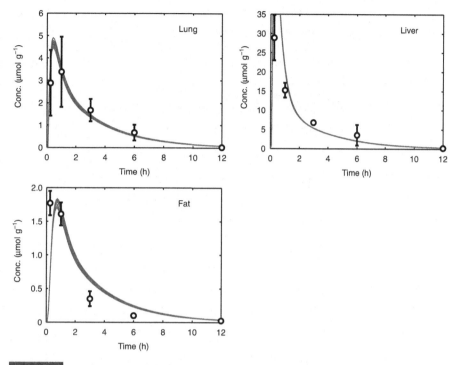

Figure 3.11

(continued) Optimized-fit flow-limited time course of distribution of matrine in rat following an oral dose for Parameter Class 1. Rats were given an oral dose of 15 mg per kg body weight at $t = 0$. Measured data are plotted as mean value with error bars corresponding to standard error. The simulated concentration time courses are plotted as solid lines. Twelve different simulated time courses are plotted, corresponding to 12 different sets of estimated parameters. Each of the 12 parameter sets results in simulations that match the data equally well. The means plus/minus standard deviations of the estimated parameter values are reported in Table 3.4. Data are from Gao and Law [19] and are plotted as mean value plus/minus standard error.

by the kidney is much faster. Parameter Class 2 predicts a much faster absorption and much slower filtration. The fast absorption in Parameter Class 2 is associated with sharper peaks in concentration in the concentration time courses than are predicted by Parameter Class 1.

Furthermore, while the transport models for all organs invoke fundamentally different assumptions in the flow limited versus the permeability limited case, the predicted concentration time courses for the two models are much the same for all organs other than liver. So we are left with a number of questions. Which of the versions of the model (flow limited or permeability limited) is a better match to reality? Are some organs flow limited with respect to matrine while others are permeability limited? Which parameter value range (Parameter Class 1 or 2) more closely matches the real situation? The answer to these questions is that, given the

Table 3.4: Parameter values associated with Parameter Class 1 used in the flow-limited PBPK model for matrine distribution in rat. Means plus/minus standard deviation are reported for λ, k_a, and k_f. (The flow listed for the liver is the hepatic flow; the total flow to the liver is the sum of the hepatic, spleen, stomach, small intestine, and large intestine flows.)

Organ	V_2 (ml)	V_1 (ml)	Flow (% of F_o)	PS (ml s^{-1})	λ
1. Heart	1.13	0.0451	5	1.106 ± 0.029	
2. Thyroid	0.66	0.0264	0.1	1.165 ± 1.007	
3. Brain	1.39	0.0558	1.5	1.725 ± 0.027	
4. Muscle	144.1	5.7656	35	1.487 ± 0.030	
5. Bone	24.9	0.9946	6	7.389 ± 0.336	
6. Fat	40.4	1.6144	2	0.352 ± 0.011	
7. Skin	57.7	2.3064	6	3.135 ± 0.122	
8. Kidney	2.77	0.1106	19	11.58 ± 0.295	
9. Other	35.67	1.4260	7.7	6.523 ± 0.271	
10. Spleen	0.838	0.0335	1.7	4.190 ± 0.080	
11. Stomach	1.822	0.0729	1.4	1.022 ± 0.647	
12. Sm. int.	6.78	0.2710	10	0.221 ± 0.074	
13. La. int.	1.66	0.0665	3	1.091 ± 0.841	
14. Liver	13.9	0.5564	1.6	3.061 ± 0.030	
15. Lung	1.432	0.0573	100	0.819 ± 0.027	
16. Large vessels	—	15.60	100	—	

$$k_f = 0.234 \pm 0.007 \quad \text{ml s}^{-1}$$
$$k_a = 7.75 \times 10^{-4} \pm 2 \times 10^{-5} \quad \text{ml s}^{-1}$$

data we have considered so far, we simply do not know. Yet, however unsatisfactory that answer may seem, the modeling of the transport kinetics behind matrine distribution in rat does yield valuable information. After all, our analysis determined a relatively small number of qualitatively different putative model behaviors. Specifically, we have generated two different competing hypotheses (quantified by the two different parameter classes and associated model predictions) to explain the available data. Furthermore, we can use the PBPK model analysis to determine an experiment necessary to disprove one (or perhaps both) versions of the model. Analysis of such an experiment is the subject of the next section.

3.4.4   Model validation and discrimination

As we have seen, based on the experiment of Gao and Law we are not able to distinguish between two distinct classes of PBPK model parameter values.

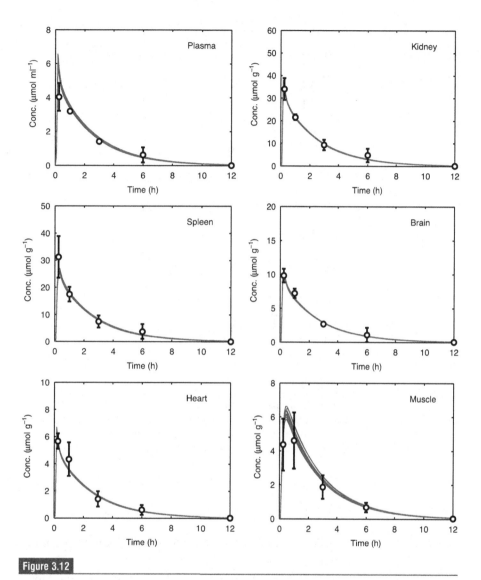

Figure 3.12

Optimized-fit flow-limited time course of distribution of matrine in rat following an oral dose for
Parameter Class 2. (Figure and legend continued on next page.)

These parameter value classes predict drastically different absorption and filtration
kinetics. Parameter Class 1 has relatively rapid filtration and slow absorption, while
Parameter Class 2 has relatively rapid absorption and slow filtration. In both cases
the estimated absorption and filtration constants combine to give approximately
the same overall decay rate in plasma for the experiment of Gao and Law.

A simple experiment to distinguish the behaviors predicted by the two parameter
classes would be to inject the drug intravenously, bypassing the absorption by gut

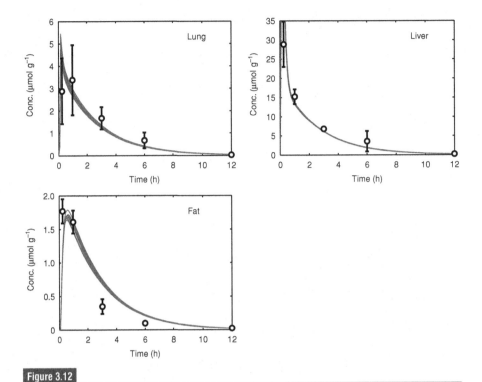

Figure 3.12

(continued) Optimized-fit flow-limited time course of distribution of matrine in rat following an oral dose for Parameter Class 2. Rats were given an oral dose of 15 mg per kg body weight at $t = 0$. Measured data are plotted as mean value with error bars corresponding to standard error. The simulated concentration time courses are plotted as solid lines. Twelve different simulated time courses are plotted, corresponding to 12 different sets of estimated parameters. Each of the 12 parameter sets results in simulations that match the data equally well. The means plus/minus standard deviations of the estimated parameter values are reported in Table 3.5. Data are from Gao and Law [19] and are plotted as mean value plus/minus standard error.

and allowing us to estimate k_f without interference from the absorption process. Fortuitously, such an experiment was performed by Wu *et al.* [67]. Figure 3.13 plots measurements from Wu *et al.* on plasma concentration following (a) an oral dose and (b) an intravenous injection.

Model simulations for the two parameter classes are plotted along with the measured data in the figure.[8] Here these model predictions are true predictions in the sense that no parameter estimation or model fitting is done. Clearly this experiment allows us to distinguish between the two models. Since Wu *et al.* sampled more data points on the early part of the time course, we are able to see that Parameter Class 1 matches the oral dose experiment better than Parameter

---

[8] The intravenous dose experiment is modeled by setting the initial condition in the large-vessel blood pool to the appropriate value.

Table 3.5: Parameter values associated with Parameter Class 2 used in the flow-limited PBPK model for matrine distribution in rat. Means plus/minus standard deviation are reported for λ, k_a, and k_f. (The flow listed for the liver is the hepatic flow; the total flow to the liver is the sum of the hepatic, spleen, stomach, small intestine, and large intestine flows.)

Organ	V_2 (ml)	V_1 (ml)	Flow (% of F_o)	λ
1. Heart	1.13	0.0451	5	1.073 ± 0.026
2. Thyroid	0.66	0.0264	0.1	1.607 ± 0.875
3. Brain	1.39	0.0558	1.5	1.893 ± 0.053
4. Muscle	144.1	5.7656	35	1.432 ± 0.070
5. Bone	24.9	0.9946	6	1.964 ± 0.462
6. Fat	40.4	1.6144	2	0.427 ± 0.017
7. Skin	57.7	2.3064	6	0.889 ± 0.213
8. Kidney	2.77	0.1106	19	6.480 ± 0.135
9. Other	35.67	1.4260	7.7	1.690 ± 0.356
10. Spleen	0.838	0.0335	1.7	5.195 ± 0.140
11. Stomach	1.822	0.0729	1.4	1.254 ± 1.114
12. Sm. int.	6.78	0.2710	10	0.340 ± 0.154
13. La. int.	1.66	0.0665	3	1.123 ± 0.412
14. Liver	13.9	0.5564	1.6	5.461 ± 0.154
15. Lung	1.432	0.0573	100	0.852 ± 0.035
16. Large vessels	—	15.60	100	—

$$k_f = 2.86 \times 10^{-3} \pm 1.4 \times 10^{-3} \quad \mathrm{ml\,s^{-1}}$$
$$k_a = 0.1849 \pm 0.047 \quad \mathrm{ml\,s^{-1}}$$

Class 2. When the drug is injected intravenously, the differences between the sets of parameter estimates becomes more apparent, and Parameter Class 1 clearly wins the comparison.

3.5 Recapitulation and lessons learned

In this chapter we have developed a general framework for whole-body physiologically based pharmacokinetic (PBPK) modeling and applied the framework to analyze transport of a pharmaceutical compound in rat.

An important lesson that emerged from this analysis is that a large-scale model such as the one created here typically involves a large number of adjustable parameters, not all of which can be estimated from the available data. We tackled this issue using two strategies. First, we searched computationally for different sets

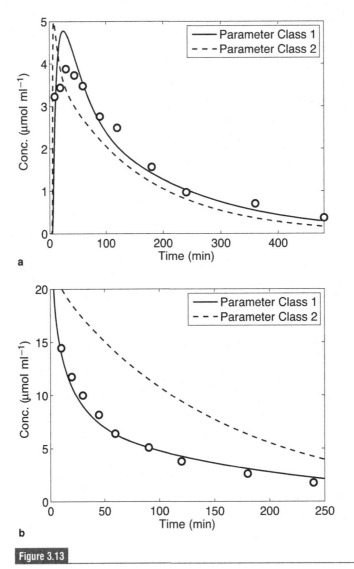

Figure 3.13

Time course of matrine in plasma following (a) oral and (b) intravenous doses in rat. Parameter values are set to the mean values from Table 3.4 and 3.5 for Parameter Class 1 and 2, respectively. Data are from Wu *et al.* [67].

of estimates for the 32 adjustable parameters in our whole-body model for matrine transport in the rat. Two key insights arose from that search. First, there are (at least) two general classes of parameter estimate sets that may equally effectively explain the PBPK data set of Gao and Law on matrine transport throughout the rat following oral dose of the drug. Second, we noted that the model predictions in comparison with the data from that experiment are largely insensitive to the estimated permeability–surface area products (PS's) for the various organs/tissues of the body.

The insensitivity of model predictions to the PS values allowed us to generate a reduced flow-limited model, where the PS values are effectively infinite. The ability of the flow-limited model to match the data effectively allowed us to reduce the number of adjustable parameters necessary to explain the data from 32 to 17. (Keep in mind that the permeability-limited model could match the data as well, so our analysis does not prove or disprove either case.)

While invoking the flow-limited assumption did drastically reduce the number of adjustable parameters, we were still left with the two classes of estimates for the remaining parameter estimates (most notably for the absorption and filtration rates) that were associated with qualitatively different predicted transport behavior. Based on an additional set of measured data (Figure 3.13) we were able to choose one particular set of parameters over the other and arrive at a single version of the model and associated parameters that can match both data sets.

Problems

3.1   Linear distributed modeling. Consider the spatially distributed transport model of Eq. (3.2) with first-order consumption of solute in the extra-capillary space:

$$\frac{\partial c_1}{\partial t} = -v\frac{\partial c_1}{\partial x} - \frac{PS}{V_1}(c_1 - c_2)$$

$$\frac{\partial c_2}{\partial t} = +\frac{PS}{V_2}\frac{V_1}{V_2}(c_1 - c_2) - ac_2.$$

What is the steady-state solution? Show that the spatial averaging $\bar{c}(t) = \frac{1}{L}\int_0^L c(x, t)dx$ leads to ordinary differential equations equivalent to Eq. (3.4) for this linear model.

3.2   Lumped compartmental versus distributed modeling. What are the solutions to Eqs (3.2) and (3.4) in the limit $PS/F \to 0$, given $c_A(t) = c_o(1 - e^{-kt})$ and the zero-concentration initial conditions used in Section 3.1.2? [Hint: See Section 9.5 for how to solve the advection equation.]

3.3   Quasi-steady model reduction 1. Show that Eq. (3.10) is a solution to Eq. (3.9). Plot the short-timescale solution for $c_1(t)$ predicted by Eq. (3.10) on the same plot as $c_1(t)$ obtained from a numerical integration of the two-compartment model of Eq. (3.6). How well and where does the short-timescale approximation match the solution to the full second-order system?

3.4 Quasi-steady model reduction 2. Derive the equation for dc_2/dt in Eq. (3.13) for the case where $c_{out} = 0$. Similarly, derive the equation for dc_2/dt in Eq. (3.22) for both $c_{out} = 0$ and $c_{out} > 0$.

3.5 Quasi-steady model reduction 3. Derive Eq. (3.20) using the passive absorption mass flux term defined in Eq. (3.19).

3.6 Flow-limited compartmental transport. Derive Eq. (3.22) from the stated assumptions. Given a constant input concentration, compare solutions to the systems of Eq. (3.5) (the two-compartment model), Eq. (3.8) (the quasi-steady case of the two-compartment model), and Eq. (3.22) (the flow-limited case of the two-compartment model). [You may choose to express the solutions analytically or use computer simulation.]

3.7 Open-ended computer exercise. Determine reasonable values for the human for the volumes and flows of the compartments of the whole-body PBPK model. Assuming that the λ's for matrine are the same in the human as in the rat, build a model to simulate whole-body transport kinetics of matrine in humans. Simulate the concentration time course driven by a fixed (constant) oral dose every 6 h. How long does it take for a periodic steady state to be achieved? What is the ratio between maximum and minimum plasma concentration in the periodic steady state? Does it make a big difference whether or not you assume values from Parameter Class 1 or 2? Flow-limited or non-flow-limited?

4 Cardiovascular systems simulation

Overview

This chapter is dedicated to studying and simulating blood pressures and flows in the circulatory system. We have already seen how transport phenomena are central to the operation of biological systems. In the previous chapter we saw how the pumping of the heart is responsible for driving blood flow to transport solutes throughout the body. Here we focus on the mechanics of the heart and circulatory system themselves.

Pumping of the heart and flow of blood throughout the circulatory system represent a critical life-support system in man. Malfunction of the heart and/or the circulatory system is associated with a great number of diseases and patho-physiological conditions. For example, hypertension – chronic systemic high blood pressure – puts stress on the heart that can ultimately lead to its failure. Here we will see that the functioning and malfunctioning of the circulatory system are best understood in terms of mathematical models that capture the key mechanistic underpinnings of its anatomy and physiology.

Our modeling and analysis in this chapter will rely on lumped parameter circuit models, analogous to electrical circuits made up of resistors, capacitors, and inductors. Readers not familiar with simple circuit analysis may choose to review Section 9.7 of the Appendices, which provides a short background on the subject, before undertaking this chapter.

We will begin our study of the circulatory system with an analysis of the main pump responsible for moving blood through the circuit described in Section 3.2 of the previous chapter.

4.1 The Frank–Starling mechanism of heart

We owe a great deal of our understanding of the mechanical operation of the heart to Ernest H. Starling (1866–1927), a British physiologist, and Otto Frank (1865–1944), a German physiologist, who made their major contributions around the end of the nineteenth and beginning of the twentieth centuries. The so-called *Frank–Starling mechanism*,[1] which stated one way says that the greater the volume of blood that fills the heart during diastole (the filling phase of the cardiac cycle) the greater the contraction and subsequent ejection of blood from the heart during systole (the contraction phase), is central to our understanding of how blood flow in the circulation is regulated.

Starling's laboratory developed a *heart–lung preparation* and conducted critical experiments to elucidate "the connection between venous inflow, venous pressure and ventricular output" [48]. In their heart–lung preparation, the systemic circulation is removed from the circuit in an experimental animal, as illustrated in Figure 4.1, leaving the pulmonary circulation and the heart intact. The systemic circulation is replaced by a system of tubes and reservoirs, allowing the input pressure and flow to the right atrium and output pressure and flow from the left ventricle to be manipulated and measured in an animal with the heart and lungs intact.

Results from one set of experiments using this preparation are shown in Figure 4.1(b). Cardiac output (blood flow out of the left ventricle into the aorta) is plotted against pressure measured at the entrance to the right atrium. The data in this figure were obtained from two different protocols for varying venous pressure and venous outflow: one where the height of the reservoir of blood connected to the right atrium was adjusted, and another where the input resistance was altered by adjusting the clip (labeled "A") on the input tubing. Both protocols yield similar results: cardiac output increases with increasing venous pressure up to some maximal point. Beyond the maximal point achieved, high venous pressures (over 120 mmH$_2$O, or approximately 10 mmHg) the cardiac output either levels off or decreases with further increases in venous pressure.

Starling's apparatus for measuring pressures and flows in mammalian hearts *in situ* was revolutionary at the time, allowing him and his colleagues to demonstrate the relationship between the pumping power of the heart and the amount that the heart muscle is stretched by filling the heart with blood. The essential observation that relates the venous pressures to the Frank–Starling mechanism is that when it is relaxed (in diastole) the filling of the heart is driven by the pressures in the great veins: the vena cava on the right side and the pulmonary vein on the left side. Thus the Frank–Starling mechanism predicts that as the pressure in the vena cava and

---

[1] The Frank–Starling mechanism is also called *Starling's law of the heart* and the *Frank–Starling law of the heart*. Apparently the mechanism that bears their names was appreciated even before either Starling or Frank was born [32]. Starling's papers of the early twentieth century credit a number of earlier works, including Frank's description of the mechanism in frog hearts (1895). While neither Starling nor Frank discovered the Frank–Starling mechanism, they both made advances critical to its dissemination and understanding. Starling, in particular, in 1918 proposed a molecular biochemical mechanism [32] that largely agrees with the modern molecular theory of the mechanism.

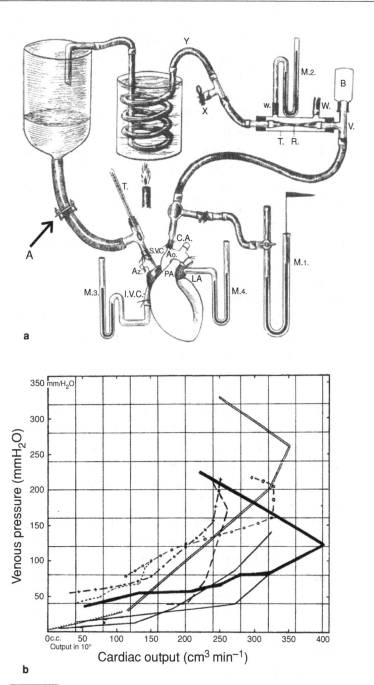

a

b

Figure 4.1

Heart–lung preparation and associated data of Starling and co-workers: (a) a diagram of their experimental setup; (b) data on cardiac output plotted against venous pressure at the inlet to the right atrium. Reprinted from Patterson and Starling [48] with permission from Wiley-Blackwell (publisher).

the pulmonary vein increases, the rate and amount of filling of the heart increase, as consequently does the output of the ventricles.[2]

Since there is normally no appreciable drop in pressure from the great veins to the atria during the filling phase of the cardiac cycle, the atrial pressures and the venous pressures can be equivalently thought of as the *filling pressures* driving blood flow into the left and right ventricles during diastole. The relationships between output of the heart and the filling pressures are conventionally plotted as cardiac output versus the atrial pressures, as illustrated in Figure 4.2. These data (from Uemura *et al.* [63]) were obtained with modern laboratory equipment in animals where the pressures and flows were varied by changing the total volume of blood in the circulation.[3] Left atrial and right atrial pressure (P_{LA} and P_{RA}, respectively) and cardiac output here represent the mean values of these variables averaged over several cardiac cycles.

Figure 4.2 shows data both from normal individual animals (in this case dogs) and from individuals following inducement of heart failure (failure of the heart to adequately pump blood to the body) by cutting off blood flow to the heart muscle. The data in the figure are effectively fitted by an empirical function used by Uemura *et al.* [63], and illustrated by the solid lines in the graphs:

$$F = F_o \left[\ln \left(P - P_o \right) + a \right],\tag{4.1}$$

where P represents either P_{LA} or P_{RA}, and F_o, P_o, and a are fitting parameters. The parameter values used to match the data in the figure are listed in the legend and indicated in the plots. Note that for either set of cardiac output curves we are able to effectively match the data by holding P_o and a constant between the different individuals and varying only F_o to mimic observed differences in cardiac contractility.

From analysis of data such as these, we can construct a family of curves representing cardiac output as a function of P_{RA} (or P_{LA}) under different circumstances, as is plotted in Figure 4.3 illustrating the cardiac output curve (F as a function of P_{RA}) under different levels of sympathetic and parasympathetic stimulation. These curves are described by Eq. (4.1) and obtained by varying the value of F_o, as indicated in the figure.

Of course, a number of physiological determinants (such as the state of health or disease of the heart and neurohumoral input to the heart) influence the shape of the cardiac output curve. The following section presents a simple analysis of the physiological regulation of cardiac output that adopts the

---

[2] In the century since Starling and Frank described the Frank–Starling mechanism, the molecular-level mechanism responsible has been studied in some detail. Essentially, our modern understanding is rooted in the sliding filament theory that explains how parallel arrangements of polymeric biomolecules in the muscle cells of the heart generate force when calcium is released into the cells and interacts with the polymer filaments. The more the muscle cells are stretched, the more sites on the filaments are exposed for binding to calcium, increasing the contractile capacity of the tissue.

[3] Nervous and endocrine influences on cardiac function were removed from the system by isolating the carotid sinuses at constant pressure (see Section 4.4.1) and by cutting the efferent vagosympathetic nerve trunk.

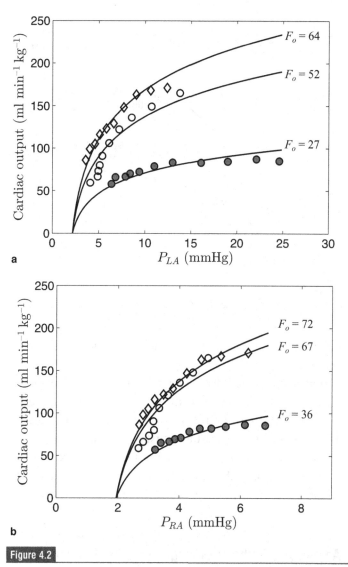

Figure 4.2

Cardiac output curves. Cardiac output is plotted as functions of (a) left atrial pressure (P_{LA}) and (b) right atrial pressure (P_{RA}). Data are shown for two healthy individuals (open circles and diamonds) and for one individual under left-ventricular heart failure (filled circles). Solid lines are fits obtained from Eq. (4.1), with $P_o = 1.59$ mmHg and $a = 0.51$ for cardiac output as a function of P_{LA}, and $P_o = 1.59$ mmHg and $a = 1.04$ for cardiac output as a function of P_{RA}. The values of F_o are varied between the individuals, as indicated in the plots. (Parameter F_o is reported in units of ml min^{-1} per unit body mass.) Data are obtained from Uemura *et al.* [63].

cardiac output curve – modeled by Eq. (4.1) – as a simple model of the heart that captures the Frank–Starling mechanism. This model is not concerned with any particular detailed mechanisms of stimulation or suppression of the intrinsic contractility of the heart, nor is it concerned with the mechanistic underpinnings of the Frank–Starling mechanism. In contrast, the model is simply an empirically backed

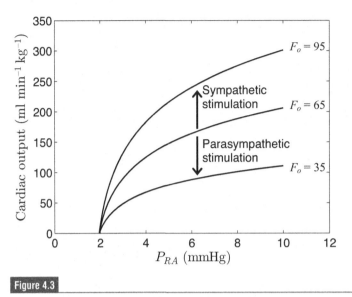

Figure 4.3

Effect of sympathetic/parasympathetic stimulation on cardiac output curves.

family of curves that captures cardiac output as a function of the right atrial pressure.

It should be noted that by adopting this model – expressing cardiac output as a function of right-atrial pressure – we are ignoring the pulmonary part of the circuit and treating "both sides of the heart and the lung as a single unit" [23]. This is reasonable, at least as a first approximation under normal circumstances, because in the steady-state conditions that we will explore, the flow out of the right ventricle is equal to the flow out of the left ventricle: thus the filling pressures P_{RA} and P_{LA} must tend to increase and decrease in synchrony.

4.2 An analysis of the physiological factors that control cardiac output

4.2.1 Guyton's model of the systemic circulation

Missing so far from our analysis of flow through the circuitry of Figure 3.5 is a treatment of the mechanical properties of the *peripheral circulation* – that is, the components of circuitry outside of the heart. Figure 4.4 shows the systemic and pulmonary circulations in a diagram that illustrates the relative volumes and distensibilities (or compliances) of the systemic arterial and venous and pulmonary arterial and venous parts. The diagram shows that the majority of blood at any time is in the systemic circulation and that the majority of systemic blood is in

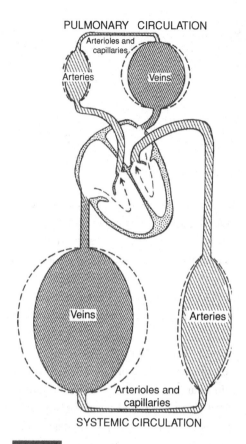

PULMONARY CIRCULATION

SYSTEMIC CIRCULATION

Figure 4.4

Diagram illustrating the relative volumes and pressures in the systemic and pulmonary circulations. Figure reprinted with permission from Guyton *Circulatory Physiology: Cardiac Output and its Regulation* [23], Chapter 7, page 137, Figure 56, copyright Elsevier.

the systemic veins. Similarly, the greatest capacity for distention, or increase in volume, is in the systemic veins.

In a landmark study Guyton *et al.* [26] analyzed the mechanical properties of the systemic circulation based on a model that discretized the systemic circulation into three compartments (representing large arteries, capillaries and venules, and large veins) to illustrate the relationships between right atrial pressure, blood flow, and the relative resistances and compliances in the arterial and venous compartments of the circulation. In a later study [24] the three-compartment model was simplified to the two-compartment model illustrated in Figure 4.5. This circuit model, which we investigate here, represents the simplest model that allows us to illustrate the major concepts of how pressure and volume are distributed around a network of compliant vessels, such as the systemic circulation.

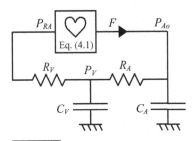

Figure 4.5

Two-compartment (four-element circuit) model of the systemic circulation.

Here the circulation is modeled as a series of discrete (lumped) ideal resistors and capacitors.[4] Specifically, the systemic circulation is treated as two serial lumped arterial and venous compartments. An arterial resistance and compliance, R_A and C_A, are associated with the arterial compartment; venous resistance and compliance, R_V and C_V, are associated with the venous compartment. (Resistors are used to mimic the viscous resistance to flow associated with different compartments of the circulation. Capacitors mimic the distensibility of the different compartments.)

The pressure of the blood leaving the left side of the heart is labeled P_{Ao} (for aortic pressure) and the right-side filling pressure P_{RA} (for right atrial pressure). It is assumed that the pulmonary circulation is noncompliant (nondistensible), and therefore, ignoring the relatively small temporal variations in volume of blood in the heart, the total volumes in both circulations (systemic and pulmonary) remain constant. Thus interactions between the pulmonary circulation and the heart are ignored in this model. Furthermore, our analysis (like that of Guyton *et al.*) treats steady-state situations, where volumes are not changing with time, and the flows into and out of each compartment in the circuit are equal and denoted F.

The elements labeled C_A and C_V in Figure 4.5 are compliant elements that model the elastic distensibility of the arterial and venous compartments of the circulation. Guyton's model invokes a simple linear assumption that the volume of blood in each compartment is proportional to its compliance value (C_A and C_V) multiplied by the pressure drop from the inside of the compartment to the outside of the network. A linear Ohm's-law relationship is associated with the resistance elements labeled R_A and R_V. This means that the pressure drop across the resistance element can be computed as flow multiplied by resistance. For example, the pressure labeled P_V is computed $P_V = P_{RA} + FR_V$, where F is the flow through the resistor. Similarly, summing the pressure drop across both resistances, $P_{Ao} = P_{RA} + FR_V + FR_A$.

---

[4] See Section 9.7 of the Appendices for an overview of circuit analysis, including mathematical modeling with ideal resistors and capacitors.

Combining these equations for the pressures with the linear volume–pressure relationship, we have

$$V_A = P_{Ao}C_A = [P_{RA} + F(R_A + R_V)]C_A$$
$$V_V = P_V C_V = (P_{RA} + FR_V)C_V \tag{4.2}$$

for the volumes in the arterial and venous compartments, respectively.

Guyton *et al.* introduced a variable known as the mean systemic filling pressure, P_{MF}, which is defined as the total volume of the circuit divided by the total compliance:

$$P_{MF} = \frac{\sum V}{\sum C} = \frac{[P_{RA} + F(R_A + R_V)]C_A + (P_{RA} + FR_V)C_V}{C_T}, \tag{4.3}$$

where $C_T = C_A + C_V$ is the total compliance of the network. To see why it is useful to introduce the quantity P_{MF}, an algebraic manipulation of Eq. (4.3) yields

$$P_{RA} = P_{MF} - FR_{eq}, \tag{4.4}$$

where we have defined $R_{eq} = R_V + R_A C_A/C_T$, which is called the *equivalent resistance* or *resistance to venous return*. One of the insights from Guyton's analysis[5] is that the relative importance of R_V in determining R_{eq} is magnified by the fact that C_A is substantially smaller than C_T. Thus changes in R_V have a relatively greater impact on the equivalent resistance than they have on the total resistance $R_T = R_A + R_V$.

Figure 4.6 shows data obtained by Guyton [22] on F as a function of P_{RA} along with model fits predicted by Eq. (4.4) for $P_{RA} > 0$. These data were obtained in an animal preparation where "in place of the heart a perfusion pump was connected from the right atrium to the aorta." The flow through the systemic circulation was varied by varying the speed of the pump. The different data sets (indicated by different symbols) were obtained "by increasing or decreasing the total quantity of blood in the circulatory system," effectively varying the quantity P_{MF}. Here we can see that the simple linear model holds reasonably well for positive right-atrial pressures: as the flow is increased, P_{RA} decreases linearly with F, as predicted.

The data demonstrate that the model breaks down when P_{RA} drops below zero. Indeed, the linear model of Figure 4.5 makes no physical sense for negative values of P_{RA}, which would imply a negative venous volume. Experimentally, the flow tends to plateau at some maximal value that is associated with $P_{RA} \rightarrow 0$. At low

---

[5] Guyton's concepts of venous return and resistance to venous return have generated a good deal of controversy and confusion in the physiology literature. In a paper on this subject [6] a colleague and I have concluded that "Because Guyton's venous return curves have generated much confusion and little clarity, we suggest that the concept and previous interpretations of venous return be removed from educational materials." Here I am granting myself the indulgence of disobeying that advice. Section 4.2.3 details exactly why Guyton's analysis has generated so much confusion.

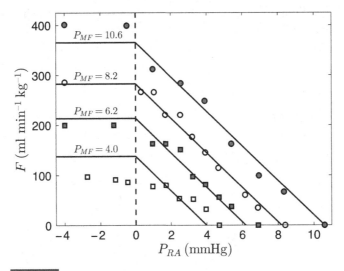

Figure 4.6

Relationship between right atrial pressure (P_{RA}) and flow (F). The data are obtained from a sacrificed animal with a blood flow driven by a mechanical pump. The different data sets are obtained for different blood volumes, effectively varying the mean filling pressure P_{MF}. Solid lines are fits to the simple circuit model described by Eq. (4.4) with $R_{eq} = 0.029$ mmHg min kg ml^{-1} and P_{MF} values indicated in the graph. Data are from Guyton [22], scaled assuming animal weight of 15 kg.

right-atrial (and venous) pressures the veins are observed to collapse, effectively limiting the flow through the circuit.

4.2.2 What the model tells us about the relationship between flow and right atrial pressure

The simple model of Figure 4.5 can be used to illustrate a number of concepts. In particular, Eq. (4.4) demonstrates the relationship between right atrial pressure and flow in a compliant circuit. As long as the total volume and individual resistances and compliances remain constant in the network, P_{MF} and R_{eq} remain constant, showing that an increase in flow results in a decrease in right-atrial pressure.

The change in right atrial pressure associated with a change in flow can be computed from Eq. (4.4),

$$\Delta P_{RA} = -R_{eq} \Delta F, \tag{4.5}$$

and the change in venous volume from Eq. (4.2),

$$\Delta V_V = \Delta F \left(-R_{eq} + R_V \right) C_V. \tag{4.6}$$

Since by definition $R_{eq} > R_V$, the change in venous volume associated with an increase in flow is negative. Since total volume is conserved, an increase in flow shifts blood from the venous compartment to the arterial. Similarly, a reduction in flow shifts volume from the arterial to the venous. This result makes intuitive sense if one thinks of the heart as a pump. When pumping rate decreases, blood pools at the inlet to the pump; when pumping increases, less blood pools at the inlet side, and therefore blood is shifted to the outlet side.

During exercise, when peripheral resistance drops, the balance between a resistance-driven increase in P_{RA} and a flow-driven decrease in P_{RA} depends on the intrinsic dependence of the heart's pumping of blood on P_{RA}. This balance is easily visualized using a graphical method introduced by Guyton [23]. Guyton *et al.* [26] famously wrote Eq. (4.4) as

$$F = \frac{P_{MF} - P_{RA}}{R_{eq}}, \tag{4.7}$$

implying a functional dependence of flow on right atrial pressure, and termed the predicted relationship between F and P_{RA} the "venous return curve." For reasons discussed below the term "venous return curve" is misleading, and here we use the term "vascular function curve" [38]. Exercise 4.3 asks the reader to show that there exists an equivalent expression for flow as an Ohm's-law relationship between aortic pressure and an equivalent resistance we could term the *equivalent resistance to cardiac output*. The functional utility of expressing flow in terms of P_{RA} in Eq. (4.4) or (4.7) is that the relationship can be plotted on the same axes as a model describing the Frank–Starling law of the heart, as demonstrated in Figure 4.7.

Figure 4.7 shows that the vascular function curve and Frank–Starling curve intersect at the steady-state operating point. In Figure 4.7 the operating point is at approximately $P_{RA} = 4.05$ mmHg and $F = 136$ ml min^{-1} kg^{-1}. Curves such as these allow us to graphically illustrate the behavior predicted by Eq. (4.5). For example, Figure 4.8(a) shows what happens when cardiac contractility increases while the parameters of the systemic circulation are held constant. Here the vascular function curve remains the same while the Frank–Starling cardiac output curve is shifted up. Specifically, this graph illustrates the impact of a doubling in contractility, resulting in an increase in flow to $F = 178$ ml min^{-1} kg^{-1} and a drop in right atrial pressure to $P_{RA} = 2.85$ mmHg. Thus a 100% increase in contractility results in a 30% increase in flow and a 30% drop in P_{RA}.

Similarly, we can hold cardiac function fixed while lowering the resistance of the systemic circulation (for example by reducing R_A, R_V, or both). The resulting drop in R_{eq} shifts the vascular function curve up, as illustrated in Figure 4.8(b). Here we have reduced R_{eq} by 50% (equivalent to doubling the slope $1/R_{eq}$), resulting

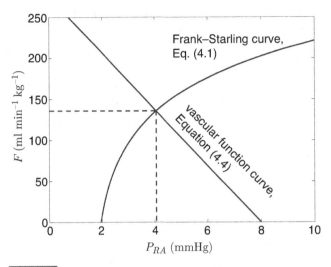

Figure 4.7

Vascular function and cardiac output curves. The linear vascular function curve of Eq. (4.4) intersects with the Frank–Starling cardiac output curve of Eq. (4.1) at the values of flow and right atrial pressure predicted by the steady-state circuit model. Parameter values for the vascular function curve are $R_{eq} = 0.029$ mmHg min kg ml^{-1} (from Figure 4.6) and $P_{MF} = 8$ mmHg. Parameter values for the Frank–Starling cardiac output curve are $P_0 = 1.59$ mmHg, $a = 1.04$, and $F_0 = 70$ ml min^{-1} kg^{-1}.

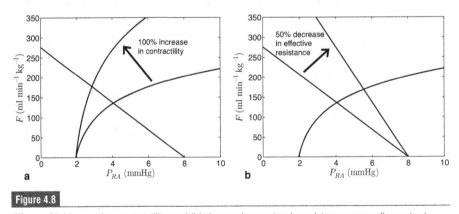

Figure 4.8

Effects of (a) increasing contractility and (b) decreasing systemic resistance on cardiac output.

in an increase in flow to $F = 169$ ml min^{-1} kg^{-1} and an *increase* in right atrial pressure to $P_{RA} = 5.55$ mmHg. Thus a 100% increase in effective conductivity ($1/R_{eq}$) results in a 24% increase in flow and a 34% increase in P_{RA}, given the parameters used in Figures 4.7 and 4.8.

The examples of Figure 4.8 are particularly instructive, because they show that a physiological increase in cardiac output could be associated with increasing, decreasing, or even constant, right atrial pressure. In exercise, the effective resistance decreases (increasing atrial pressure) and cardiac contractility increases (decreasing atrial pressure). Both the increase in contractility and the decrease in resistance cause an increase in flow; both tend to have opposing effects on right atrial pressure. Similarly, if contractility were to decrease, for example in left heart failure, the reduction in contractility would cause a decrease in flow and an increase in right atrial pressure.

Because of their simplicity and utility, diagrams such as those of Figures 4.7 and 4.8 are widely used teaching tools in graduate and medical physiology courses. They can illustrate how changes in flow and right atrial pressure are related to changes in contractility and effective resistance. However, there is some danger in relying on graphical methods such as these without understanding the underlying physical model. For one thing, since the volume shift represented by Eq. (4.6) for the linear circuit model is not explicitly apparent in this graphical analysis, these plots do tend to obscure an important phenomenon that underlies all of these changes, as pointed out by Brenglemann [11]. More importantly, textbook descriptions that accompany the analyses of these simple models are often misleading, and sometimes incorrect, and have led to widespread misunderstanding of the concept of "venous return" and its relationship to cardiac output, as discussed in the next section.

But first, let us see how we can use the Guyton circuit analysis to illustrate the stability imparted in the system by the Frank–Starling mechanism. To simulate how the system responds to a perturbation, we make our model of the system time dependent by assuming that the heart responds to changes in filling pressure with a response function defined by the Frank–Starling mechanism with time constant τ:

$$\frac{dF}{dt} = [F_{FS}(P_{RA}) - F]/\tau, \tag{4.8}$$

where $F_{FS}(P_{RA})$ is the steady-state cardiac output as a function of right atrial pressure. We may use the empirical function of Uemura et $al.$ [63] of Eq. (4.1) to define the steady-state Frank–Starling curve: $F_{FS}(P_{RA}) = F_o[\ln(P - P_o) + a]$. With a time-dependent pump, the circuit model of Figure 4.5 is governed by the equations

$$P_{RA} = P_V - F R_V$$
$$\frac{dP_A}{dt} = \frac{1}{C_A}[F - (P_A - P_V)/R_A]$$
$$\frac{dP_V}{dt} = \frac{1}{C_V}[(P_A - P_V)/R_A - (P_V - P_{RA})/R_V]. \tag{4.9}$$

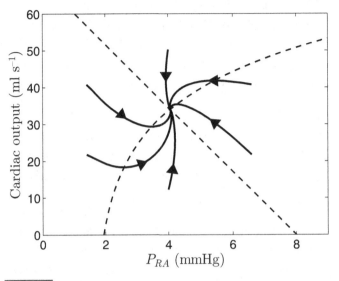

Figure 4.9

Simulation of Eqs (4.8) and (4.9). Parameter values for the circulation are $C_A = 1.838$ ml mmHg^{-1}, $C_V = 125$ ml mmHg^{-1}, $R_A = 2.453$ mmHg s ml^{-1}, and $R_V = 0.08$ mmHg s ml^{-1}. Parameter values for the heart are $\tau = 4$ s, $P_0 = 1.59$ mmHg, $a = 1.04$, and $F_0 = 17.5$ ml s^{-1}.

Figure 4.9 plots flow out of the heart versus right atrial pressure, showing what happens when the system is perturbed away from the steady-state point. (The Frank–Starling and vascular function curves are plotted as dashed lines.) Solutions governed by Eqs (4.8) and (4.9) for different initial conditions are illustrated as trajectories in the F–P_{RA} plane. Regardless of initial condition, the system relaxes to the single stable steady-state point associated with the intersection of the Frank–Starling and vascular function curves.[6]

How does the Frank–Starling mechanism impart stability? For one thing, the above equations predict only one steady-state point, and therefore multiple steady states are not possible. At a given flow, right atrial pressure is always driven toward the vascular function curve; at a given right atrial pressure, the flow is always driven toward the Frank–Starling curve. Anywhere in the F–P_{RA} plane, the system state is driven toward the stable point.

To illustrate an extreme counterexample, consider the reversed Frank–Starling curve illustrated in Figure 4.10. Here we are simulating a heart that obeys a reverse Frank–Starling mechanism where cardiac output decreases with increasing filling pressure. The "reversed" Frank–Starling curve is simulated here by reversing the

[6] See Exercise 4.4.

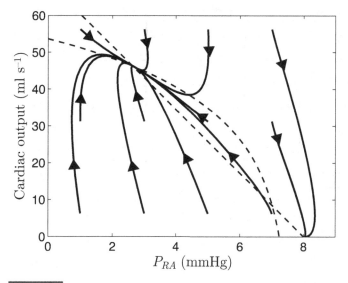

Simulation of Eqs (4.8) and (4.9), with the reversed Frank–Starling mechanism given by Eq. (4.10). Parameter values of the circulation are the same as for Figure 4.9. Parameter values for the heart are $\tau = 2$ s, $P_o = 7.59$ mmHg, $a = 1.04$, and $F_o = 17.5$ ml s^{-1}.

pressure dependence in the Uemura *et al.* equation:

$$F_{FS}(P_{RA}) = \begin{cases} F_o[\ln(P_o - P_{RA}) + a], & P_{RA} \le P_o - e^{-a} \\ 0, & P_{RA} > P_o - e^{-a}. \end{cases} \tag{4.10}$$

In this case, given the parameter values reported in the legend of Figure 4.10, there are two points where the Frank–Starling and vascular function curves intersect. Simulated trajectories reveal that only one of the intersection points is a stable steady state. Furthermore, not all initial conditions result in trajectories to that stable point. Certain trajectories lead to zero flow, with $P_{RA} = P_{MF}$. Therefore the reverse Frank–Starling mechanism would not be a particularly useful way for nature to design a heart, if indeed such a design were even physically feasible! The standard Frank–Starling heart coupled to a compliant circulation is a physiologically stable system, for which those of us with beating hearts can be grateful.

The control system associated with this compliant circuit model is illustrated in Figure 4.11. Right atrial pressure is the input to the Frank–Starling mechanism: the higher the pressure, the higher the flow. Feedback via the circulation provides a self-regulating mechanism, where an increase flow leads to a decrease in right atrial pressure, and thus restoration of flow to the stable steady-state value.

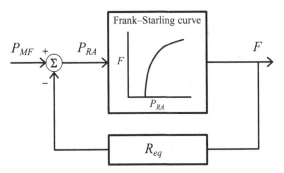

Figure 4.11

Control system of associated with Figure 4.5. The set point for the flow is $F_{FS}(P_{MF})$, the cardiac output that would be generated by a filling pressure of P_{MF}. The error function is the set point minus the feedback signal, $P_{MF} - F R_{eq}$, which is equal to P_{RA}. The output, F, is equal to $F_{FS}(P_{MF} - F R_{eq})$.

4.2.3 How the simple Guyton model is commonly misinterpreted

While we have seen throughout the examples presented so far in this book that simple mathematical models can be critically useful in capturing the important phenomena underlying a particular biological process, misinterpretation and mis-use of mathematical models can be dangerous. For example, Guyton's concept of venous return expressed by Eq. (4.7), intended to help explain in simple terms a mechanism behind the physiological regulation of cardiac output, has unintention-ally lead to widespread confusion on the subject.

Consider the statement "the amount of blood pumped by the heart each minute is determined by the rate of blood flow into the heart from the veins, which is called venous return," which expresses a common sentiment [25]. Guyton's models of the circulatory system support the idea that venous return to the heart determines cardiac output no more than the idea that cardiac output determines venous return. Indeed, in Guyton's mathematical models, and in any real circulatory system in steady state, venous return is simply a synonym for cardiac output. Thus it would be no more meaningful (and no less misleading) to point out that the amount of blood returning to the heart each minute is determined by cardiac output. Indeed, in the passage quoted above, the authors go on: "The heart in turn automatically pumps this incoming blood into the systemic arteries, so that it can flow around the circuit again." At the very least, this passage encourages the misinterpretation that the heart's role is secondary, that venous return is driven by the difference between mean filling pressure and vena cava pressure working against the resistance to venous return, and that the heart simply ejects whatever venous-determined venous return is fed into it.

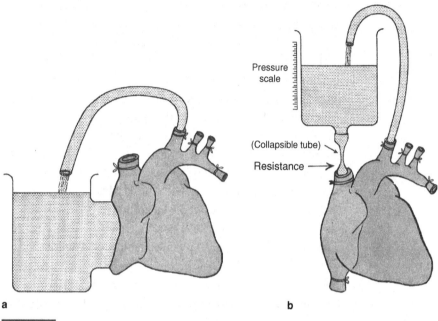

Pressure scale

(Collapsible tube)

Resistance →

a   b

Figure 4.12

Competing hypotheses explaining the physiological regulation of cardiac output. The left diagram (a) illustrates the concept that cardiac output is controlled by the heart. The circulation is thought of as a "central reservoir with blood always available to the right atrium" [23]. The right diagram (b) illustrates the concept that cardiac output is controlled by the periphery. Flow of blood to the right atrium is restricted by the circulation, and "the heart is capable of pumping any amount of blood that flows into the right atrium" [23]. Reprinted with permission from Guyton *Circulatory Physiology: Cardiac Output and its Regulation* [23], Chapter 7, Pages 128 and 130, Figures 52 and 53, Copyright Elsevier.

This confusion can be traced to Guyton, who discussed two competing models for the regulation of cardiac output: "regulation of cardiac output by the heart" and "regulation of cardiac output by the peripheral circulation." These two competing ideas are illustrated in Figure 4.12, taken from Guyton [23].

Guyton reconciled these two competing ideas as follows:

Under normal resting conditions and during light exercise, but not during very severe exercise, the heart has so much reserve pumping power that . . . it is almost entirely peripheral factors, such as peripheral resistance and degree of filling of the peripheral vessels, that control the output of the heart . . . When the heart is in a state of failure, a tremendous amount of blood is often dammed behind the heart because the heart is incapable of pumping the blood rapidly enough to keep the veins semicollapsed, which is their normal state. Under these conditions, it is obviously the heart that is primarily controlling the cardiac output.

Yet, even though Guyton asserted that in a state of heart failure the heart is most responsible for determining cardiac output, the idea that cardiac output "is determined by" venous return has been extended even to describe what is happening in heart failure: "[the venous return] curve shows that when heart pumping fails and causes the right atrial pressure to rise, the backward force of the rising atrial pressure on the veins of the systemic circulation decreases venous return of the blood to the heart" [25]. The first part of this statement, that decreasing cardiac contractility results in an increase in right atrial pressure, is justified by the Guyton model, and illustrated by the graphical analysis of Figure 4.8. However, the cause-and-effect of the second part is backwards, as illustrated in Figure 4.11. It is the decrease in venous return (equal to cardiac output) that causes the right atrial pressure to rise, not the other way around!

To sum up, what we have here is a confusion of contradictory statements, all apparently backed up by the same simple model and associated data sets. So which is it: "regulation of cardiac output by the heart" or "regulation of cardiac output by the peripheral circulation"? An understanding of the simple model of Figure 4.5 reveals of course that neither is meaningful under any reasonable circumstance. At a given level of contractility (given the Frank–Starling cardiac output curve) and given fixed mechanical properties of the circulation (fixed vascular function curve), the cardiac output is determined by the combined *system* of the heart pumping into and being supplied by blood flow through the circulation.[7]

4.3 Pulsatile mechanics of the circulatory system

In the Guyton analysis of the circulation, which ignores the pulsatile nature of cardiac pumping and circulatory blood flow, flow and pressure variables represent mean values averaged over several cardiac cycles. That simplification is valid and appropriate in investigating certain phenomena, such as the interaction between mean pressures, flows, and volumes in the arteries and veins. Yet we appreciate the heartbeat as a fundamental sign of life in higher animals. Analyzing the rhythmic pulsatile nature of the heartbeat yields insights not available from time-independent analysis of the function of the circulation.

Figure 4.13 illustrates how pressures in the left and right atria and ventricles, the aorta, and the pulmonary artery vary over the course of a heartbeat. Two cardiac cycles (heartbeats) are illustrated, with right-heart (pulmonary) pressures

[7] Personally, I believe that the fundamental root of all of this confusion is the choice of expressing Eq. (4.4) as (4.7), falsely implying that cardiac output is a variable *determined* by right atrial pressure and R_{eq}. If you have the opportunity, ask a physician, "What determines cardiac output?" His or her answer might tell you whether he or she thinks in terms of Figure 4.12(a) or Figure 4.12(b) – or equivalently in terms of Eq. (4.4) or Eq. (4.7).

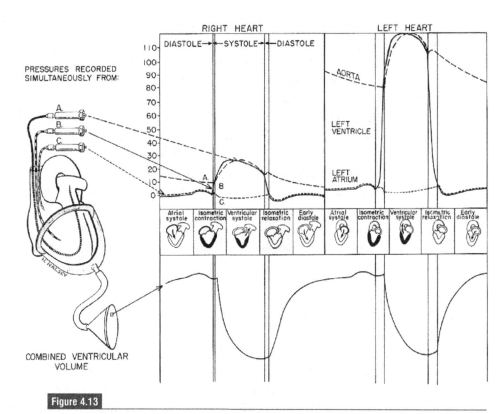

PRESSURES RECORDED
SIMULTANEOUSLY FROM:

COMBINED VENTRICULAR
VOLUME

Figure 4.13

Cardiac pressures and volumes over the cardiac cycle. Reprinted from Rushmer [52] with permission.

---

illustrated in the first cycle and left-heart (systemic) pressures illustrated in the second. The lower panel shows how total (right plus left) ventricular volume changes over the cardiac cycle.

Let us first consider the isometric (isovolemic) phase of the cardiac cycle. During this phase the ventricles are filled with blood and the muscles of the ventricle walls contract, building up pressure in the chamber. The contraction remains isometric as long as the ventricular pressure stays below the pressure of the major artery that the ventricle feeds. For example, in the left heart the isometric contraction phase lasts until the pressure in the chamber exceeds that in the aorta, at which point the aortic valve opens, and blood flows from the ventricle through the aorta and into the systemic circulation. In Figure 4.13, the phase of the cardiac cycle associated with ventricular contraction pumping blood out of the ventricle is identified as *ventricular systole*. It is also commonly called the *ejection phase*. (The whole period of ventricular contraction, including the isometric contraction and the ejection phase, is called *systole*.) The ejection phase (or ventricular systole) proceeds until the ventricle relaxes, the aortic valve closes, and ventricular pressure drops rapidly and drastically – from values in the neighborhood of 100 mmHg to close to the left atrial pressure of around 5–10 mmHg.

Because of compliance in the aorta and the systemic arteries, the aortic pressure does not drop as rapidly as ventricular pressure during ventricular relaxation. During this period, called *diastole*, when the aortic valve is closed and the ventricles remain relaxed, the aortic pressure gradually decreases from approximately 110 mmHg to 80 mmHg. In diastole the ventricular pressure drops below the atrial pressure, allowing the ventricles to fill, preparing the heart for the next isometric contraction and the next period of systole.

The time course of ventricular volume changes over the cardiac cycle is illustrated in the lower panel of Figure 4.13. Approximately 50–60% of the maximal diastolic ventricular volume is ejected from the left ventricle during systole in the hearts of large mammals, such as humans. The *ejection fraction* can be as high as 80–90% in rats and mice.

4.3.1 Time-varying elastance models of the heart

One simple effective approach to simulating the time-dependent pumping of the heart is to treat the chambers of the heart as having time-varying *elastances* that generate the pressures that drive the blood flow. (Elastance, or elasticity, is the inverse of compliance.) Pressure in, for example, the left ventricle, may be simulated as

$$P_{LV}(t) = E_{LV}(t) \cdot [V_{LV}(t) - V_o], \qquad (4.11)$$

where $E_{LV}(t)$ is the left ventricular elastance, $V_{LV}(t)$ is the volume of blood in the ventricle, and V_o is the unstressed volume. The varying elastance of the left ventricle is modeled as a smooth function of time that rises in amplitude as the heart develops pressure in systole and decreases as the heart relaxes.

Ellwein [18] describes a simple and effective varying elastance model for the left ventricle that captures time-dependent pumping in a four-parameter function:

$$E_{LV}(\theta) = \begin{cases} \dfrac{(E_{max} - E_{min})}{2}\left[1 - \cos\left(\dfrac{\pi\theta}{T_M}\right)\right] + E_{min}, & 0 \le \theta \le T_M \\[2ex] \dfrac{(E_{max} - E_{min})}{2}\left[\cos\left(\dfrac{\pi(\theta - T_M)}{T_R}\right) + 1\right] + E_{min}, & T_M \le \theta \le T_M + T_R \\[2ex] E_{min}, & T_M + T_R \le \theta \le T, \end{cases}$$

$$(4.12)$$

where T is the period of the cardiac cycle and θ is the elapsed time within a given cardiac cycle, $\theta \in (0, T)$. The parameters E_{max} and E_{min} represent the maximal systolic and minimal diastolic elastances, respectively; T_M is the time

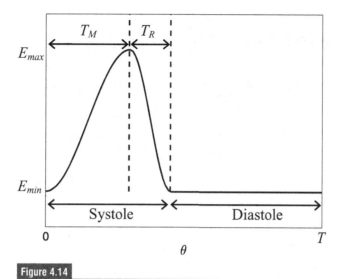

Figure 4.14

$E_{LV}(\theta)$ predicted by the varying elastance model of Ellwein [18].

to reach maximal elastance; and T_R is the relaxation time from peak elastance to reach minimal elastance. The shape of the curve predicted by Eq. (4.12) and the contraction and relaxation times are illustrated in Figure 4.14. We can see that this phenomenological model represents the contractility of the heart as a continuously differentiable function that increases to a maximum during systole and relaxes to a constant value (E_{min}) during diastole.

The behavior of this model for left-ventricular pumping may be explored by incorporating it in a circuit model for the systemic circulation, as illustrated in Figure 4.15. Here we have invoked a three-element model of the circulation (two resistors R_{out} and R_s, and one capacitor C_s), hooked up to a model of the left ventricle based on Eq. (4.12). The left-ventricle model includes a fixed pressure source (P_{LA}), which feeds into the ventricle; the left ventricle is represented by the variable capacitor in the figure. The mitral valve, separating the atrium from the ventricle, is treated as an ideal diode – an element that allows flow only in the direction indicated by the arrow. An ideal diode is also used to model the aortic valve.

The circuit of Figure 4.15 is governed by a system of ordinary differential equations derived as follows. The variable θ, which represents the time elapsed in the cardiac cycle, and varies between 0 and T, is computed from the modulus of time t divided by the period T: $\theta = \text{modulus}(t, T)$. The modulus operator (invoked with the syntax $\text{mod}(t, T)$ in MATLAB) returns the remainder of the division of t by T. Given a value of θ and V_{LV}, the left ventricular pressure $P_{LV}(t, V_{LV})$ in this model is computed from Eq. (4.12).

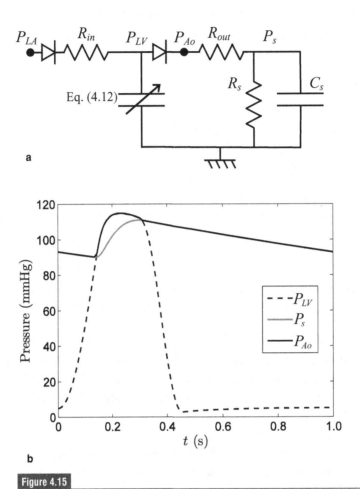

Figure 4.15

(a) A simple model of the circulation connected to a varying elastance model of the left ventricle. The governing equations for the model are given by Eqs (4.12)–(4.15). (b) Model-predicted pressures in the circuit. Parameter values for the circuit model are $R_{in} = 0.01$ mmHg s ml^{-1}, $R_{out} = 0.05$ mmHg s ml^{-1}, $R_s = 4.0$ mmHg s ml^{-1}, and $C_s = 1.0$ mmHg ml^{-1}. Left atrial pressure is set to $P_{LA} = 5$ mmHg. Parameter values for the left-ventricle model are $T = 1$ s, $T_M = 0.3$ s, $T_R = 0.15$ s, $V_o = 4.85$ ml, $E_{max} = 3.34$ mmHg ml^{-1}, and $E_{min} = 0.085$ mmHg ml^{-1}.

The differential equation for ventricular volume is a simple statement of mass conservation:

$$\frac{dV_{LV}}{dt} = F_{LA\_to\_LV} - F_{LV\_to\_S}, \tag{4.13}$$

where $F_{LA\_to\_LV}$ is the flow from the atrium to the ventricle and $F_{LV\_to\_S}$ is the flow from the ventricle to the systemic circulation. These flows are computed, accounting for the one-way valves in the circuit:

$$F_{LA\_to\_LV} = \max \left[\frac{P_{LA} - P_{LV}(t)}{R_{in}}, 0 \right]$$

$$F_{LV\_to\_S} = \max \left[\frac{P_{LV}(t) - P_s(t)}{R_{out}}, 0 \right], \qquad (4.14)$$

where P_s is the pressure at the node indicated in the figure. The pressure P_s is governed by Eq. (9.24), where the flow into the capacitor C_s is $F_{LV\_to\_S} - P_s/R_s$:

$$\frac{dP_s}{dt} = (F_{LV\_to\_S} - P_s/R_s)/C_s. \qquad (4.15)$$

Finally, we may compute the model-predicted aortic pressure as $P_{Ao}(t) = P_s(t) + F_{LV\_to\_S}R_{out}$.

The circuit model of Figure 4.15 (given the parameter values listed in the legend) predicts that aortic pressure varies from a peak of approximately 115 mmHg during systole to a minimum of approximately 90 mmHg. The aortic pressure follows the left ventricular pressure wave when left ventricular pressure exceeds systemic pressure. Once left ventricular pressure drops below systemic pressure, the aortic valve closes and the left ventricular pressure rapidly drops while P_{Ao} and P_s are maintained by the volume stored in the capacitor C_s. The systemic pressures P_{Ao} and P_s gradually decline during diastole as volume flows out of the the capacitor through the systemic resistances R_s.

The pressure time courses plotted in Figure 4.15 reproduce many features of the aortic and left ventricular pressures illustrated in Figure 4.13. Features not captured by the simple model include the small variations in left atrial pressure that are seen in Figure 4.13. In fact, here the model assumes that P_{LA} remains constant. Another feature not captured by the model is the observed transient increase in aortic pressure that occurs following the closing of the aortic value (which occurs when the left ventricular pressure drops below aortic pressure). This transient increase in aortic pressure while ventricular pressure is dropping results in a small notch (called the *dicrotic notch*) in the aortic pressure waveform. Pressure increases immediately following the closing of the aortic valve owing to inertial wave propagation in the aorta, a phenomenon that is explored in the following section.

4.3.2 Simulation of the aortic pressure waveform

To simulate the systemic pressure in a more realistic fashion than that afforded by the simple model of Figure 4.15 we can represent the circulatory system using a

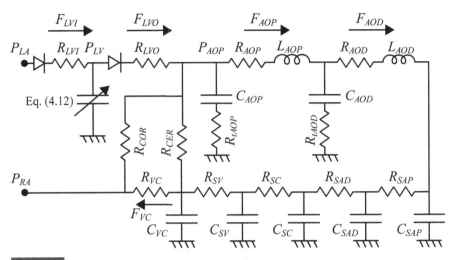

Figure 4.16

Model of the systemic circulation adapted from Olansen *et al.* [44].

circuit that breaks the system down into a model that contains more representative discrete elements. The circuit of Figure 4.16, adapted from Olansen *et al.* [44], breaks the systemic circulation down into nine resistive blocks associated with the proximal and distal aorta, proximal and distal systemic arteries, systemic capillaries, systemic veins, vena cava, coronary circulation, and cerebral circulation. The subscript notation used in the circuit diagram and for the variables in the model is defined in Table 4.1.

Here there is a capacitance associated with each of the nine circulatory compartments, with the exceptions of the cerebral and coronary circulations. (Since relatively little volume is stored in the cerebral and coronary circulations, the capacitances of these components do not significantly influence the whole-network behavior, and thus may be ignored for the purpose of simulating whole-network function.)

Transmural resistances (R_{tAOP} and R_{tAOD}) are associated with the proximal and distal aortic compartments. These resistances model the effective resistance of these large vessels to transmural flow – that is, flow that increases or decreases the volume of the vessel compartment. Transmural resistances are ignored in the remaining circulatory compartments.

This model assumes that blood flowing in the aortic vessel compartments has finite momentum, modeled using the inductor elements L_{AOP} and L_{AOD}. Pressure drops across these elements are proportional to the rate of change in flow through them, according to Eq. (9.25). The momentum (or inductance or *inertance*) associated with these elements transmits the inertial wave that is responsible for the dicrotic notch in the aortic pressure transient.

Table 4.1: Parameter values for circuit elements to simulate the aortic pressure wave and flows in the systemic circulation. Values are expressed in pressure, volume, and time units of mmHg, ml, and seconds.

Identifier	Description	Resistance	Capacitance	Inductance
LVI	LV input	0.005	–	–
LVO	LV output	0.01	–	–
AOP	Proximal aorta	0.0409	0.4255	0.0044
tAOP	Transmural AOP	0.0717	–	–
AOD	Distal aorta	0.0470	0.3146	0.0081
tAOD	Transmural AOD	0.0518	–	–
SAP	Proximal arteries	0.9089	0.25	–
SAD	Distal arteries	0.8144	0.50	–
SC	Systemic capillaries	0.6180	0.01	–
SV	Systemic veins	0.0727	100	–
VC	Vena cava	0.0364	250	–
COR	Coronary vessels	31	–	–
CER	Cerebral circulation	15.6	–	–

The parameter values associated with the model circuit element are listed in Table 4.1. All values are expressed in units of milliliters, mmHg, and seconds.

The network looks complicated, but simulation and analysis requires nothing more than a systematic application of the rules outlined in Section 9.7 of the Appendices. There actually exist a number of alterative ways to express the governing equations for this model. For example, differential equations for pressure and volume of a capacitive element are equivalent and redundant. Assuming that the pressure variables are computed explicitly (from the differential and algebraic equations developed below), the following flows may be computed directly:

$$F_{LVI} = \max \left(\frac{P_{LA} - P_{LV}}{R_{LVI}}, 0 \right)$$

$$F_{LVO} = \max \left(\frac{P_{LV} - P_{AOP}}{R_{LVO}}, 0 \right)$$

$$F_{COR} = (P_{AOP} - P_{RA})/R_{COR}$$

$$F_{CER} = (P_{AOP} - P_{VC})/R_{CER}$$

$$F_{VC} = (P_{VC} - P_{RA})/R_{VC}. \tag{4.16}$$

Here the subscript in a flow variable indicates the resistive element with which it is associated. For example, the flow into the left ventricle (F_{LVI}) is computed from the pressure drop across the input resistor, divided by the input resistance. As in the previous model, here $P_{LV}(t)$ is modeled using the simple varying elastance model of Eq. (4.12). Note that the coronary flow drains into the coronary venous sinus, downstream of the main trunk of the vena cava, while the cerebral circulation drains upstream of the vena cava element.

The transmural flows may be computed from the application of Kirchhoff's current law:

$$F_{tAOD} = F_{AOP} - F_{AOD}$$

$$F_{tAOP} = F_{LVO} - F_{AOP} - F_{COR} - F_{CER}. \tag{4.17}$$

These expressions are simple statements of mass conservation that arise from the network topology.

The rate of change of pressure in each capacitor in the network is governed by the flow into (or out of) the element. For example, the rate of volume change of the vena cava (dV_{VC}/dt) is equal to the flow in from the systemic veins and the cerebral circulation ($(P_{SV} - P_{VC})/R_{SV} + F_{CER}$) minus the flow out of the vena cava (F_{Vc}). Dynamic mass balances such as this yield ordinary differential equations for pressures in the vena cava, systemic veins, systemic capillaries, distal arteries, and proximal arteries:

$$\frac{dP_{VC}}{dt} = [(P_{SV} - P_{VC})/R_{SV} + F_{CER} - F_{VC})]/C_{VC}$$

$$\frac{dP_{SV}}{dt} = [(P_{SC} - P_{SV})/R_{SC} - (P_{SV} - P_{VC})/R_{SV})]/C_{SV}$$

$$\frac{dP_{SC}}{dt} = [(P_{SAD} - P_{SC})/R_{SAD} - (P_{SC} - P_{SV})/R_{SC})]/C_{SC}$$

$$\frac{dP_{SAD}}{dt} = [(P_{SAP} - P_{SAD})/R_{SAP} - (P_{SAD} - P_{SC})/R_{SAD}]/C_{SAD}$$

$$\frac{dP_{SAP}}{dt} = [F_{AOD} - (P_{SAP} - P_{SAD})/R_{SAP}]/C_{SAP}. \tag{4.18}$$

For the left ventricle and aortic compartments we express the differential equations in terms of the volumes based on the following dynamic mass balances:

$$\frac{dV_{LV}}{dt} = F_{LVI} - F_{LVO}$$

$$\frac{dV_{AOD}}{dt} = F_{tAOD}$$

$$\frac{dV_{AOP}}{dt} = F_{tAOP}. \tag{4.19}$$

Derivation of differential equations for the aortic flows follows from application of Kirchhoff's voltage law:

$$P_{AOD} = P_{SAP} + L_{AOD}\frac{dF_{AOD}}{dt} + R_{AOD}F_{AOD}$$

$$P_{AOP} = P_{AOD} + L_{AOP}\frac{dF_{AOP}}{dt} + R_{AOP}F_{AOP}. \tag{4.20}$$

Here P_{AOD} is computed from the downstream pressure P_{SAP} plus the pressure drops across the downstream inductor and resistor. Likewise, P_{AOP} is computed from the downstream pressure P_{AOD} plus the pressure drops across the downstream elements. Rearranging these expression yields

$$\frac{dF_{AOD}}{dt} = (P_{AOD} - P_{SAP} - R_{AOD}F_{AOD})/L_{AOD}$$

$$\frac{dF_{AOP}}{dt} = (P_{AOP} - P_{AOD} - R_{AOP}F_{AOP})/L_{AOP}. \tag{4.21}$$

Since we have an equation for V_{AOD}, we may compute P_{AOD} directly from

$$P_{AOD} = V_{AOD}/C_{AOD} + F_{tAOD}R_{tAOD}, \tag{4.22}$$

which gives the pressure as the sum of the pressure drops across the distal aortic capacitor and the transmural resistor.

These expressions provide an explicit system of differential equations for five pressures, three volumes, and two flows, for a total of ten state variables. All that is missing is an expression for P_{AOP} to apply in Eqs (4.16) and (4.21). This equation for P_{AOP} is a little bit trickier to obtain than that for P_{AOD}. We cannot directly use the expression

$$P_{AOP} = V_{AOP}/C_{AOP} + F_{tAOP}R_{tAOP},$$

because our expression for F_{tAOP} depends on F_{LVO}, which depends on P_{AOP}. So we need another independent equation to find an expression for P_{AOP}. To obtain that, we write the following set of equations:

$$R_{COR}F_{COR} = P_{AOP} - P_{RA}$$

$$R_{CER}F_{CER} = P_{AOP} - P_{VC}$$

$$F_{tAOP} = F_{LVO} - F_{AOP} - F_{COR} - F_{CER}$$

$$P_{AOP} = V_{AOP}/C_{AOP} + R_{tAOP}F_{tAOP}$$

$$F_{LVO} = \max\left(\frac{P_{LV} - P_{AOP}}{R_{LVO}}, 0\right). \tag{4.23}$$

These five equations can be solved for the five unknowns F_{COR}, F_{CER}, F_{tAOP}, F_{LVO}, and P_{AOP}. Thus Eq. (4.23) may be manipulated to give P_{AOP} as a

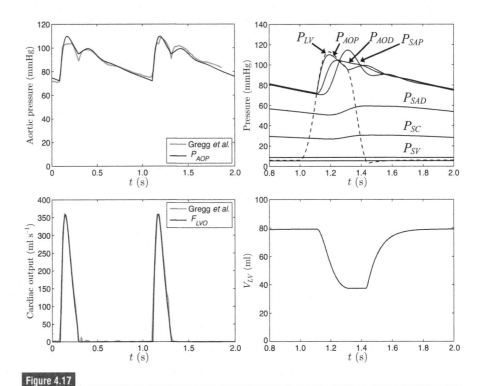

Figure 4.17

Simulation of heart beat using varying elastance model of the left ventricle and circuit model of Figure 4.16. The data on the aortic pressure and the cardiac outflow are from Gregg $et\ al.$ [21]. Parameter values used are listed in Table 4.1 for the circuit, and for the heart: $H = 0.98\ \text{s}^{-1}$, $T_M = 0.3\ \text{s}$, $T_R = 0.15\ \text{s}$, $V_0 = 5.04\ \text{ml}$, $E_{max} = 3.0\ \text{mmHg ml}^{-1}$, and $E_{min} = 0.081\ \text{mmHg ml}^{-1}$.

function of the flow and volume variables, for which we have independent explicit equations:

$$P_{AOP} = \begin{cases} \dfrac{\dfrac{V_{AOP}}{C_{AOP}} + R_{tAOP}\left[\dfrac{P_{LV}}{R_{LVO}} - F_{AOP} + \dfrac{P_{RA}}{R_{COR}} + \dfrac{P_{VC}}{R_{CER}}\right]}{1 + R_{tAOP}\left(\dfrac{1}{R_{LVO}} + \dfrac{1}{R_{COR}} + \dfrac{1}{R_{CER}}\right)} & P_{LV} > P_{AOP} \\[20pt] \dfrac{\dfrac{V_{AOP}}{C_{AOP}} + R_{tAOP}\left[-F_{AOP} + \dfrac{P_{RA}}{R_{COR}} + \dfrac{P_{VC}}{R_{CER}}\right]}{1 + R_{tAOP}\left(\dfrac{1}{R_{COR}} + \dfrac{1}{R_{CER}}\right)} & P_{LV} \le P_{AOP}. \end{cases}$$

(4.24)

With this equation, we have a complete set of equations to simulate the circuit of Figure 4.16 as a system of ten state variables.

Simulations from the model are plotted in Figure 4.17 and compared with data from Gregg $et\ al.$ [21]. The data on aortic pressure and cardiac output from Gregg $et\ al.$ [21] were measured in a conscious dog, and the model parameters were adjusted here to match the observed data. (Parameter values for the circuit model are listed in Table 4.1; parameter values for the varying elastance left ventricle are listed in the figure legend. The left and right atrial pressures are set at $P_{LA} = 6$ mmHg and $P_{RA} = 3$ mmHg.)

The ventricular volume curve follows a shape similar to that shown in Figure 4.13 with a predicted ejection fraction of 54%. The aortic valve is open for only 21% of the cardiac cycle, resulting in the short bursts of F_{LVO} illustrated in Figure 4.17. The magnitude of variability in the pressure wave decreases from the aorta to the small vessels. The aortic pressure varies over a range of approximately 40 mmHg from peak systolic to minimum diastolic pressure, while the pressure at the inlets of the capillaries (P_{SC}) varies over a range of approximately 7 mmHg. The predicted capillary flow is proportional to $P_{SC} - P_{SV}$ and is essentially steady compared with the time dependence F_{LVO}. Likewise, the variation in flow in upstream vessels is increasingly buffered by capacitance in the circuit as the pressure wave moves downstream.

The dicrotic notch in the aortic pressure wave arises because the shutting of the aortic valve causes the flow supplied to the proximal aorta from the left ventricle to be suddenly cut off while the inductance of the aortic compartments maintains flow. The relatively stiff proximal aorta is filled to its maximal volume of the cardiac cycle and starts to empty before the valve closes. While the valve remains open, the upstream (proximal) aortic pressure drops with the dropping left ventricular pressure at the end of the ejection phase, as demonstrated in the pressure traces in Figure 4.17. Owing to inertia and compliance in the system, the downstream (distal) aortic pressure does not drop as fast as the proximal pressure, causing P_{AOD} to be temporarily greater than P_{AOP}. This pressure difference leads to a temporary reversal in flow following the shutting of the valve and a rebound in proximal aortic pressure.

Pressure wave interference and interaction through the circuit lead to a small increase in the magnitude of the pressure wave when it reaches the upstream ends of the arteries (P_{SAP}), illustrated in Figure 4.18. This figure shows how the pressure wave is magnified at the bottom of the aorta, in P_{SAP}, which is the upstream end of the systemic arteries, before largely disappearing downstream of P_{SAP}.

4.4   Dynamic changes in blood pressures and flows

Now that we understand the basic mechanics of the circulatory system, we may use the models we have developed to explore how the system responds to perturbations such as dynamic changes in filling pressure P_{LA} and heart rate H. For this purpose, we may use the simple model of Section 4.3.1 and Figure 4.15. (Using the simple noninductive model, the effects of inertial wave propagation are ignored. Repeating

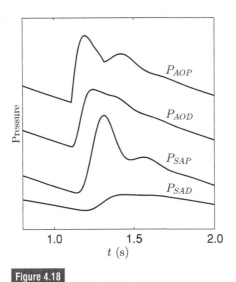

Figure 4.18

Illustration of arterial pressure wave propagation in the systemic circulation. The pressure waves are arbitrarily shifted on the pressure axis for visualization.

the following analysis with the more complex model of Figure 4.16 would yield results qualitatively similar to what we will find with the simpler model.)

First, let us simulate what happens when the filling pressure is increased and maintained at a new value. Figure 4.19(a) shows the predicted aortic pressure wave and cardiac output when the filling pressure is increased from 5 to 7 mmHg at time $t = 10$ s. The system starts in a periodic steady state with the pressure wave of that in Figure 4.15. Within a few heartbeats following the increase in filling pressure, a new steady state with a peak systolic value of over 160 mmHg is reached. The cardiac output increases commensurately with aortic pressure. (Model parameters for this simulation are the same as those in Figure 4.15.)

Figure 4.19(b) plots predicted average flow over the cardiac cycle as a function of the filling pressure. The amount the ventricle fills during diastole increases linearly with the filling pressure; and with the ejection fraction remaining nearly constant, the cardiac output is proportional to the ventricular volume. Thus the model predicts a linear relationship between P_{LA} and cardiac output. Since pressures in this model are measured relative to the ground pressure (0 mmHg) in Figure 4.15, the cardiac output is zero at $P_{LA} = 0$. Since the right atrial pressure – the output of the systemic circulation – is set to ground in Figure 4.15, the left atrial pressure in this model is measured relative to right atrial pressure.

The model-predicted linear relationship between cardiac output and filling pressure is not consistent with the nonlinear cardiac output curves plotted in Figure 4.2.

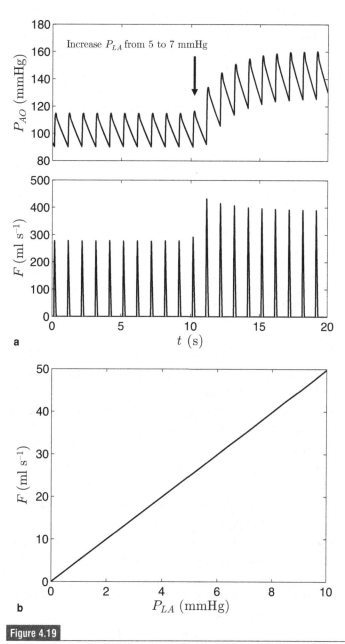

Figure 4.19

Response in aortic pressure and cardiac output to changes in filling pressure: (a) model-predicted aortic pressure and cardiac output following a sudden increase (from 5 to 7 mmHg) in left atrial pressure; (b) mean steady-state cardiac output as a function of left atrial pressure.

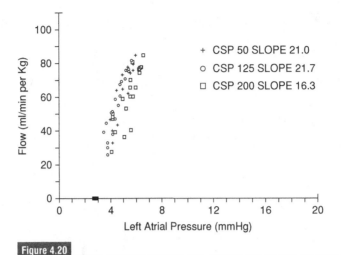

Figure 4.20

Cardiac output curve data measured in the canine model of Greene and Shoukas [20]. (CSP stands for carotic sinus pressure, which was varied independently in these experiments.) Figure reprinted from Greene and Shoukas [20], with permission.

Yet the predicted linear cardiac output curve is consistent with other experimental observations. See, for example, data from Greene and Shoukas [20] plotted in Figure 4.20, which are consistent with a linear relationship. The exact nature of the observed cardiac output curve depends on the experimental prep used to obtain the measurements. Uemura *et al.* [63] obtained the data in Figure 4.2 by varying the total blood volume: left and right atrial pressures (and indeed pressure throughout the circulation) increase in proportion to blood volume. Thus in Uemura *et al.*'s experiments the heart is working against an increasing pressure head as the right and left atrial pressures increase. Greene and Shoukas [20] obtained the data in Figure 4.20 using a pump to vary the flow while maintaining constant blood volume. In Figure 4.21 we are simulating the system of Figure 4.15 that drains to ground. Thus there is no influence of output flow on the filling pressure.

Next, let us simulate what happens when the heart rate H is increased and maintained at a new value. Figure 4.21(a) shows the predicted aortic pressure wave and cardiac output when H is increased from 1 to 1.5 s^{-1} at time $t = 10$ s. The system starts in the same periodic steady state observed in the initial period in the previous simulation (Figure 4.19). Following the step increase in heart rate, the pressure goes up while the cardiac output per heartbeat decreases slightly. (Again, model parameters for this simulation are the same as those in Figure 4.15.) The flow per heartbeat decreases with increasing H because as H increases the shorter diastolic period allows less time for filling, resulting in less blood in the ventricle at the beginning of the ejection phase. As a result, less blood is pumped

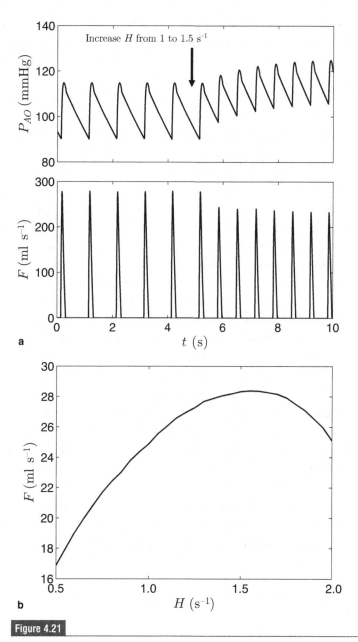

Figure 4.21

Response in aortic pressure and cardiac output to changes in heart rate: (a) show model-predicted aortic pressure and cardiac output following a sudden increase in H (from 1 to 1.5 s^{-1}); (b) mean steady-state cardiac output as a function of H.

per heartbeat when $H = 1.5$ s^{-1}, compared with when $H = 1$ s^{-1}. Yet the higher heart rate, multiplied by a slightly lower volume of blood pumped per beat, results in an increased average cardiac output, as illustrated in Figure 4.21(b).

Figure 4.21(b) plots average cardiac output predicted by the model as a function of H. We can see that there is a maximum in the plot: as H increases from 1 to about 1.6 s^{-1}, cardiac output increases with H. As H is increased even more, the cardiac output starts to drop, as the filling and ejection of the ventricle cannot keep up with the cardiac cycle.

These results imply that cardiac output is essentially the same at heart rates of 60 and 120 beats per minute. Yet one expects that when the pulse doubles, the cardiac output ought to increase substantially, not remain constant. Indeed, cardiac output does increase in mammals over a heart-rate range of more than threefold in some species (including in physically fit humans). So what is our model missing? At least two major additional phenomena are necessary to account for exercise-induced increases in cardiac output. First, our model assumes that the contraction and relaxation times are constant. For the simulations presented above, $T_M = 0.3$ s, $T_R = 0.15$ s, resulting in a total systolic time of 0.45 seconds. Thus at $H = 2$ s^{-1} the diastolic phase is only 0.05 s – not nearly enough time to effectively fill the left ventricle. More accurately capturing effects related to exercise would require accounting for decreases in T_M and T_R with increasing H.

Second, the simulations described above assume that cardiac contractility (represented by E_{max} and E_{min}) remains constant. In exercise, cardiac contractility increases and systemic resistance decreases, helping to boost cardiac output to transport oxygen more rapidly to the exercising muscles. Changes in heart rate, contraction times, and cardiac contractility in exercise are effected by a number of physiological mechanisms, including stimulation of the heart by sympathetic nerves that occurs with exercise. The stimulation of sympathetic (and parasympathetic) nerves is influenced by negative-feedback control systems, such as the baroreflex system described in the next section.

4.4.1 Baroreceptor control of systemic pressure

Arterial baroreceptors are sensory cells located in the aorta and the carotid arteries that respond to changes in pressure in the large systemic arteries. Stimulation of baroreceptors effects an autonomic nervous system response that leads to decreased blood pressure through lowered heart rate, contractility, and arterial tone.

Olufsen *et al.* [45] employ a model of this system to analyze data from human subjects performing the simple procedure of rising to stand from a sitting position. The physiology of the baroreflex system and the Olufsen *et al.* model are conveniently described in tandem. In the model, the baroreceptors are assumed to respond to a time-averaged mean pressure $\bar{p}(t)$, which in terms of our simple three-element model for the circulation of Figure 4.15 is computed

$$\frac{d\bar{p}}{dt} = \alpha(P_{Ao} - \bar{p}), \tag{4.25}$$

where the rate constant α is a model parameter. The baroreceptors fire nerve impulses at a rate $n = n_l + N$, where n_l is the rate of firing of long-term responding cells and N is the baseline firing rate.[8] The variable n_l is governed by the equation

$$\frac{dn_l}{dt} = k_l \frac{d\bar{p}}{dt} \frac{n(M-n)}{(M/2)^2} - \frac{n_l}{\tau_l}, \tag{4.26}$$

where M is the maximal firing rate, and the rate constant k_l and time constant τ_l are model parameters.

Via Eq. (4.26), increases in pressure result in increases in n; decreases in pressure decrease n. In the absence of pressure changes, the n_l decays to zero and n stays at the baseline rate of N. The parasympathetic tone is assumed proportional to baroreceptor firing rate. In the Olufsen *et al.* model, the nondimensional parameter T_{para} represents the parasympathetic tone:

$$T_{para} = n/M. \tag{4.27}$$

The influence of firing rate on the sympathetic tone (T_{symp}) is assumed to be somewhat more complex. In the Olufsen model the sympathetic tone depends on the firing rate delayed by τ_d:

$$T_{sym} = \frac{1 - n(t - \tau_d)/M}{1 + \beta T_{para}(n)}, \tag{4.28}$$

where β is another model parameter that represents the damping influence of the parasympathetic tone on the sympathetic system. Instead of using a delayed function, we implement an approximate version of this model by introducing a new variable to represent the delayed firing rate:

$$\frac{dn_{delay}}{dt} = \left(n(t) - n_{delay}(t)\right)/\tau_d \tag{4.29}$$

and modeling T_{sym} as

$$T_{sym} = \frac{1 - n_{delay}/M}{1 + \beta T_{para}(n)}. \tag{4.30}$$

The model implements linear kinetics of the unitless sympathetic and parasympathetic concentrations of neurotransmitters norepinephrine and acetylcholine,

---

[8] Here we are ignoring the contributions from short- and medium-term responding cells.

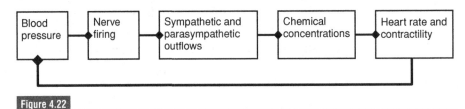

Figure 4.22

Diagram of baroreflex system.

C_{nor} and C_{ach}, respectively:

$$\frac{dC_{nor}}{dt} = \left(-C_{nor} + T_{sym}\right)/\tau_{nor}$$

$$\frac{dC_{ach}}{dt} = \left(-C_{ach} + T_{para}\right)/\tau_{ach}. \tag{4.31}$$

These concentrations are assumed to influence heart rate and contractility through the following relation:

$$f = 1 + M_s C_{nor} - M_p C_{ach}$$

$$H = f H_o$$

$$E_{max} = f E_{max,o}, \tag{4.32}$$

where f is a scaling factor that depends on C_{nor} and C_{ach}. (The Olufsen *et al.* model does not explicitly simulate cardiac contractility; here we have assumed the same scaling factor multiplying heart rate and contractility. Influences of C_{nor} and C_{ach} on arterial resistance are ignored.) Here, E_{max} is the maximal elastance of our simple varying elastance model for the left ventricle, and H is the heart rate. Since heart rate is variable in this model, we introduce an equation for θ, the time variable that tracks progress within a cardiac cycle. Adopting the approach of Olufsen *et al.*, the differential equation for θ is

$$\frac{d\theta}{dt} = H, \tag{4.33}$$

and the value of θ is reset to 0 every time it reaches 1 [45].

The overall operation of the baroreflex system is illustrated in Figure 4.22. The blood pressure time course affects baroreceptor firing rate (Eq. (4.26) in the model), which affects sympathetic and parasympathetic tones (Eqs (4.27) and (4.30)), which affect neurotransmitter concentrations (Eq. (4.31)), which affect heart rate and contractility (Eq. (4.32)).

The influence of the baroreflex system on modulating blood pressure is illustrated in Figure 4.23, where the simulation experiment of Figure 4.19 is repeated. When the filling pressure is increased from 5 to 7 mmHg, the blood pressure increases acutely, but then returns closer to the original range within a few

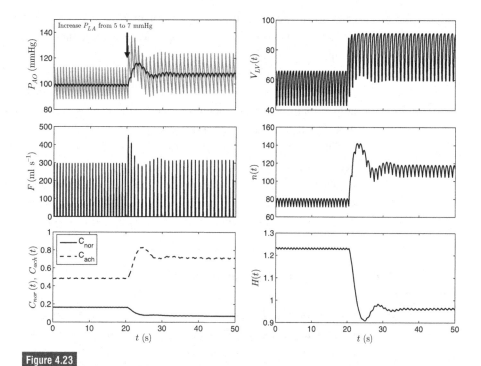

Figure 4.23

Response in aortic pressure and cardiac output to changes in filling pressure with baroreflex system. Parameter values for the baroreflex system model are: $\alpha = 0.78$ s^{-1}, $N = 80$ s^{-1}, $M = 160$ s^{-1}, $\tau_l = 350$ s, $\tau_d = 6.12$ s, $\tau_{nor} = 0.72$ s, $\tau_{ach} = 1.32$ s, $\beta = 4.48$, $k_l = 5$, $M_s = 1.222$, $M_p = 0.444$, $R_{in} = 0.01$ mmHg s ml^{-1}, $R_{out} = 0.05$ mmHg s ml^{-1}, $R_s = 3.5$ mmHg s ml^{-1}, $C_s = 1.0$ mmHg ml^{-1}. Parameter values for the left-ventricle model are: $H_0 = 1.25$ s^{-1}, $T_M = 0.3$ s, $T_R = 0.15$ s, $V_0 = 4.0$ ml, $E_{max,o} = 2.8$ mmHg ml^{-1}, $E_{min} = 0.08$ mmHg ml^{-1}.

seconds. The reduction in the blood pressure compared with the transient response with no baroreflex system (see Figure 4.19) is accomplished through a reduction in the heart rate and cardiac contractility.

Recall that in the simulation without the baroreflex system, the aortic pressure increased by more than 40 mmHg in response to the step increase in P_{LA}. Here the mean pressure (plotted as a thick black line in the upper left panel of Figure 4.23) increases from approximately 99 mmH at $P_{LA} = 5$ mmHg to a maximum of approximately 115 mmHg before returning to a final value of approximately 107 mmHg at $P_{LA} = 7$ mmHg.

4.5 Mechanisms of hypertension

Hypertension is the disease condition of chronic high blood pressure: that is, blood pressure that is higher than the physiological normal healthy range. Potential causes of hypertension include an impaired ability of the kidneys to excrete salt (leading to increased blood volume), increased vascular resistance, and overactive sympathetic tone. Both increased systemic resistance and blood volume lead to an increase in blood in the venous circulation, increasing filling pressure, and leading to increased blood pressure. Thus while the models developed in this chapter may help explain how these potential underlying causes lead to high blood pressure, they do not reveal the underlying causes.

The baroreflex system is unable to reverse chronic hypertension effectively because the effective gain of the system (represented by the parameter k_l in the model of the previous section) becomes attenuated in adaptation to hypertension. This attenuation of the reflex gain is observed in analysis of cardiovascular dynamics in hypertensive animals using a related model [13].

4.6 Recapitulation and lessons learned

This chapter, which began with relatively simple descriptions of the phenomena of cardiac pumping and circulatory mechanics, closed with the presentation of a relatively complex model accounting for pulsatile mechanics of the heart and the circulatory system as well as regulation via the baroreflex system. Thus a lesson from previous chapters was repeated: the same system may be simulated and analyzed at many different levels of detail and complexity. The appropriate level of detail depends on what is to be asked of a given model. The simple constant-pressure model of Section 4.2.1 revealed how a heart governed by the Frank–Starling mechanism coupled to a compliant circulation maintains a steady, stable mean cardiac output. That simple model elucidated the pathophysiological mechanism of increased venous pressure with diminished cardiac contractility. It also illustrates how an increase in blood volume leads to an increase in blood pressure. In addition, in Section 4.2.2 (mainly as an exercise in simulation), we modified the constant-pressure Frank–Starling–Guyton model of Guyton to be time dependent and explored the behavior of an imaginary "reverse" Frank–Starling heart.

While the crude time-dependent analysis of the Frank–Starling–Guyton model yields some insight into the physiological control of cardiac output, much more realistic simulation of the time-dependent nature of cardiac pumping was presented in Section 4.3. Armed with time-dependent models of cardiac pumping and lumped-parameter models of the systemic circulation, we were able to simulate the aortic pressure wave in some detail, including a realistic reproduction of the dicrotic notch. These models allowed us to investigate how the circulatory system responds to dynamic changes, such as in cardiac filling pressure.

A model that realistically captures the response to changes in filling pressure (illustrated in Figure 4.23) involves 11 parameters to describe the baroreflex system, four parameters to describe the simple circuit of Figure 4.15, and six parameters to describe the varying elastance heart model. Thus, even without invoking a complex circuit model such as that of Figure 4.16, simulations of the circulatory system can require the introduction of more adjustable parameters than may be practically possible to identify.

Problems

4.1   Circuit analysis 1. The governing equations for the circuit of Figure 9.6 are expressed in Section 9.7 using the flow variable F_1 as the dependent variable. Write down equivalent governing equations using F_2 and pressure at the inlet node as the dependent variables.

4.2   Circuit analysis 2. The resistor–inductor–capacitor network of Figure 9.5 is shown in Section 9.7 of the Appendices to be governed by Eq. (9.37). Show that this second-order system may be equivalently expressed as a system of two first-order differential equations for $F_1(t)$ and $F_2(t)$. Does the solution expressed in Eq. (9.38), with frequency of oscillation

$$\omega = \frac{\sqrt{4CL(1 + R_1/R_2) - (CR_1 + L/R_2)^2}}{2CL},$$

require the condition $4CL(1 + R_1/R_2) \geq (CR_1 + L/R_2)^2$? Why or why not?

4.3   Venous return curves and resistance to venous return. Show that the flow F in the circuit of Figure 9.7 can equally be expressed as the aortic pressure minus the mean filling pressure divided by an equivalent resistance to cardiac output, R_{co}, defined

$$F = \frac{P_{Ao} - P_{MF}}{R_{co}}.$$

Which is the more significant contribution to the resistance to cardiac output, R_A or R_V?

4.4   Computer exercise. Dynamic analysis of the Frank–Starling mechanism. The trajectories plotted in Figures 4.9 and 4.10 are obtained by setting the initial F and P_{RA} to arbitrary non-steady-state values. To obtain initial values of P_A and P_V for use in Eq. (4.9) that are consistent with a constant mean filling pressure, we set $P_V(0) = P_{RA}(0) + F(0)R_V$ and $P_A(0) = (P_{MF}(C_A + C_V) - P_V(0)C_V)/C_A$. Use these equations for the initial conditions and the parameter values reported in the legends to reproduce the trajectories plotted in the figures.

4.5 Computer exercise. One practical check on the correctness of the code used to simulate the model in Section 4.3.2 is to verify that, in the steady state, the model-predicted mean flow into the circuit is equal to the mean flow out of the circuit:

$$\frac{1}{T}\int_o^T F_{LVO}(t)\,dt = \frac{1}{T}\int_o^T (F_{VC}(t) + F_{COR})\,dt.$$

Implement a numerical simulation of the model and verify that this steady-state mass conservation statement holds true.

4.6 Computer exercise. The linear cardiac output curve predicted in Figure 4.19 is related to the linear left-ventricular pressure–volume relationship in the model. Replace the linear relationship with a second-order expansion

$$P_{LV}(t) = E_{LV}(t) \cdot V_o\left[(V_{LV}(t)/V_o - 1) + \beta\,(V_{LV}(t)/V_o - 1)^2\right], \quad (4.34)$$

and estimate values of E_{max}, E_{min} with $\beta = 0.2$ that produce a reasonable steady-state aortic pressure waveform. Repeat the simulations in Figure 4.19 and determine the predicted nonlinear cardiac output curve.

5 Chemical reaction systems: thermodynamics and chemical equilibrium

Overview

This and the following two chapters are focused on analyzing and simulating chemical systems. These chapters will introduce basic concepts of thermodynamics and kinetics for application to biochemical systems, such as biochemical synthesis, cellular metabolism and signaling processes, and gene regulatory networks. Although we have seen examples of chemical kinetics in previous chapters, notably in Sections 2.3 and 2.4, in those examples we developed the expressions governing the chemistry more from intuition than from a physical theory. One of the primary goals here will be to develop a formal physical/chemical foundation for analyzing and simulating complex biochemical systems.

As is our practice throughout this book, these concepts will be applied to analyze real data (and understand the behavior of real systems) later in this chapter and elsewhere. Yet, because the rules governing the behavior of biochemical systems are grounded in thermodynamics, we must begin our investigation into chemical systems by establishing some fundamental concepts in chemical thermodynamics. The concept of *free energy* is particularly crucial to understanding thermodynamic driving forces in chemistry. We will see that both a physical definition and an intuitive understanding of free energy require physical definitions and intuitive understandings of temperature and entropy. All of this means that this chapter will begin with some abstract thought experiments and derivations of physical concepts.

5.1 Temperature, pressure, and entropy

5.1.1 Microstates and macrostates

All thermodynamic theory arises from the fact that physical systems composed of many atoms and/or molecules attain a large number (often a practically infinite number) of microstates under defined macroscopic conditions, such as temperature, pressure, and volume. Any physical object familiar within our everyday experience – such as a cup of coffee or a living organism – is made up of a fantastic number of particles moving and interacting on time and space scales that are much smaller than we may observe with our unaided senses. Regardless of its complexity, the laws of quantum mechanics predict that in the steady state the number of microstates associated with an isolated system (a system with fixed dimensions that does not exchange mass or energy with its surroundings) is in principle finite and fixed under fixed macroscopic conditions. (The macroscopic conditions define the thermodynamic state, also called the *macrostate*, which is the macroscopically observed state of the system.) For a given system made up of particles all of the same type, the thermodynamic state is determined by the number of particles (N), the volume of the container (V), and the total internal energy (E). (For systems made up of more than one type of particle, specification of the thermodynamic state requires the number of each type of particle in the system to be specified.)

Thus at nonzero temperature there are an extremely large number of microstates associated with a defined thermodynamic state. We call the number of microstates associated with a macrostate the *NVE partition function* or *microcanonical partition function*[1] and denote this number by the symbol Ω. Even though it is usually practically impossible to actually specify the number Ω for a given system, thermodynamic theory hangs on a concept related to Ω: whether or not we can enumerate the number of microstates for a given system, it is postulated that isolated systems naturally adopt a macrostate that is associated with the largest possible value of Ω. In fact, this fundamental postulate of statistical thermodynamics is a consequence of statistical likelihood.[2] Left to their own devices (i.e., in isolation) systems attain a thermodynamic state (macrostate) that is associated with the largest possible Ω.

5.1.2   Example: a simple two-state system

Imagine an isolated system composed of N noninteracting particles, each of which can adopt one of two states, identified as state A and state B. Assume that at a given instant there are N_A particles in state A and N_B particles in state B. Further assume that the particles can freely interconvert between the states A and B, and

[1] This cumbersome term tells us that Ω is the partition function associated with the microcanonical ensemble – that is, an ensemble of particles making up an isolated system. We will meet other ensembles in Section 5.4.

[2] More precisely, systems are assumed to sample all possible microstates with equal probability. Thus a microstate with larger Ω has a higher probability than one with lower Ω. In thermodynamic systems with astronomically large numbers of states the difference in Ω between any two different macrostates is so large that the state with smaller Ω is never observed.

that there is no energy difference between the two states. Finally assume that individual particles of state A and state B access identical state spaces. This means that each type has the same single-particle entropy. For given values of N_A and N (or equivalently N_A and N_B), the microcanonical partition function – the sum of all states obtainable by the system – is equal to the number of states accessible to an individual particle times the number of ways N_A particles may be selected from a set of N total particles. Ignoring the common factor associated with the single-particle entropy,

$$\Omega = \frac{N!}{N_A! \, N_B!} = \frac{N!}{N_A! \, (N - N_A)!}. \tag{5.1}$$

Equation (5.1) is a simple enumeration of the number of ways N_A out of a total of N particles can be in state A. For example, if $N_A = 1$, then

$$\Omega = \frac{N!}{1! \, (N - 1)!} = N,$$

indicating that since each of the N particles may be the one particle in state A, there are N possible states. In other words, there are N ways for N_A to equal 1. Similarly, for $N_A = 2$, $\Omega = N \cdot (N - 1)$ and so on.

Equation (5.1) may also be interpreted as follows. If, instead of there being only two unique states, each of the N particles in the system were to attain one of N unique states, then there would be $\Omega = N!$ possible ways to arrange the particles amongst the states. In Eq. (5.1) it is assumed that the N_A state-A particles are indistinguishable from one another (having identical chemical states) and that the state-B particles are also indistinguishable. The terms $N_A!$ and $N_B!$ in the denominator of Eq. (5.1) are correction factors applied to ensure that indistinguishable identical states are not counted more than once in the partition function.

Equation (5.1) has a maximum at $N_A = N/2$ (for even N). Thus in the most likely state, half of the particles are in state A and half are in state B. Just how likely the most likely state is can be illustrated by plotting $\Omega(N_A)$ predicted by Eq. (5.1) for various values of N, as illustrated in Figure 5.1. The figure shows $\Omega(N_A)$ for $N = 20$ and $N = 1000$. At the lower value the distribution is relatively broad, while at $N = 1000$ the distribution approaches the limit of a narrow spike at $N_A = N/2$. Here we can see that even for modest numbers of particles the most likely state become the only likely state. When N approaches numbers on the order of 10^{23} there is essentially no chance of attaining a state significantly different from $N_A = N/2$ in equilibrium.[3]

---

[3] See Exercise 5.1 for more on this example.

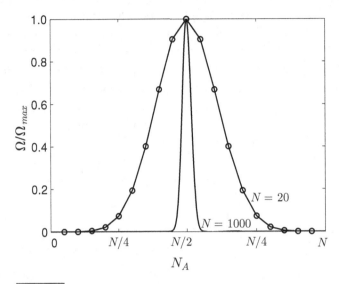

Microcanonical partition function for the example of Section 5.1.2, predicted by Eq. (5.1) for $N = 20$ and 1000. The value of Ω is normalized to the maximal value occurring at $N_A = N/2$.

$$\boxed{\begin{array}{c|c} N_1,\, V_1,\, E_1 & N_2,\, V_2,\, E_2 \end{array}}$$

Two NVE systems in thermal contact.

5.1.3 Relationship between temperature and entropy

To derive a physical definition for temperature, let us consider a simple thought experiment where two systems are put into thermal contact.[4] Thermal contact means that energy (in the form of heat) is allowed to flow between the two systems. We assume that the numbers of particles and the volumes of the individual systems remain fixed. Even though the two systems are not isolated, in thermal equilibrium there will be no net exchange of energy between them. The equilibrium state will be associated with microcanonical partition functions $\Omega_1(N_1, V_1, E_1)$ and $\Omega_2(N_2, V_2, E_2)$, where N_1 and N_2 are the numbers of particles in systems 1 and 2, V_1 and V_2 are the volumes of the two systems, and E_1 and E_2 are the energies.

We further assume that the composite system, made up of systems 1 and 2 in contact, is an isolated system with $E = E_1 + E_2$ constant. The composite system is illustrated in Figure 5.2. At a given macrostate, system 2 independently accesses

---

[4] This thought experiment is a starting point for nearly all expositions on statistical thermodynamics.

Ω_2 microstates for every microstate that system 1 accesses. Thus there are

$$\Omega = \Omega_1 \Omega_2 \tag{5.2}$$

microstates associated with the composite system. Since energy is freely exchanged between the two systems, E_1 and E_2 will become distributed in such a way that Ω is maximized in thermal equilibrium. This implies

$$\frac{\partial \Omega}{\partial E_1} = \Omega_1 \frac{\partial \Omega_2}{\partial E_1} + \Omega_2 \frac{\partial \Omega_1}{\partial E_1} = 0$$

$$= -\Omega_1 \frac{\partial \Omega_2}{\partial E_2} + \Omega_2 \frac{\partial \Omega_1}{\partial E_1} = 0 \tag{5.3}$$

or

$$\frac{1}{\Omega_1} \frac{\partial \Omega_1}{\partial E_1} = \frac{1}{\Omega_2} \frac{\partial \Omega_2}{\partial E_2}$$

$$\frac{\partial \ln \Omega_1}{\partial E_1} = \frac{\partial \ln \Omega_2}{\partial E_2} \tag{5.4}$$

(where we have applied the identity $\frac{\partial}{\partial E_1} = \frac{dE_2}{dE_1}\frac{\partial}{\partial E_2} = -\frac{\partial}{\partial E_2}$). In other words, two systems at the same temperature T have the same value of $\partial \ln \Omega / \partial E$. In fact, the quantity temperature is defined to be proportional to $(\partial \ln \Omega / \partial E)^{-1}$:

$$k_B T = \left(\frac{\partial E}{\partial \ln \Omega} \right),$$

where the constant k_B is called *Boltzmann's constant*. Furthermore, in thermodynamics we introduce a variable called *entropy*, which is represented using the symbol S and is defined $S = k_B \ln \Omega$. Therefore temperature is defined as a function of entropy:

$$T = \left(\frac{\partial E}{\partial S} \right)_{N,V}, \tag{5.5}$$

where the notation $(\cdot)_{N,V}$ reminds us that the partial derivative is taken holding other state variables constant.

5.1.4   Relationship between pressure and entropy

Through a thought experiment similar to that of the previous section, we may derive a thermodynamic definition for pressure as a function of entropy. In this case, we imagine that the two systems of Figure 5.2 are allowed to exchange volume through some flexible coupling, such that any pressure difference between the two systems disappears and total volume $V = V_1 + V_2$ is preserved. Again, starting with Eq. (5.2) and assuming that free exchange of volume between the two systems is associated with a macrostate that maximizes Ω for the composite system,

$$\frac{\partial \Omega}{\partial V_1} = \Omega_1 \frac{\partial \Omega_2}{\partial V_1} + \Omega_2 \frac{\partial \Omega_1}{\partial V_1} = 0$$

$$= -\Omega_1 \frac{\partial \Omega_2}{\partial V_2} + \Omega_2 \frac{\partial \Omega_1}{\partial V_1} = 0 \tag{5.6}$$

or equivalently

$$\frac{\partial \ln \Omega_1}{\partial V_1} = \frac{\partial \ln \Omega_2}{\partial V_2}. \tag{5.7}$$

Since the free exchange of volume between the two systems implies pressure equilibration, systems at the same temperature and pressure[5] have the same value of $\partial S / \partial V$.

More specifically, since pressure P is inversely proportional to volume,

$$\frac{P}{T} = \left(\frac{\partial S}{\partial V} \right)_{N,E}. \tag{5.8}$$

5.2 Free energy under constant-temperature and constant-volume conditions

The relations (5.5) and (5.8) were derived from the postulate that isolated systems move to equilibrium states that maximize the number of associated microstates, or equivalently maximize their entropy. Since chemical and biological systems are rarely isolated, the principles governing the behavior of isolated systems do not always provide the most convenient basis for our studies. For example, imagine a system in which heat flows in and out so that the temperature is maintained at a constant value. Such a system is illustrated in Figure 5.3.

Here we have a gas-filled piston attached to a spring with spring constant α. The position of the piston is indicated by x, where x_o is the unstressed position of the spring, so that the mechanical energy stored in the spring is given by $\frac{\alpha}{2}(x - x_o)^2$. Entropy in the piston increases as the gas expands and the configuration space accessible to the gas molecules increases. However, gas expansion is resisted by the spring, and intuition tells us that in equilibrium there will be a trade-off

---

[5] Thus we have assumed that these systems are in thermal as well as mechanical equilibrium. If the pressures are allowed to equilibrate without allowing heat to exchange then *thermal* equilibrium in general will not be achieved. See Exercise 5.3.

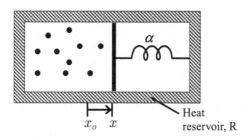

Figure 5.3

Diagram illustrating the system considered in Section 5.2. The system is composed of a gas-filled piston connected to a spring with spring constant α and held at constant temperature.

between an entropic driving force favoring expansion and a mechanical driving force resisting expansion. This is easy to show by adopting the ideal gas law, which states that

$$PV = Nk_BT$$

in the piston. Equating the pressure force acting on the piston with the spring force in equilibrium, we have

$$P = \frac{\alpha(x - x_o)}{A} = \frac{Nk_BT}{V} = \frac{Nk_BT}{Ax}, \tag{5.9}$$

where A is the surface area of the piston. Solving for the equilibrium position, we have

$$x = \frac{x_o + \sqrt{x_o^2 + 4Nk_BT/\alpha}}{2}. \tag{5.10}$$

The predicted behavior confirms our intuition. If the spring constant is zero, then the solution is unbounded, because gas expansion in unhindered. If the spring is infinitely stiff ($\alpha \to \infty$) then the volume is fixed and $x = x_o$.

Clearly, for the case of a finite spring constant the entropy of the system composed of the piston and the spring is not maximized. Yet the fundamental postulate that *isolated* systems maximize entropy still holds. Therefore, to investigate the thermodynamics of the constant-temperature system of Figure 5.3, we need to consider an isolated system that encapsulates the piston and the spring. Specifically, let us consider the composite system made up of the system 1 (the piston and the spring) plus the heat reservoir R. We assume that the heat reservoir is large enough to maintain constant temperature. The entropy of the composite system is $S_{tot} = S_1 + S_R$; a perturbation to the total entropy is computed

$$\delta S_{tot} = \delta S_1 + \delta S_R. \tag{5.11}$$

Since temperature is constant, from Eq. (5.5) a change in entropy of the reservoir is associated with an energy flow into the reservoir: $T\delta S_R = \delta E_R$. Since the composite system is isolated, $\delta E_1 = -\delta E_R$. Therefore we have

$$T\delta S_{tot} = T\delta S_1 - \delta E_1, \tag{5.12}$$

in which the total entropy change is expressed as a function only of changes in the properties of system 1.

From this equation, at the thermodynamic state that maximizes total entropy,

$$T\delta S_1 = \delta E_1.$$

The right-hand side of this equation can be expressed in terms of change in position δx:

$$E_1 = \frac{\alpha}{2}(x - x_o)^2$$

$$\delta E_1 = \alpha(x - x_o)\delta x. \tag{5.13}$$

Expressing the term $T\delta S_1$ in terms of δx requires a relationship between S_1 and the dimensions of the piston. Invoking the assumption of an ideal gas of noninteracting particles, Ω_1 is proportional to V^N, and thus entropy is given by[6]

$$S_1 = k_B \ln \Omega_1 = N k_B \ln V + C, \tag{5.14}$$

where C is a constant that does not depend on the volume of the gas, V. Therefore

$$T\delta S_1 = N k_B T \cdot \frac{\partial V}{\partial x} \cdot \frac{d \ln V}{dV} \cdot \delta x = N k_B T \frac{\delta x}{x}. \tag{5.15}$$

Equating Eqs (5.13) and (5.15) we arrive at

$$x(x - x_o) = \frac{N k_B T}{\alpha}, \tag{5.16}$$

which is the same result as obtained above in Eq. (5.9).

Thus for system 1, the quantity $S_1 - E_1/T$ is maximized rather than simply the entropy S_1. Changes to this quantity represent changes to the total entropy of system 1 plus that of its surroundings. For systems like this it is conventional to define the quantity

$$F = E - TS, \tag{5.17}$$

which is called the *Helmholtz free energy*. This quantity is minimized in equilibrium in isothermal (constant-temperature), isovolemic (constant-volume) systems. (Note that although the volume of the gas-filled piston was allowed to vary in this experiment, the total volume of system 1, piston plus spring, was assumed constant.)

---

[6] The ideal gas law can be obtained from applying Eq. (5.8) to Eq. (5.14).

This expression is intuitively satisfying, because at high temperatures the free energy is dominated by the entropy term, while at low temperatures it is dominated by internal energy. At absolute zero temperature, the equilibrium minimization of F implies a minimum in internal energy.

5.3 Free energy under constant-temperature and constant-pressure conditions

To obtain the appropriate free energy for a constant-pressure system, we can extend the above analysis by allowing system 1 to exchange volume with the reservoir while maintaining constant pressure. Again, the entropy of the composite system is assumed to be maximized in equilibrium.

From Eqs (5.5) and (5.8) we have

$$T\delta S_{tot} = T\delta S_1 - \delta E_1 - P\delta V_1, \qquad (5.18)$$

where again the total entropy change is expressed as a function only of changes in the properties of system 1. (In this equation δV is the change in volume of system 1, composed of the piston and the spring.) For systems like this we define the quantity

$$G = E - TS + PV, \qquad (5.19)$$

which is called the *Gibbs free energy* and is minimized in equilibrium in isothermal isobaric (constant-pressure) systems.

5.4 Thermodynamic ensembles, partition functions, and driving forces

Section 5.1.1 introduced Ω, the partition function of the microcanonical ensemble, an ensemble of particles held at constant N, V, and E, representing an isolated system. In this system, also called the *NVE ensemble*, entropy is the thermodynamic driving force: NVE systems move toward equilibrium defined by maximizing entropy. As we have demonstrated above, and as summarized in Table 5.1, different thermodynamic driving forces apply to different systems. A closed isothermal isovolemic (NVT) system naturally adopts a macrostate that minimizes Helmholtz free energy; a closed isothermal isobaric (NPT) system naturally adopts a macrostate that minimizes Gibbs free energy.

In the following section we will make use of several identities related to partition functions of these various ensembles. The partition function associated with the NVT ensemble is called the *canonical partition function*, and is defined

$$Q = \sum_i e^{-\beta E_i}, \qquad (5.20)$$

Table 5.1: Thermodynamic driving forces in three systems.

System	Thermodynamic driving force
NVE	Maximize entropy S
NVT	Minimize Helmholtz free energy $E - TS$
NPT	Minimize Gibbs free energy $E + PV - TS$

where the sum is over all possible microstates of the system, E_i is the internal energy of state i, and the factor β is equal to $(k_B T)^{-1}$. Thus Q is a weighted sum over all states, where the weighting of each state is determined by its energy. Furthermore, it is a fundamental property of NVT systems that the distribution of states follows Boltzmann statistics:

$$P_i = \frac{e^{-\beta E_i}}{Q}, \tag{5.21}$$

where P_i is the probability of state i. A further identity related to the canonical ensemble that will be useful to us relates the Helmholtz free energy to the partition function:

$$F = -k_B T \ln Q. \tag{5.22}$$

For the NPT ensemble we have the partition function:

$$Z = \sum_i e^{-\beta H_i}, \tag{5.23}$$

where $H_i = E_i + PV_i$ is the *enthalpy* of state i, the statistical distribution of states

$$P_i = \frac{e^{-\beta H_i}}{Z}, \tag{5.24}$$

and the Gibbs free energy

$$G = -k_B T \ln Z. \tag{5.25}$$

(See, for example, Beard and Qian [8] for more details and derivations of Eqs (5.20)–(5.25).)

5.5   Chemical reactions, stoichiometry, and equilibrium constants

Armed with some basic theoretical concepts of thermodynamics, we are now prepared to develop some practical formulae for applications in chemistry. Namely, we wish to be able to compute the free energy changes associated with the turnover of chemical reactions, and to predict and understand chemical reaction equilibria. To do so, we require expressions for how the appropriate partition function (or the appropriate free energy) varies with the numbers or concentrations of particles in a reacting system. We start by defining the single-particle partition function for a given chemical species in given system. The single-particle NVT partition function is

$$Q_1 = \sum_i e^{-\beta E_i},\qquad(5.26)$$

where this sum is over all states accessible to the single particle. Similarly the single-particle NPT partition function is

$$Z_1 = \sum_i e^{-\beta H_i}.\qquad(5.27)$$

If an NVT system consists of two independent molecules of the same type, each with single-particle partition functions Q_1, then the two-molecule NVT partition function is

$$Q_2 = \frac{1}{2}\sum_i \sum_j e^{-\beta(E_i+E_j)} \approx \frac{1}{2}Q_1^2,\qquad(5.28)$$

where, again, the sums are over all states of the individual particles. The approximation $Q_2 \approx Q_1^2/2$ is valid as long as the likelihood that both molecules adopt the same state simultaneously is vanishingly small. For the general N-molecule case:

$$Q_N = \frac{1}{N!}Q_1^N.\qquad(5.29)$$

Equations (5.28) and (5.29) assume that the molecules do not interact in such a way that the energy associated with a given conformation of one molecule is affected by the conformation of any other molecule in the system. This dilute-system limit is broadly valid for chemical reactions in aqueous media under many circumstances.

Recall that we have encountered the $N!$ factor that corrects for double counting of indistinguishable states in Eq. (5.1). In fact, the same sort of argument that we have applied to the NVT ensemble applies to other ensembles as well. For example, for the NVE ensemble,

$$\Omega_N = \frac{\Omega_1^N}{N!}.$$

In an NVE system consisting of N_A particles of type A and N_B particles of type B, then the dilute-system partition function would take the form

$$\Omega = \Omega_A\Omega_B = \frac{\Omega_{A,1}^{N_A}\,\Omega_{B,1}^{N_B}}{N_A!\,N_B!}.\qquad(5.30)$$

Furthermore, if state-A particles and state-B particles have the same single-particle entropy, then

$$\Omega = \Omega_A \Omega_B = \frac{\Omega_{A,1}^N}{N_A! \, N_B!} = \frac{\Omega_{A,1}^N}{N_A! \, (N - N_A)!}. \tag{5.31}$$

This result is different from Eq. (5.1), because in the case of Eq. (5.1) we assumed that all of the particles in a given state (A or B) were in the same indistinguishable state. Here, we have assumed that a particle in each state can access a large number of different microstates, and that the chance of two particles having exactly the same state at exactly the same time is zero. (Regardless, Eqs (5.1) and (5.31) both predict the same relative probability $\Omega(N_A)/\Omega_{max}$ distribution, as illustrated in Figure 5.1.)

From Eqs (5.22) and (5.29) we have

$$\begin{aligned} F &= -k_B T \ln(Q_N) = k_B T \, (\ln N! - N \ln Q_1) \\ &= k_B T \, (N \ln N - N - N \ln Q_1), \end{aligned} \tag{5.32}$$

where we have invoked the Stirling approximation that $\ln N! \approx N \ln N - N$ for large N. Defining the chemical potential μ as the free energy change per molecule added to the system,

$$\mu = (\partial F/\partial N)_{V,T},$$

yields

$$\mu = k_B T (\ln N - \ln Q_1). \tag{5.33}$$

By introducing the constant μ^o, which does not depend on the number (or concentration) of molecules, we can obtain

$$\mu = k_B T \ln(N/V/C_o) - k_B T \ln(Q_1/V/C_o)$$
$$\mu = \mu^o + k_B T \ln(C/C_o), \tag{5.34}$$

where μ^o depends on the molecular species, C is the concentration of molecules of that species, and C_o is a reference concentration that is typically taken as $1 \text{ mol} \, l^{-1}$.

Analogously, we can show that in the NPT ensemble

$$\begin{aligned} \mu &= (\partial G/\partial N)_{P,T} \\ &= \mu^o + k_B T \ln(C/C_o), \end{aligned} \tag{5.35}$$

where $\mu^o = -k_B T \ln(Z_1/V/c_o)$. In a system composed of molecules of a number of different species, the definition $\partial G/\partial N = \mu$ yields

$$(\delta G)_{P,T} = \sum_{i=1}^{N_s} \mu_i \delta N_i, \tag{5.36}$$

where the sum is over contributions from N_s different kinds of species, which are indexed by i. This equation sums up all contributions to Gibbs energy from small changes in the numbers of molecules of all different species in a system, where each species is present at its individual chemical potential μ_i.

We can apply equation chemical reaction defined by the general expression

$$0 \rightleftharpoons \sum_{i=1}^{N_s} \nu_i A_i, \tag{5.37}$$

where $\{A_i\}$ represent chemical species and $\{\nu_i\}$ are the stoichiometric coefficients,[7] as follows. When the reaction of Eq. (5.37) proceeds an infinitesimal amount in the forward (left-to-right) direction, the free energy of the system changes by an amount given by

$$(\delta G)_{P,T} = \sum_{i=1}^{N_s} \mu_i \nu_i \delta \phi, \tag{5.38}$$

where the variable ϕ represents the turnover or the number of times the reaction has progressed. (In other words, the reaction stoichiometry constrains the changes in the numbers of the individual species: $\delta N_i = \nu_i \delta \phi$.) It follows that

$$\left(\frac{dG}{d\phi}\right)_{P,T} = \sum_{i=1}^{N_s} \mu_i \nu_i. \tag{5.39}$$

Chemists use the notation $\Delta_r G = (dG/d\phi)_{P,T}$ to denote the change in free energy of a chemical system associated with the turnover of a given chemical reaction. Substituting the expression for μ_i from Eq. (5.35), we have

$$\Delta_r G = \sum_{i=1}^{N_s} \nu_i \left[\mu_i^o + k_B T \ln(C_i/C_o)\right]$$

$$= \Delta_r G^o + k_B T \sum_{i=1}^{N_s} \nu_i \ln(C_i/C_o), \tag{5.40}$$

where $\Delta_r G^o = \sum_{i=1}^{N_s} \nu_i \mu_i^o$, which is called the *standard reference Gibbs free energy* of the reaction.

---

[7] For example, for the reaction $A_1 + 2A_2 \rightleftharpoons A_3$ the stoichiometric coefficients are $\nu_1 = -1$, $\nu_2 = -2$, and $\nu_3 = +1$.

If we express $\Delta_r G^o$ and $\Delta_r G$ in units of energy per unit mole of flux through a reaction (rather than as the free energy change associated with a single turnover of the reaction), they become

$$\Delta_r G = \Delta_r G^o + RT \sum_{i=1}^{N_s} v_i \ln(C_i/C_o)$$

$$\Delta_r G^o = \sum_{i=1}^{N_s} v_i \Delta_f G_i^o = \sum_{i=1}^{N_s} v_i \mu_i^o, \qquad (5.41)$$

where $\Delta_f G_i^o$ (called the *free energy of formation* of species i) is another notation used for μ_i^o under isothermal isobaric conditions, R is equal to $N_A k_B$, and $N_A \approx 6.022 \times 10^{23}$ is the Avogadro constant. The value of $\Delta_f G_i^o$ for a given species depends on the environmental conditions, most notably temperature, pressure, and the ionic solution strength; it does not depend on the concentration of molecular species in a dilute solution.

Chemical equilibrium is achieved when the driving force for the reaction $\Delta_r G$ goes to zero. Equilibrium yields

$$-\frac{\Delta_r G^o}{RT} = \sum_{i=1}^{N_s} v_i \ln(C_i/C_o) = \ln \prod_{i=1}^{N_s} (C_i/C_o)^{v_i}$$

$$e^{-\Delta_r G^o/RT} = K_{eq} = \prod_{i=1}^{N_s} (C_i/C_o)^{v_i}, \qquad (5.42)$$

where K_{eq} is the equilibrium constant for the reaction. Like $\Delta_r G^o$, the equilibrium constant does not depend on the concentrations of the reactants. For example, the simple reaction

$$A \rightleftharpoons B$$

has a reference free energy change

$$\Delta_r G^o = \mu_B^o - \mu_A^o$$
$$= RT \left(\ln Z_{A,1} - \ln Z_{B,1} \right), \qquad (5.43)$$

where $Z_{A,1}$ and $Z_{B,1}$ are the single-particle NPT partition functions for states A and B, respectively. If $Z_{A,1} = Z_{B,1}$, then $\Delta_r G^o = 0$, $K_{eq} = 1$, and the equilibrium mass-action ratio is

$$K_{eq} = e^{-\Delta_r G^o/RT} = (C_B/C_A)_{eq} = 1. \qquad (5.44)$$

5.6   Acids, bases, and buffers

Equation (5.42) provides the basis for computing equilibrium concentrations in reaction systems, given knowledge of the reference free energies. It also provides the basis for predicting the direction in which reactions not in equilibrium will proceed. In biological systems certain reactions tend to proceed so quickly that under most practical circumstances they may be treated as maintained in chemical equilibrium. A familiar example is the dissociation of water into hydroxyl and hydrogen ions,

$$H_2O \rightleftharpoons OH^- + H^+,$$

with equilibrium constant

$$K_w = [OH^-] \cdot [H^+] \approx 10^{-14} \ M^2. \tag{5.45}$$

(By convention the concentration of water is usually not included in chemical thermodynamic expressions for reactions in aqueous media, since the concentration/density of water remains constant for reactions in dilute solution at constant temperature.) Although the standard reference free energy change, which is computed

$$\Delta_r G^o = \mu^o_{OH^-} + \mu^o_{H^+} - \mu^o_{H_2O}, \tag{5.46}$$

depends on conditions such as the temperature and ionic strength of the solution, the equilibrium constant $K_w = e^{-\Delta_r G^o / RT}$ remains roughly constant in the temperature range 25–37 °C; thus in neutral solution $[OH^-] = [H^+] \approx 10^{-7}$.

A weak acid is a chemical substance that can donate a hydrogen ion into solution when dissolved in water. A general expression for the hydrogen ion association/dissociation reaction is

$$AH \rightleftharpoons A^- + H^+. \tag{5.47}$$

The equilibrium constant for this reaction is called a *dissociation constant* and is often denoted K_d:

$$K_d = 10^{-pK} = \left(\frac{[H^+] \cdot [A^-]}{[AH]} \right)_{eq}, \tag{5.48}$$

where the so-called pK value is $-\log_{10} K_d$. In a solution where one weak acid (or *buffer*) is present, the total exchangeable hydrogen ion concentration is given by the sum of the free plus bound ions

$$
\begin{aligned}
H_o &= [H^+] + [AH] + W - [OH^-] \\
&= [H^+] + [H^+][A^-]/K_d + W - [OH^-] \\
&= [H^+] + [H^+][A^-]/K_d + W - K_w/[H^+], \tag{5.49}
\end{aligned}
$$

where the term W is a constant equal to $[H_2O] + [OH^-]$, which is the total concentration of water. The concentration of exchangeable hydrogen ions present in water molecules is equal to $W - [OH^-]$. The concentration of unprotonated buffer $[A^-]$ can be obtained from

$$A_{tot} = [A^-] + [AH]$$
$$= [A^-]\left(1 + [H^+]/K_d\right),\qquad(5.50)$$

yielding

$$[A^-] = A_{tot}\frac{1}{1 + [H^+]/K_d}.\qquad(5.51)$$

Combining Eqs (5.49) and (5.51) gives

$$H_o = \left([H^+] + \frac{A_{tot}[H^+]}{[H^+] + K_d}\right) + W - \frac{K_w}{[H^+]}.\qquad(5.52)$$

Plotting $H_o - W$ versus $[H^+]$ in Figure 5.4 illustrates how weak acids act as buffers to changes in free hydrogen ion concentration. In the neighborhood of pH $\approx pK$ it takes a relatively large amount of added (or subtracted) hydrogen ion to affect a change in pH. The higher the buffer concentration, A_{tot}, the less sensitive pH is to H_o, particularly at pH values near the pK of the buffer. It is possible to measure the pK of a buffer by titrating in a strong acid such as HCl that fully dissociates and plotting total hydrogen added versus measured pH, constructing a *titration curve*, like the one plotted in Figure 5.4. The buffer pK value may be estimated from the inflection point on the measured curve, as illustrated in the figure.

The buffering ability of the solution may be quantified by evaluating $d[H^+]/dH_o$ from Eq. (5.52):

$$\frac{d[H^+]}{dH_o} = \left(\frac{dH_o}{d[H^+]}\right)^{-1} = \left(1 + \frac{A_{tot}K_d}{([H^+] + K_d)^2} + \frac{K_w}{[H^+]^2}\right)^{-1},\qquad(5.53)$$

or

$$\delta[H^+] = \frac{\delta H_o}{\left(1 + \frac{A_{tot}K_d}{([H^+]+K_d)^2} + \frac{K_w}{[H^+]^2}\right)},\qquad(5.54)$$

or

$$\delta\text{pH} = -\frac{\delta H_o}{\beta},\quad \beta = 2.303[H^+]\left(1 + \frac{A_{tot}K_d}{([H^+] + K_d)^2} + \frac{K_w}{[H^+]^2}\right),\qquad(5.55)$$

where β is called the *buffer capacity* of the solution. The buffer capacity for a weak acid with a $pK = 7.5$ is plotted as a function of pH in Figure 5.5. It is apparent

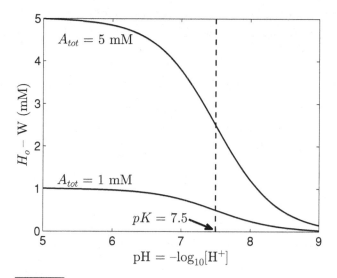

Figure 5.4

Relationship between H_o and $[H^+]$ predicted by Eq. (5.52), for two different values of A_{tot} and with $pK = 7.5$.

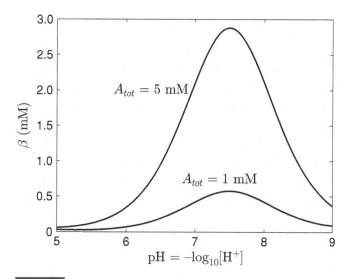

Figure 5.5

Buffer capacity β predicted by Eq. (5.55), for two different values of A_{tot} and with $pK = 7.5$.

that the maximal buffering capacity occurs at pH value equal to the pK of the buffer.

The buffer capacity is a measure of the rate at which pH changes in response to the addition of strong acid – an acid such as HCl that completely dissociates.

This is illustrated by assuming that a concentration C_a of HCl is added to a buffer solution. Since the HCl fully dissociates into H^+ and Cl^-, conservation of charge predicts that

$$[H^+] = [OH^-] + [Cl^-] + [A^-]$$

$$= \frac{K_w}{[H^+]} + C_a + \frac{A_{tot} K_d}{K_d + [H^+]}. \qquad (5.56)$$

Rearranging yields

$$C_a = [H^+] - \frac{A_{tot} K_d}{K_d + [H^+]} - \frac{K_w}{[H^+]},$$

or

$$\frac{\partial C_a}{\partial [H^+]} = 1 + \frac{A_{tot} K_d}{([H^+] + K_d)^2} + \frac{K_w}{[H^+]^2},$$

or

$$\frac{\partial C_a}{\partial \text{pH}} = -\beta.$$

5.7 Analysis of biochemical reactions

In the final section of this chapter we explore how the concepts of chemical equilibrium derived above can be used to analyze data from experiments designed to probe the thermodynamic properties of biochemical reactions. Specifically, we will explore how the standard free energy and enthalpy of a biochemical reaction may be estimated from equilibrium data. In these examples, we will see that typical experimental solutions are a complex mixture of a large number of interconverting ionic species; analysis of data from these experiments requires us to account for the complex distribution of biochemical reactants into many related species in solution.

5.7.1 Example: equilibrium of a biochemical reaction

As an example of how the principles of equilibrium in the isothermal isobaric ensemble are applied in biochemistry, let us analyze data on the creatine kinase reaction, published in 1979 by Lawson and Veech [36]. The enzyme creatine kinase catalyzes the chemical reaction

$$\text{ADP}^{3-} + \text{creatine-P}^{2-} + \text{H}^+ \rightleftharpoons \text{ATP}^{4-} + \text{creatine}^0, \qquad (5.57)$$

where the species ADP^{3-}, creatine-P^{2-}, ATP^{4-} and creatine^0 represent adenosine diphosphate, creatine phosphate, adenosine triphosphate, and creatine, respectively. The valences on these species indicate the proton and metal-ion binding state. Specifically, ADP^{3-}, for example, indicates adenosine diphosphate with no freely dissociateable cations such as H^+, K^+, or Mg^{2+} bound. Indeed, the negatively charged species in Eq. (5.57) all tend to associate with a number of different cations in solution, and the total quantity of these *reactants*[8] in a solution is given by the sum of all cation-bound states:

$$[\Sigma ADP] = [ADP^{3-}] + [H.ADP^{2-}] + [K.ADP^{2-}] + [Mg.ADP^-] + \ldots$$

$$[\Sigma creatine-P] = [creatine-P^{2-}] + [H.creatine-P^-] + [Mg.creatine-P^0] + \ldots$$

$$[\Sigma ATP] = [ATP^{4-}] + [H.ATP^{3-}] + [K.ATP^{3-}] + [Mg.ATP^{2-}] + \ldots$$

$$[\Sigma creatine] = [creatine^0]. \tag{5.58}$$

The neutrally charged species creatine is assumed to not associate with dissolved cations. Which cation-bound states on the right-hand side of Eq. (5.58) are present in significant quantities depends on the solution conditions. For example, under particularly acidic conditions, a two-proton bound state of ADP, H_2ADP^-, may make a relatively significant contribution to the total ADP, while at pH in the range of 7 and above this species does not make a significant contribution.

In fact, standard chemical assay methods provide measurement of total reactant concentrations. Thus in experiments such as those conducted by Lawson and Veech [36] to measure the equilibrium constant of reaction (5.57), the measured concentrations are reactant concentrations. More specifically, reaction solutions were mixed up under defined conditions with the creatine kinase enzyme added; the system was allowed to reach chemical equilibrium; and the final concentrations of $[\Sigma ADP]$, $[\Sigma ATP]$, $[\Sigma creatine-P]$, and $[\Sigma creatine]$ were measured. Raw data on these measurements reported by Lawson and Veech are provided in Table 5.2.

The second column of Table 5.2 reports the total Mg^{2+} ion added (in the form of $MgCl_2$ or Mg-acetate) to the reaction solution for each measurement. Lawson and Veech also report their estimate of free (unbound) $[Mg^{2+}]$, which is also provided in Table 5.2. Lawson and Veech also report adding KCl to achieve a final ionic strength of $I = 0.25$ M. The ionic strength is a measure of the ionic screening strength of dissolved ions in solution, and is defined

$$I = \frac{1}{2} \sum_{i=1}^{N_i} z_i^2 C_i, \tag{5.59}$$

---

[8] A chemical species is a chemical entity with defined molecular structure, mass and charge. The term *biochemical reactant* is used to refer to entities that may be composed of a number of rapidly interconverting species. For example, ADP^{3-} represents a defined species, while the abbreviation ADP is commonly used to refer to a biochemical reactant.

Table 5.2: Reported data from Lawson and Veech [36] on creatine kinase equilibrium. Concentrations are given in units of mM.

	Total [Mg]	[ΣATP]	[Σcreatine]	[ΣADP]	[Σcreatine-P]	Free [Mg^{2+}]	pH
1.	0.0026	0.194	10.2	0.269	0.266	0.00056	7.067
2.	0.0026	0.192	10.2	0.223	0.268	0.00089	7.038
3.	0.0026	0.174	10.3	0.287	0.101	0.00051	7.042
4.	0.0026	0.168	10.2	0.244	0.202	0.00064	7.010
5.	0.0057	0.0544	10.4	0.145	0.126	0.0019	6.988
6.	0.0057	0.0893	10.4	0.106	0.260	0.0018	6.991
7.	0.0057	0.0874	10.4	0.302	0.0919	0.0015	7.005
8.	0.0057	0.151	10.4	0.241	0.202	0.0019	7.023
9.	0.10	3.10	5.61	0.664	0.905	0.0034	7.117
10.	0.10	3.14	5.44	0.685	0.877	0.0035	7.111
11.	0.10	3.12	5.56	0.687	0.907	0.0035	7.112
12.	0.10	2.34	6.49	0.708	0.557	0.0056	6.998
13.	0.10	2.47	6.56	0.685	0.513	0.0054	7.005
14.	0.10	2.36	6.35	0.728	0.522	0.0055	7.004
15.	1.00	3.10	5.59	0.723	0.812	0.048	7.140
16.	1.00	3.18	5.60	0.703	0.857	0.046	7.145
17.	1.00	3.14	5.54	0.726	0.831	0.047	7.143
18.	1.00	2.44	6.46	0.578	0.470	0.073	7.007
19.	1.00	2.50	6.48	0.598	0.486	0.071	7.005
20.	1.00	2.46	6.58	0.618	0.476	0.072	7.000
21.	2.30	0.303	10.0	0.148	0.161	0.82	7.065
22.	2.30	0.124	10.0	0.0702	0.168	0.88	7.103
23.	2.30	0.112	9.9	0.127	0.0681	0.87	7.100
24.	2.30	0.243	9.8	0.150	0.122	0.83	7.080
25.	2.30	0.092	10.0	0.0962	0.0773	0.89	7.093
26.	2.30	0.131	10.2	0.0559	0.203	0.86	7.113
27.	2.30	0.130	10.3	0.237	0.0458	0.84	7.117
28.	2.30	0.120	10.1	0.0702	0.135	0.87	7.115
29.	8.40	0.572	10.01	0.181	0.182	3.36	7.110
30.	8.40	0.243	9.51	0.156	0.090	3.53	7.111
31.	8.40	0.254	9.47	0.098	0.160	3.48	7.160
32.	8.40	0.516	9.14	0.187	0.152	3.42	7.090
33.	8.40	0.522	10.3	0.203	0.164	3.32	7.161
34.	8.40	0.319	10.2	0.428	0.0463	3.38	7.120
35.	8.40	0.323	9.92	0.0567	0.368	3.42	7.190
36.	8.40	0.211	9.51	0.0617	0.217	3.53	7.140
37.	24.0	0.442	11.53	0.189	0.143	11.9	7.317
38.	24.0	0.217	9.39	0.094	0.147	12.1	7.328
39.	24.0	0.202	9.62	0.152	0.083	12.0	7.332

Table 5.2: (cont.)

	Total [Mg]	[ΣATP]	[Σcreatine]	[ΣADP]	[Σcreatine-P]	Free [Mg^{2+}]	pH
40.	24.0	0.504	8.84	0.188	0.179	11.8	7.340
41.	24.0	0.465	9.21	0.208	0.155	11.7	7.392
42.	24.0	0.302	9.89	0.064	0.324	11.9	7.365
43.	24.0	0.293	10.0	0.416	0.043	11.7	7.378
44.	24.0	0.192	10.0	0.068	0.196	12.0	7.350

Here we use the notation $[\Sigma ADP]$ to make it clear that this concentration refers to the sum of contributions of all species that make up ADP.

where N_i is the number of different types of ions present, and z_i and C_i are the valance and concentration of the ith ion. Although Lawson and Veech did not report the quantity of KCl added to each measurement, it may be estimated from the reported data:

$$[KCl]_{added} = 0.25 \text{ M} - \frac{1}{2} \sum_{i \notin [KCl]_{added}} z_i^2 C_i, \qquad (5.60)$$

where the summation in this equation is over all ionic species, excluding the KCl added to achieve the final I of 0.25 M.

The complete enumeration of substances present and their breakdown into ionic species for experiments like this one can become lengthy. Here, let us go through the detailed steps for the first experimental entry in Table 5.2. In this case, the solution contained a 25 mM phosphate buffer, which indicates that

$$[\Sigma phosphate] = [HPO_4^{2-}] + [H_2PO_4^-] + [Mg.HPO_4^0] = 25 \text{ mM}.$$

(Other ionic species of phosphate are possible, but were not considered by Lawson and Veech. Therefore we will not consider other ionic species in calculating how much KCl was added to achieve the reported I.) Assuming rapid equilibrium assocation/dissociation, we can show that[9]

$$\left[HPO_4^{2-}\right] = [\Sigma phosphate] \left(\frac{1}{1 + [H^+]/K_{H,Pi} + [Mg^{2+}]/K_{Mg,Pi}} \right) = 16.81 \text{ mM}$$

$$\left[H_2PO_4^-\right] = [\Sigma phosphate] \left(\frac{[H^+]/K_{H,Pi}}{1 + [H^+]/K_{H,Pi} + [Mg^{2+}]/K_{Mg,Pi}} \right) = 8.19 \text{ mM}$$

$$\left[Mg.HPO_4^0\right] = [\Sigma phosphate] \left(\frac{[Mg^{2+}]/K_{Mg,Pi}}{1 + [H^+]/K_{H,Pi} + [Mg^{2+}]/K_{Mg,Pi}} \right) = 0.88 \text{ μM},$$

$$(5.61)$$

[9] See Exercise 5.5.

Table 5.3: Ion dissociation constants for calculations in Section 5.7.1 and 5.7.2.

Reaction	Diss. const. from [36]	Diss. consts from [39]
$H.ATP^{3-} \rightleftharpoons H^+ + ATP^{4-}$	$K_{H,ATP} = 1.08 \times 10^{-7}$ M,	3.62×10^{-7} M
$H.ADP^{2-} \rightleftharpoons H^+ + ADP^{3-}$	$K_{H,ADP} = 1.20 \times 10^{-7}$ M,	5.03×10^{-7} M
$H_2PO_4^- \rightleftharpoons H^+ + HPO4^{2-}$	$K_{H,Pi} = 1.76 \times 10^{-7}$ M,	2.49×10^{-7} M
$H.creatine-P^- \rightleftharpoons H^+ + creatine-P^{2-}$	$K_{H,CrP} = 3.16 \times 10^{-5}$ M,	3.53×10^{-5} M[a]
$Mg.ATP^{2-} \rightleftharpoons Mg^{2+} + ATP^{4-}$	$K_{Mg,ATP} = 7.19 \times 10^{-5}$ M,	1.43×10^{-4} M
$Mg.ADP^{1-} \rightleftharpoons Mg^{2+} + ADP^{3-}$	$K_{Mg,ADP} = 7.58 \times 10^{-4}$ M,	1.04×10^{-3} M
$Mg.HPO4^0 \rightleftharpoons Mg^{2+} + HPO4^{2-}$	$K_{Mg,Pi} = 0.0107$ M,	2.46×10^{-2} M
$Mg.creatine-P^0 \rightleftharpoons Mg^{2+} + creatine-P^{2-}$	$K_{Mg,CrP} = 0.050$ M,	0.0544 M[a]
$K.ATP^{3-} \rightleftharpoons K^+ + ATP^{4-}$	$K_{K,ATP} = $ NA,	0.128 M
$K.ADP^{2-} \rightleftharpoons K^+ + ADP^{3-}$	$K_{K,ADP} = $ NA,	0.163 M
$KHPO_4^- \rightleftharpoons K^+ + HPO4^{2-}$	$K_{K,Pi} = $ NA,	0.438 M
$K.creatine-P^- \rightleftharpoons K^+ + creatine-P^{2-}$	$K_{K,CrP} = $ NA,	NA
$Mg.H.ATP^- \rightleftharpoons Mg^{2+} + H.ATP^{3-}$	$K_{Mg,HATP} = 0.0282$ M,	3.95×10^{-2} M
$Mg.H.ADP^0 \rightleftharpoons Mg^{2+} + H.ADP^{2-}$	$K_{Mg,HADP} = 0.0309$ M,	8.46×10^{-2} M
$H.Tris^+ \rightleftharpoons H^+ + Tris$	$K_{H,Tris} = $ NA,	8.47×10^{-9} M

Values from Li *et al.* [39] adjusted to $T = 37°$ C and $I = 0.25$ M.
[a] From [58]

where we have used the values of dissociation constants used by Lawson and Veech. (See Table 5.3).

Since, according to Lawson and Veech, creatine phosphate, ATP, and ADP were added as mono-sodium salts, $[Na^+]$ can be calculated

$$[Na^+] = [\Sigma ATP] + [\Sigma ADP] + [\Sigma creatine-P]_{initial} = 0.551 \text{ mM} \qquad (5.62)$$

for data entry 1. (We will assume that the sodium ions are all unbound. This approximation is valid because, as we will see, the free K^+ concentration is more than 100 times that of Na^+. Thus sodium binding may be ignored.) Here $[\Sigma creatine-P]_{initial} = 0.088$ mM is the initial creatine phosphate added by Lawson and Veech (not shown in Table 5.2). Phosphate was added as the potassium salt, and so

$$[K^+]_{Pi} = [\Sigma phosphate] = 25 \text{ mM}. \qquad (5.63)$$

Here the subscript "Pi" is used to distinguish this concentration as the potassium ion that is added with the phosphate salt. Thus $[K^+]_{Pi}$ contributes to the sum on the right-hand side of Eq. (5.60).

Magnesium ion was added as either $MgCl_2$ or $Mg(acetate)_2$. Without affecting the results, let us make the arbitrary assumption that the magnesium salt is $MgCl_2$

in all cases, yielding

$$[\text{Cl}^-]_{MgCl_2} = 2\,[\text{Mg}] = 0.0052 \text{ mM} \tag{5.64}$$

for data entry 1.

Given the reported free $[\text{Mg}^{2+}]$ and pH, we can compute the estimated breakdown of ATP, ADP, creatine, and creatine phosphate into ionic species:

$$[\text{ATP}^{4-}] = [\Sigma\text{ATP}]/P_{ATP} = 0.108 \text{ mM}$$

$$[\text{H.ATP}^{3-}] = [\Sigma\text{ATP}]\left([\text{H}^+]/K_{H,ATP}\right)/P_{ATP} = 0.0854 \text{ mM}$$

$$[\text{Mg.ATP}^{2-}] = [\Sigma\text{ATP}]\left([\text{Mg}^{2+}]/K_{Mg,ATP}\right)/P_{ATP} = 0.001 \text{ mM}$$

$$[\text{Mg.H.ATP}^-] = [\Sigma\text{ATP}]\left([\text{H}^+][\text{Mg}^{2+}]/(K_{H,ATP}K_{Mg,HATP})\right)/P_{ATP} \approx 0 \text{ mM}$$

$$P_{ATP} = 1 + [\text{H}^+]/K_{H,ATP} + [\text{Mg}^{2+}]/K_{Mg,ATP}$$

$$+ [\text{H}^+][\text{Mg}^{2+}]/(K_{H,ATP}K_{Mg,HATP})$$

$$[\text{ADP}^{3-}] = [\Sigma\text{ADP}]/P_{ADP} = 0.156 \text{ mM}$$

$$[\text{H.ADP}^{2-}] = [\Sigma\text{ADP}]\left([\text{H}^+]/K_{H,ADP}\right)/P_{ADP} = 0.112 \text{ mM}$$

$$[\text{Mg.ADP}^{1-}] = [\Sigma\text{ADP}]\left([\text{Mg}^{2+}]/K_{Mg,ADP}\right)/P_{ADP} = 0.001 \text{ mM}$$

$$[\text{Mg.H.ADP}^0] = [\Sigma\text{ADP}]\left([\text{H}^+][\text{Mg}^{2+}]/(K_{H,ADP}K_{Mg,HADP})\right)/P_{ADP} \approx 0 \text{ mM}$$

$$P_{ADP} = 1 + [\text{H}^+]/K_{H,ADP} + [\text{Mg}^{2+}]/K_{Mg,ADP}$$

$$+ [\text{H}^+][\text{Mg}^{2+}]/(K_{H,ADP}K_{Mg,HADP})$$

$$[\text{creatine-P}^{2-}] = [\Sigma\text{creatine-P}]/P_{CrP} = 0.265 \text{ mM}$$

$$[\text{H.creatine-P}^-] = [\Sigma\text{creatine-P}]\left([\text{H}^+]/K_{H,CrP}\right)/P_{CrP} = 0.001 \text{ mM}$$

$$[\text{Mg.creatine-P}^0] = [\Sigma\text{creatine-P}]\left([\text{Mg}^{2+}]/K_{Mg,CrP}\right)/P_{CrP} \approx 0 \text{ mM}$$

$$P_{CrP} = 1 + [\text{H}^+]/K_{H,CrP} + [\text{Mg}^{2+}]/K_{Mg,CrP}, \tag{5.65}$$

where all the dissociation constants used by Lawson and Veech are listed in Table 5.3. The polynomials P_{ATP}, P_{ADP}, and P_{CrP} are called *binding polynomials*; the binding polynomial for phosphate appears in the denominator of Eq. (5.61).

Finally, we note that a certain amount of HCl or KOH must be added to the reaction media to achieve the near-neutral pH values indicated for the experiments. We can estimate the amount of HCl or KOH added based on the net H^+/OH^- imbalance associated with the chemicals in the buffer media as follows. First, recall that phosphate was added as KH_2PO_4. Thus at the given pH there is a net surplus of approximately $[\Sigma\text{phosphate}] - [\text{H}_2\text{PO}_4^-]$ hydrogen ion to be accounted for. The monosodium ATP and ADP (Na·H$_3$·ATP and Na·H$_2$·ADP) can dissociate hydrogen ions in solution as well. Thus at a given pH and ionic conditions, the net addition of hydrogen ion due to binding to ATP in solutions is

$$3\,[\Sigma\text{ATP}] - \big([\text{H.ATP}^{3-}] + [\text{Mg.H.ATP}^-]\big)\,.$$

Similarly the net addition due to dissociation from ADP is

$$2\,[\Sigma\text{ADP}] - \big([\text{H.ADP}^{2-}] + [\text{Mg.H.ADP}^0]\big)\,.$$

Creatine phosphate, added as monosodium salt, can also dissociate free hydrogen ion; the net addition of hydrogen ion into solution from addition of creatine phosphate is

$$[\Sigma\text{creatine-P}] - [\text{H.creatine-P}^-]\,.$$

In some experiments of Lawson and Veech, Tris (tris(hydroxymethyl) aminomethane) was added as an additional buffer. The net loss of hydrogen ion due to binding to Tris is computed

$$[\text{H.Tris}^+]\,.$$

(For data entry 1 in Table 5.2 there is no Tris present, so this quantity does not contribute.) The total proton surplus/deficit that must be accounted for by the addition of HCl or KOH is computed

$$
\begin{aligned}
H_{surplus} ={}& [\Sigma\text{phosphate}] - [\text{H}_2\text{PO}_4^-] \\
&+ 3\,[\Sigma\text{ATP}] - \big([\text{H.ATP}^{3-}] + [\text{Mg.H.ATP}^-]\big) \\
&+ 2\,[\Sigma\text{ADP}] - \big([\text{H.ADP}^{2-}] + [\text{Mg.H.ADP}^0]\big) \\
&+ [\Sigma\text{creatine-P}] - [\text{H.creatine-P}^-] \\
&- [\text{H.Tris}^+] \\
={}& 18.27 \text{ mM}.
\end{aligned}
\tag{5.66}
$$

Since the value of $H_{surplus}$ computed above is positive, it must be balanced by an equal addition of a strong base, in this case KOH. The potassium ion associated with this is[10]

$$[\text{K}^+]_{KOH} = H_{surplus}.\tag{5.67}$$

(If the computed $H_{surplus}$ were negative, it would be balanced by an equal addition of HCl.)

---

[10] This rough calculation ignores binding of other cations such as K^+ and Na^+ to the biochemical reactants in the buffer, and ignores the negligible contribution from self-buffering of water through Eq. (5.45).

Summing up all ionic species contributions from Eqs (5.61) to (5.67), we find that 0.1887 M of KCl was added to achieve a final $I = 0.25$ M. The total estimated added KCl for each data entry in Table 5.2 is listed in Table 5.4, along with the estimated total [K]. Given values for total potassium and magnesium ion concentrations, we can use mass balance equations to re-estimate the free $[Mg^{2+}]$ and to estimate the free $[K^+]$ in these experiments, as follows. (These calculations differ from the original calculations of Lawson and Veech in that here K^+ binding is accounted for and we use different values of the binding constants, as described below.)

Accounting for K^+ binding, the binding polynomials in Eqs (5.61) and (5.65) are modified to include a K^+-dependent term:

$$[K.ATP^{3-}] = [\Sigma ATP] \left([K^+]/K_{K,ATP}\right)/P_{ATP}$$

$$P_{ATP} = 1 + [H^+]/K_{H,ATP} + [Mg^{2+}]/K_{Mg,ATP}$$
$$+ [H^+][Mg^{2+}]/(K_{H,ATP}K_{Mg,HATP}) + [K^+]/K_{K,ATP}$$

$$[K.ADP^{2-}] = [\Sigma ADP] \left([K^+]/K_{K,ADP}\right)/P_{ADP}$$

$$P_{ADP} = 1 + [H^+]/K_{H,ADP} + [Mg^{2+}]/K_{Mg,ADP}$$
$$+ [H^+][Mg^{2+}]/(K_{H,ADP}K_{Mg,HADP}) + [K^+]/K_{K,ADP}$$

$$[K.creatine\text{-}P^-] = [\Sigma creatine\text{-}P] \left([K^+]/K_{K,CrP}\right)/P_{CrP}$$

$$P_{CrP} = 1 + [H^+]/K_{H,CrP} + [Mg^{2+}]/K_{Mg,CrP} + [K^+]/K_{K,CrP}$$

$$[K.HPO_4^-] = [\Sigma phosphate] \left([K^+]/K_{K,Pi}\right)/P_{Pi}$$

$$P_{Pi} = 1 + [H^+]/K_{H,Pi} + [Mg^{2+}]/K_{Mg,Pi} + [K^+]/K_{K,Pi}. \tag{5.68}$$

Expressing the conservation of total magnesium and potassium ions, we have

$$\text{total } [Mg] = [Mg^{2+}] + [Mg.ATP^{2-}] + [Mg.H.ATP^-] + [Mg.ADP^-]$$
$$+ [Mg.H.ADP^0] + [Mg.HPO_4^0] + [Mg.creatine\text{-}P^0]$$
$$\text{total } [K] = [K^+] + [K.ATP^{3-}] + [K.ADP^{2-}] + [K.creatine\text{-}P^-] + [K.HPO_4^-].$$

$$\tag{5.69}$$

Equation (5.69) is a set of nonlinear equations that may be solved for free $[Mg^{2+}]$ and $[K^+]$ for every experiment represented in Tables 5.2 and 5.4. The values of these quantities in Table 5.4 correspond to solutions to these equations obtained using the dissociation constants from Li *et al.* [39] listed in Table 5.3.

Table 5.4: Estimated total added KCl, total potassium ion, and free magnesium and potassium ion concentrations for the experimental measurements on creatine kinase equilibrium from Lawson and Veech [36].

	Added [KCl] (M)	Total [K] (M)	$[Mg^{2+}]$ (mM)	$[K^+]$ (M)
1.	0.187743	0.231013	0.00122892	0.223872
2.	0.188617	0.231387	0.00124821	0.22435
3.	0.189075	0.23185	0.00126032	0.224767
4.	0.189835	0.232159	0.00128504	0.225194
5.	0.192114	0.233528	0.00333679	0.226724
6.	0.191245	0.232822	0.00320432	0.226021
7.	0.191055	0.233147	0.00308633	0.226194
8.	0.189578	0.232026	0.00287499	0.225032
9.	0.154351	0.164351	0.00987239	0.160306
10.	0.154047	0.164047	0.0097563	0.159985
11.	0.1542	0.1642	0.00981816	0.160135
12.	0.164862	0.174862	0.0131746	0.171187
13.	0.163617	0.173617	0.0125555	0.169893
14.	0.164707	0.174707	0.013045	0.171006
15.	0.160773	0.170773	0.125363	0.166889
16.	0.159848	0.169848	0.121897	0.16594
17.	0.160331	0.170331	0.123503	0.166435
18.	0.169528	0.179528	0.168876	0.17613
19.	0.168958	0.178958	0.164427	0.175519
20.	0.169328	0.179328	0.167089	0.175912
21.	0.190394	0.23404	1.3193	0.227172
22.	0.19095	0.234355	1.41587	0.227451
23.	0.191253	0.234639	1.41142	0.227717
24.	0.190597	0.234224	1.34537	0.22733
25.	0.191525	0.234702	1.42955	0.227813
26.	0.19045	0.233981	1.41289	0.227062
27.	0.190462	0.234372	1.37529	0.227363
28.	0.190731	0.23423	1.41599	0.227298
29.	0.189302	0.199302	6.41999	0.197021
30.	0.190498	0.200498	6.707	0.198232
31.	0.190656	0.200656	6.70886	0.198377
32.	0.189438	0.199438	6.47172	0.197176
33.	0.189761	0.199761	6.42904	0.197436
34.	0.190383	0.200383	6.47942	0.198061
35.	0.190059	0.200059	6.64852	0.197748
36.	0.190382	0.200382	6.76774	0.198114
37.	0.200907	0.250066	16.4597	0.24424

Table 5.4: (cont.)

	Added [KCl] (M)	Total [K] (M)	[Mg²⁺] (mM)	[K⁺] (M)
38.	0.202401	0.250694	16.682	0.244881
39.	0.202444	0.250811	16.663	0.244967
40.	0.200235	0.249734	16.3839	0.243865
41.	0.199979	0.249667	16.3574	0.243746
42.	0.200939	0.249758	16.5674	0.243895
43.	0.200823	0.250429	16.3843	0.244496
44.	0.202158	0.25046	16.6869	0.244612

The chemical equilibrium constant for reaction (5.57) determines the equilibrium ratio[11]

$$K_{eq} = \left(\frac{[\text{ATP}^{4-}][\text{creatine}^0]}{[\text{ADP}^{3-}][\text{creatine-P}^{2-}][\text{H}^+]} \right)_{eq}.$$

Expressing unbound species concentrations in terms of the reactant binding polynomials, we have

$$K_{eq} = \left(\frac{[\Sigma\text{ATP}][\Sigma\text{creatine}]}{[\Sigma\text{ADP}][\Sigma\text{creatine-P}][\text{H}^+]} \right)_{eq} \cdot \frac{P_{ADP}}{P_{ATP} \, P_{CrP}}, \qquad (5.70)$$

where all of the binding polynomials are functions of the free cation concentrations ([H⁺], [Mg²⁺], and [K⁺]) that are assumed to bind biochemical reactants in significant quantities, and are defined in Eq. (5.68). Defining the observed equilibrium ratio of biochemical reactants, we have

$$K_{obs} = \left(\frac{[\Sigma\text{ATP}][\Sigma\text{creatine}]}{[\Sigma\text{ADP}][\Sigma\text{creatine-P}]} \right)_{eq} = K_{eq} \frac{[\text{H}^+] P_{ATP} \, P_{CrP}}{P_{ADP}}. \qquad (5.71)$$

In the Lawson–Veech experiments, the magnesium ion concentration varies over several orders of magnitude, while [H⁺] and [K⁺] are maintained in a relatively narrow range. Thus the binding polynomials in Eq. (5.71) depend primarily on [Mg²⁺] over the studied range of experimental conditions, and the quantity $K_{obs}/[\text{H}^+]$ does not depend strongly on changes in [H⁺] and [K⁺] in this observed range of these variables. Figure 5.6 plots $K_{obs}/[\text{H}^+]$ versus [Mg²⁺] for all of the data points in Table 5.2. The solid curve is the theoretical prediction from Eq. (5.71), with hydrogen and potassium ion concentrations fixed at the mean values for the whole data set ([H⁺] = 7.707×10^{-8} M and [K⁺] = 0.207 M) and with $K_{eq} = 3.5 \times 10^8$, and $\Delta_r G^0 = -RT \ln K_{eq}$.

---

[11] Here, K_{eq} is a true chemical reaction equilibrium constant, in that it is the equilibrium mass-action ratio for a balanced chemical reaction.

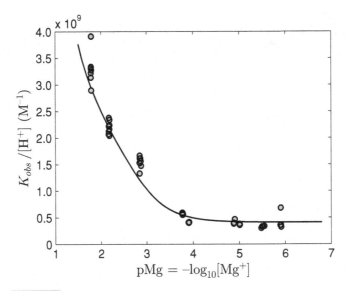

Figure 5.6

Plot of $K_{obs}/[H^+]$, defined by Eq. (5.71), versus Mg^{2+} ion concentration for the creatine kinase reaction. Data are from Lawson and Veech [36]. The solid line is the theoretical prediction associated with $K_{eq} = 3.5 \times 10^8$.

5.7.2 Example: standard enthalpy of a biochemical reaction

In 1992 Teague and Dobson [58] published a study measuring creatine kinase equilibria at different temperatures to estimate the standard enthalpy of the reaction. To see how $\Delta_r H^o$ may be estimated from equilibrium data at different temperatures, recall the definition of the Gibbs free energy

$$G = H - TS,$$

where the enthalpy H is equal to $E + PV$. For chemical reactions, the change in free energy with respect to extent of reaction ϕ is given by

$$\left(\frac{\partial G}{\partial \phi}\right)_{P,T} = \left(\frac{\partial H}{\partial \phi}\right)_{P,T} - T\left(\frac{\partial S}{\partial \phi}\right)_{P,T},$$

or in standard notation,

$$\Delta_r G = \Delta_r H - T\Delta_r S. \tag{5.72}$$

We define the standard reference enthalpy and entropy changes

$$\Delta_r G^o = \Delta_r H^o - T\Delta_r S^o.$$

Since, as we have seen in Section 5.5, the species concentrations influence the entropic contribution to the change in free energy $(\partial S/\partial \phi)_{P,T}$, the reaction entropy and enthalpy change can be expressed

$$\Delta_r S = \Delta_r S^o - RT \ln \prod_{i=1}^{N_s} (C_i/C_o)^{v_i} \qquad (5.73)$$

and

$$\Delta_r H = \Delta_r H^o. \qquad (5.74)$$

Finally, recalling that $\Delta_r G^o = -RT \ln K_{eq}$, we arrive at the following expression:

$$\ln K_{eq} = \frac{\Delta_r S^o}{R} - \frac{\Delta_r H^o}{RT}. \qquad (5.75)$$

This equation is called the *van't Hoff equation*. If $\Delta_r S^o$ and $\Delta_r H^o$ remain constant over some defined temperature range, then the van't Hoff equation predicts that $\ln K_{eq}$ varies linearly with $1/T$, and reaction $\Delta_r H^o$ may be estimated from the slope of this relationship.

Raw data associated with the study of Teague and Dobson [58] are listed in Table 5.5, which lists temperature, observed equilibrium ratio, and pH for 32 measured data points.

Recall the relationship between the observed concentration ratio K_{obs} and the observed chemical equilibrium constant defined in Eq. (5.71). To compute K_{eq} from K_{obs} requires knowledge of the free cation concentrations that contribute to the binding polynomials P_{ATP}, P_{ADP}, and P_{CrP}. Thus it is necessary to obtain estimates of free $[Mg^{2+}]$ and $[K^+]$ associated with the data points in Table 5.5 in order to obtain estimates of K_{eq}. Here Teague and Dobson report adding total magnesium of 3.25 mM and total potassium of approximately 250 mM. Applying the same conservation equation – Eq. (5.69) – used in the previous section, we obtain the estimates for free $[Mg^{2+}]$ and $[K^+]$ listed in Table 5.5. (Note that these calculations require values for the dissociation constants at the different temperatures employed in this study. Estimates for the necessary constants are listed in Tables 5.3 and 5.6.)

Given the estimated free cation concentrations, we can compute the values of K_{eq} associated with these data points.[12] Plotting the logarithm of the measured K_{eq} as a function of $1/T$ in Figure 5.7, we estimate $\Delta_r H^o = -18.93$ kJ/mol from Eq. (5.75) for this reaction (in the temperature range of 5–38° C). Note that this study provides a higher estimate for the chemical equilibrium constant at $T = 38°$ C and $I = 0.25$ M than the study of Lawson and Veech [36]: $K_{eq} = 5.18 \times 10^8$ under these condition in Figure 5.7 versus $K_{eq} = 3.5 \times 10^8$ in Figure 5.6.

[12] See Exercise 5.6.

Table 5.5: Reported data from Teague and Dobson [58] on creatine kinase equilibrium.

	$T(°C)$	K_{obs}	pH	$[Mg^{2+}]$ (mM)	$[K^+]$ (M)
1.	38	142.881	6.934	0.867042	0.236307
2.	38	137.375	6.934	0.867176	0.236307
3.	38	142.31	6.94	0.724377	0.236039
4.	38	145.708	6.94	0.720213	0.23603
5.	38	168.707	6.927	0.835308	0.236297
6.	38	163.265	6.927	0.835247	0.236297
7.	38	156.559	6.975	0.76206	0.23569
8.	38	159.002	6.975	0.766991	0.235702
9.	25	166.129	6.97	0.98186	0.236207
10.	25	180.69	6.97	0.963962	0.236184
11.	25	145.44	6.976	0.835493	0.235939
12.	25	158.477	6.976	0.82061	0.235912
13.	25	195.942	6.963	0.940063	0.236202
14.	25	192.87	6.963	0.932368	0.236174
15.	25	187.045	7.011	0.856009	0.235557
16.	25	200.117	7.011	0.85215	0.235552
17.	15	201.121	6.998	1.07653	0.236127
18.	15	191.79	6.998	1.06637	0.236113
19.	15	203.361	7.004	0.916892	0.235866
20.	15	180.272	7.004	0.927402	0.235857
21.	15	214.546	6.991	1.03457	0.236103
22.	15	214.69	6.991	1.04491	0.236118
23.	15	213.698	7.039	0.965484	0.235549
24.	15	221.672	7.039	0.967629	0.235509
25.	5	225.581	7.026	1.17486	0.236042
26.	5	226.65	7.026	1.17487	0.236042
27.	5	223.407	7.032	1.03309	0.23578
28.	5	207.588	7.032	1.02001	0.235755
29.	5	266.19	7.019	1.1441	0.236032
30.	5	257.981	7.019	1.13681	0.236018
31.	5	262.086	7.067	1.07333	0.23547
32.	5	257.264	7.067	1.07041	0.23545

5.8 Recapitulation and lessons learned

In this chapter we derived a set of thermodynamic formulae for application in chemistry. Many of these expressions, namely Eqs (5.19), (5.40), (5.42), (5.55),

Table 5.6: Ion dissociation constants for calculations in Section 5.7.2.

Reaction	Dissociation constant		
	$T = 5°\,C$	$T = 15°\,C$	$T = 25°\,C$
$H.ATP^{3-} \rightleftharpoons H^+ + ATP^{4-}$	3.83×10^{-7}	3.75×10^{-7}	3.69×10^{-7}
$H.ADP^{2-} \rightleftharpoons H^+ + ADP^{3-}$	5.37×10^{-7}	5.25×10^{-7}	5.15×10^{-7}
$H_2PO_4^- \rightleftharpoons H^+ + HPO4^{2-}$	1.98×10^{-7}	2.13×10^{-7}	2.28×10^{-7}
$H.creatine\text{-}P^- \rightleftharpoons H^+ + creatine\text{-}P^{2-}$	2.91×10^{-5}	3.09×10^{-5}	3.27×10^{-5}
$Mg.ATP^{2-} \rightleftharpoons Mg^{2+} + ATP^{4-}$	3.04×10^{-4}	2.37×10^{-4}	1.88×10^{-4}
$Mg.ADP^{1-} \rightleftharpoons Mg^{2+} + ADP^{3-}$	1.95×10^{-3}	1.58×10^{-3}	1.30×10^{-3}
$Mg.HPO4^0 \rightleftharpoons Mg^{2+} + HPO4^{2-}$	3.67×10^{-2}	3.22×10^{-2}	2.84×10^{-2}
$Mg.creatine\text{-}P^0 \rightleftharpoons Mg^{2+} + creatine\text{-}P^{2-}$	6.85×10^{-2}	6.29×10^{-2}	5.88×10^{-2}
$K.ATP^{3-} \rightleftharpoons K^+ + ATP^{4-}$	1.29×10^{-1}	1.28×10^{-1}	1.28×10^{-1}
$K.ADP^{2-} \rightleftharpoons K^+ + ADP^{3-}$	1.59×10^{-1}	1.60×10^{-1}	1.61×10^{-1}
$KHPO_4^- \rightleftharpoons K^+ + HPO4^{2-}$	4.31×10^{-1}	4.33×10^{-1}	4.35×10^{-1}
$K.creatine\text{-}P^- \rightleftharpoons K^+ + creatine\text{-}P^{2-}$	—	—	—
$Mg.H.ATP^- \rightleftharpoons Mg^{2+} + H.ATP^{3-}$	8.37×10^{-2}	6.53×10^{-2}	5.19×10^{-2}
$Mg.H.ADP^0 \rightleftharpoons Mg^{2+} + H.ADP^{2-}$	1.40×10^{-1}	1.19×10^{-1}	1.02×10^{-1}

Values for ATP, ADP, and phosphate complexes are from Li *et al.* [39] adjusted to $I = 0.25$ M. Values for creatine-P complexes are from [58].

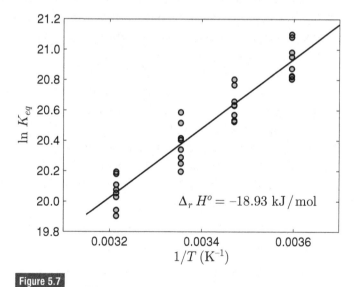

Figure 5.7

Plot of $\ln K_{eq}$, defined by Eq. (5.70), versus $1/T$ ion concentration for the creatine kinase reaction. Data are from Teague and Dobson [58]. The solid line is the fit to Eq. (5.75) with $\Delta_r H^o = -18.93$ kJ/mol. The mean value of the equilibrium constant at $T = 38°$ C is $K_{eq} = 5.18 \times 10^8$.

(5.72), and (5.75), are likely to be familiar from basic chemistry. What distinguishes the preceding exposition from that in a standard introductory chemistry text is that here these expressions were derived from fundamental principles. It is hoped that these derivations, invoking ideas regarding the statistical nature of thermodynamics, have helped the reader to develop an intuition for their origin and application.

Furthermore, these formulae for chemical thermodynamics were applied to analyze real data on the thermodynamics of a real biochemical reaction. In the examples we saw that the analysis of a system of only one biochemical reaction can require a large number of biochemical species and cation dissociation reactions to be accounted for. As a result, the total number of species accounted for in the analysis of a single reaction in vitro in Section 5.7 is much greater than the number of species occurring in the reaction of Eq. (5.57). Indeed, real systems are truly complex. Cellular biochemical systems in vivo involve large numbers of interacting reactions, some in chemical equilibrium, some not. Yet analysis and simulations of these systems rest on the same foundation of chemical thermodynamics that we have developed here.

Problems

5.1 Combinatorial entropy. For the values of N used in the example of Section 5.1.2, it is not practical to compute the factorial function exactly. Indeed, for N greater than approximately 170, $N!$ is a number greater than can be represented using standard double-precision floating-point format. One way to generate the plots shown in Figure 5.1 is to work with the logarithm of Ω, and make use of Stirling's approximation, $\ln N! = N \ln N - N$. Make use of Stirling's approximation to generate the plots in Figure 5.1. Plot the predicted curves for $N = 2000$ and $N = 10\,000$.

5.2 Ideal gas. For a system of noninteracting monatomic particles (an ideal gas) the microcanonical partition function is proportional to V^N. Based on $\Omega \sim V^N$, we can derive the state equation known as the ideal gas law:

$$\frac{P}{k_B T} = \frac{\partial \ln \Omega}{\partial V} = \frac{N}{V}.$$

Consider a gas of particles that do not interact in any way except that each particle occupies a finite volume v_o, which cannot be overlapped by other particles. What consequences does this imply for the gas law? [Hint: use the relationship $\Omega(V) \sim \int d\mathbf{x}$. You might try assuming that each particle is a solid sphere.]

5.3 Isobaric equilibrium. Assume that the two systems in Figure 5.2 are separated by a barrier that moves in response to a pressure difference but does not allow heat exchange. Therefore the system moves to an equilibrium where the pressures P_1 and P_2 on both sides of the barrier equilibrate to $P_1 = P_2$. As the barrier moves, the incremental work done by system 1 on system 2 is $\delta(P_2 V_2)$, and conservation of energy requires $\delta(P_1 V_1) = -\delta(P_2 V_2)$. Assuming the ideal gas law applies, we have $N_1 k \delta(T_1) = -N_2 k \delta(T_2)$. Show that

$$N_1(T_1 - T_1^o) = -N_2(T_2 - T_2^o),$$

where T_1^o and T_2^o are the initial temperatures in systems 1 and 2, and T_1 and T_2 are the final temperatures. Invoke a conservation argument to show that

$$N_1 T_1 < N_1 T_1^o + N_2 T_2^o.$$

5.4 Acids, bases, and buffers. Phosphate may be obtained as either a monobasic salt (e.g. Na_2HPO_4) or a dibasic salt (e.g. NaH_2PO_4). If we wish to make a buffer of 50 mM total phosphate at pH near 7, what concentration of each salt should be used? If we add 1 mM of HCl to this solution, how much will the pH change? [Hint: First compute H_o from Eq. (5.52) for the neutral buffer solution. Next, determine how much $[H^+]$ changes when H_o is increased by 1 mM.]

5.5 Ionic distribution in solution. Derive Eqs (5.61) and (5.65), and verify the calculations.

5.6 Computer exercise. Using the values total $[Mg] = 3.25$ mM and total $[K] = 250$ mM, and adopting the values of the dissociation constants from Tables 5.3 and 5.6, confirm that the free $[Mg^{2+}]$ and $[K^+]$ values listed in Table 5.5 are reasonable. Compute the estimates of K_{eq} associated with these data points, and reproduce Figure 5.7.

6 Chemical reaction systems: kinetics

Overview

The fundamental physical processes of cell biology are chemical reactions and chemical transport phenomena. In any active living cell at any given time thousands of individual chemical transformations – oxidation of primary substrates, synthesis and replication of macromolecules, reaction-driven transport of organic and inorganic substances – are occurring spontaneously and simultaneously. To study the coordinated functions of biochemical reaction *systems* it will be useful to become familiar with the basics of chemical kinetics, the rules that effectively describe the turnover of individual elementary reaction steps. (Elementary reactions are fundamental chemical transformations that are by definition not broken down into more fundamental steps.) Elementary reaction steps can be combined to model and simulate enzymatic mechanisms catalyzing biochemical reactions. Biochemical reactions are combined in biochemical systems, networks of interdependent reactions.

This chapter primarily treats the kinetics of individual reactions, elementary steps, and enzyme-mediated mechanisms for biochemical reactions.

6.1 Basic principles of kinetics

6.1.1 Mass-action kinetics

Let us begin our study of chemical kinetics with the uni-uni-molecular reaction

$$A \underset{J^-}{\overset{J^+}{\rightleftharpoons}} B,$$
(6.1)

familiar from Chapter 5. Here we denote the rate of turnover of the reaction in the left-to-right direction (or forward *flux*) J^+ and the reverse flux J^-. Therefore the net flux is given by $J = J^+ - J^-$.

If individual molecules in the system are kinetically isolated – i.e., the rate of transformation of an individual A molecule to a B molecule does not depend on interactions (direct or indirect) with other A or B molecules – then the total forward flux is proportional to the number of A molecules present:

$$J^+ = k_+ a. \tag{6.2}$$

Similarly,

$$J^- = k_- b, \tag{6.3}$$

where a and b are the concentrations of A and B molecules. In these equations the constants k_+ and k_- are called *rate constants*. The average time an individual molecule spends in state A before transforming to state B is $1/k_+$, and the average time an individual molecule spends in state B before transforming to state A is $1/k_-$. Thus k_+ and k_- have units of inverse time.

In the absence of additional reaction or transport processes, the kinetics of a and b are governed by the differential equations

$$da/dt = -k_+ a + k_- b$$
$$db/dt = +k_+ a - k_- b, \tag{6.4}$$

where $a = [A]$ and $b = [B]$. The right-hand sides of these equations are dependent, and the system may be reduced to

$$da/dt = -k_+ a + k_- (B_o + A_o - a), \tag{6.5}$$

where A_o and B_o are the initial quantities $a(0) = A_o$ and $b(0) = B_o$.

In our study of biochemical systems, we will develop models of enzyme-mediated biochemical reaction mechanisms made up of elementary reaction steps governed by mass action. We defined an elementary reaction as a reaction governed by mass action that cannot be decomposed into more fundamental steps. (In general any continuous and high-dimensional chemical process may be broken down into an arbitrary number of discrete elementary steps. However, in practice the definition of elementary steps is usually unambiguous.) For example, consider a simple proton association/dissociation reaction

$$HCOO^- + H^+ \rightleftharpoons HCOOH.$$

The size of the quantum mechanical state space adopted by the formic acid ($HCOO^-$/$HCOOH$) molecule in aqueous solution is practically infinite, even at fixed temperature, pressure, and in a confined volume. Yet at the molecular scale under biological conditions the state space is effectively collapsed into two forms: protonated ($HCOOH$) and deprotonated ($HCOO^-$). At a given fixed pH (constant H^+ chemical activity) the kinetics of transition between the two states may be characterized by the mass-action elementary reaction $A \rightleftharpoons B$.

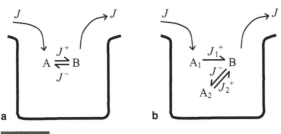

Nonequilibrium steady-state reaction system. The reaction $A \rightleftharpoons B$ is maintained in a nonequilibrium steady state by pumping A in and B out at a constant flux J.

6.1.2 Thermodynamic constraints on reaction kinetics

Equation (6.5), governing the mass-action kinetics of $A \rightleftharpoons B$, has solution

$$a(t) = (A_o + B_o)\frac{k_-}{k_+ + k_-} + \left(\frac{A_o k_+ - B_o k_-}{k_+ + k_-}\right) e^{-(k_+ + k_-)t}. \qquad (6.6)$$

In the limit $t \to \infty$,

$$\left(\frac{b}{a}\right)_\infty = \left(\frac{[B]}{[A]}\right)_\infty = \frac{k_+}{k_-}.$$

Therefore we expect k_+/k_- to equal K_{eq}, the equilibrium constant for the reaction. The values of the kinetic constants are constrained by a thermodynamic property of the reaction. This relationship arises from the fact that the forward and reverse fluxes J^+ and J^- must be equal in thermodynamic equilibrium. A more general thermodynamic relationship between forward and reverse fluxes for nonequilibrium systems may be obtained as follows.

Imagine that the reaction $A \rightleftharpoons B$ is maintained in a nonequilibrium steady state by pumping A molecules and B molecules out of a reaction system at constant flux J, as illustrated in Figure 6.1. In this system a steady state will be established where $J^+ - J^- = J$. Next, imagine that we could observe all of the individual A and B molecules in this system using a powerful imaging system. We keep track of each A molecule pumped in from the external supply by labeling it A_1. Similarly, we identify every A molecule that appears from the reverse reaction $B \to A$ with the label A_2. This labeling scheme is illustrated in Figure 6.1(b).

Since there is no transport of A_2 into or out of the system, steady-state concentrations and fluxes will obey

$$\left(\frac{[B]}{[A_2]}\right) = K_{eq} \qquad (6.7)$$

and

$$J_2^+ = J^-, \qquad (6.8)$$

where we have separated the forward flux into two components, J_1^+ and J_2^+, as illustrated in the figure.

Since A_1 and A_2 molecules differ only in their origin, they are both equally likely to undergo a reaction transition to B, regardless of the reaction mechanism.[1] Therefore

$$\frac{J_1^+}{J_2^+} = \frac{[A_1]}{[A_2]}. \tag{6.9}$$

From Eq. (6.9) we have

$$J^+ = J_1^+ + J_2^+ = \left(1 + \frac{[A_1]}{[A_2]}\right) J_2^+$$

$$= \left(\frac{[A]}{[A_2]}\right) J^- = \left(\frac{[A]}{[B]}\right) K_{eq} \, J^- \tag{6.10}$$

or

$$\left(\frac{J^+}{J^-}\right) = \left(\frac{[A]}{[B]}\right) K_{eq} = e^{-\Delta_r G / RT}. \tag{6.11}$$

This *flux-ratio theorem* has been independently derived a number of times for a number of special cases.[2]

6.1.3 Transition state theory

Recall that J^+ is the rate at which A molecules transform to B molecules, perhaps measured in units of number of transitions per unit time per unit volume of reaction medium. Similarly J^- is the rate of transition from B to A. Therefore, if over some (perhaps small) finite span of time a single state transition occurs (either $A \to B$ or $B \to A$) within some (perhaps small) finite volume, the relative probabilities of these transitions are

$$\frac{P_{A \to B}}{P_{B \to A}} = e^{-\Delta G_{AB}/(RT)}$$

$$P_{A \to B} + P_{B \to A} = 1.$$

It follows that

$$P_{A \to B} = \frac{1}{1 + e^{+\Delta G_{AB}/(RT)}}$$

$$P_{B \to A} = \frac{1}{1 + e^{-\Delta G_{AB}/(RT)}}. \tag{6.12}$$

---

[1] In other words, Eq. (6.9) holds whether the kinetics are governed by simple mass action or some more complex catalytic mechanism.

[2] See discussion in Beard and Qian [7].

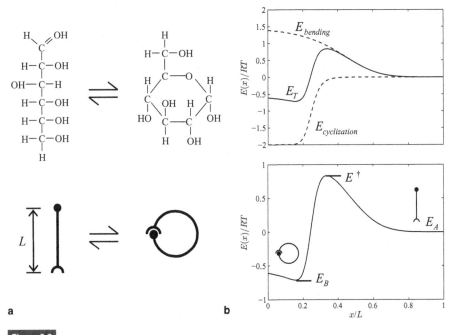

Cylicization–isomerization reaction to illustrate dependence of reaction kinetics on transition state energy: (a) two isomers of glucose, with the chemical structure shown above and schematic model below; (b) energy landscape as a function of the reaction coordinate x, the distance between the two ends of molecular chain that associate to form the cyclic isomer. The lower panel of (b) shows the total energy $E_t = E_{cyclization} + E_{bending}$ with states A and B associated with the open and closed forms, respectively.

This relationship gives us the probability of a given elementary transition, given the condition that an amount of time has passed such that exactly one transition has occurred in a given ensemble of A and B molecules. But how long do we need to wait (on average) for a transition to occur? In other words, what are the rate constants k_+ and k_-? In this section we explore the thermodynamic theory of how rate constants vary with temperature and energy barriers separating molecular states.

Thermodynamic factors affecting the kinetics of a chemical reaction such as A $\rightleftharpoons$ B may be illustrated by considering a cyclization–isomerization reaction such as is illustrated in Figure 6.2. Here we see two forms of the six-carbon carbohydrate glucose: an open chain molecule and a six-member ring. The two forms are schematically illustrated in the lower left panel of the figure. This diagram illustrates the concept that covalent binding of the two ends requires bending of the backbone of the molecule.

To develop a theory of how bonding and bending energies influence the rates of isomerization, we define a *reaction coordinate x*, which is the distance between the two ends of the molecular chain that can covalently associate. When x is near zero, the molecule is cyclic, with bonding energy, $E_{cyclization}$, at a minimum value. The bending energy, $E_{bending}$, on the other hand, is inversely proportional to x. Representative energy profiles are plotted in Figure 6.2(b). We see that the total energy $E_T = E_{cyclization} + E_{bending}$ has minima associated with the cyclic state (near $x = 0$) and with the open chain (near $x = L$). Also recall that there exist distributions of microstates associated with each of these qualitatively distinct configurational states, open and cyclic. Thus the open and cyclic states in general have different entropies that are not reflected in the energy landscape shown in the figure.

Furthermore, to transition from one state to the other, the molecule most move through an energy maximum, labeled $E^\dagger$ in the figure and called the *transition energy*. If there were a large number of molecules present in an ensemble of a given state – say the open state, for example – then we would expect that at any given instant the number of molecules adopting the transition state would be proportional to[3] $e^{-(E^\dagger - E_A)/RT}$:

$$\frac{P(E^\dagger)}{P(E_A)} \propto e^{-(E^\dagger - E_A)/RT}.$$

Here we have identified the open state as state A and the closed cyclic state as B. Furthermore we have associated a representative energy E_A with state A. This energy is formally identified to be the average internal energy of an equilibrium ensemble of molecules in conformation A. As long as the activation energy barrier is large compared with the average fluctuations in energy in state A, then this proportionality statement is reasonable. The relative probability of adopting the transition state is proportional to (rather than equal to) $e^{-(E^\dagger - E_A)/RT}$ because of differences in conformational entropy between state A and the transition state. Since in order to transition from one state to another a molecule must pass through the transition state, the average rate of transition from A to B (open to cyclic) will be proportional to $e^{-(E^\dagger - E_A)/RT}$ and the average rate of transition of a single molecule from B to A proportional to $e^{-(E^\dagger - E_B)/RT}$. Defining the transition energy barriers $\Delta E_A^\dagger = E^\dagger - E_A$ and $\Delta E_B^\dagger = E^\dagger - E_B$, we have

$$k_+ = C_+ e^{-\Delta E_A^\dagger/RT}$$

$$k_- = C_- e^{-\Delta E_B^\dagger/RT} \tag{6.13}$$

---

[3] Formally, this is true for the NVT ensemble. In NPT systems we should think in terms of transition enthalpies rather than transition energies.

for the forward and reverse rate constants for the reaction $A \rightleftharpoons B$. In these expressions C_+ and C_- are constants that are in general not equal to one another. The equations in (6.13) are called *Arrhenius equations*.

In this example we might expect the conformational entropy associated with the open chain to be greater than that associated with the closed ring. Therefore, since the open chain is identified as state A, C_- is expected to be greater than C_+, reflecting an entropic bias for the open state. More formally, we recall that k_+/k_- is equal to the equilibrium constant for this reaction. From this it is straightforward to show that, in the NVT ensemble, $C_+/C_- = e^{(S_B - S_A)/R}$ (where S_A and S_B are the single-molecule entropies associated with the two states) and to revise Eq. (6.13):

$$k_+ = k_o e^{+\Delta S_A^\ddagger/R} \cdot e^{-\Delta E_A^\ddagger/RT} = k_o e^{-\Delta F_A^\ddagger/RT}$$

$$k_- = k_o e^{+\Delta S_B^\ddagger/R} \cdot e^{-\Delta E_B^\ddagger/RT} = k_o e^{-\Delta F_B^\ddagger/RT}, \qquad (6.14)$$

where k_o is a common factor that does not depend on the energy or entropy differences between the two conformational states,[4] and the activation state entropy changes are defined analogous to the energy changes: $\Delta S_A^\ddagger = S^\dagger - S_A$ and $\Delta S_B^\ddagger = S^\dagger - S_B$. Similarly in the NPT ensemble:

$$k_+ = k_o e^{-\Delta G_A^\ddagger/RT}$$

$$k_- = k_o e^{-\Delta G_B^\ddagger/RT}, \qquad (6.15)$$

where $\Delta G_A^\ddagger$ and $\Delta G_B^\ddagger$ are the Gibbs free energies of activation for the forward and reverse transitions.

According to a variety of theories of chemical kinetics the factor k_o also shows temperature dependence. This is because, while we expect that the frequency with which a molecule adopts the transition state will be proportional to $e^{\Delta E^\ddagger/RT}$, we also expect that the rate at which molecules move out of the transition state will vary with temperature. Assuming that a molecule that is in the transition state will tend to transition out of that state at a rate proportional to the kinetic energy of its internal molecular motions, we find that the factor k_o is linearly proportional to T, as in the famous *Eyring equation*:

$$k_+ = \kappa \frac{k_B T}{h} e^{-\Delta G_A^\ddagger/RT}$$

$$k_- = \kappa \frac{k_B T}{h} e^{-\Delta G_B^\ddagger/RT}, \qquad (6.16)$$

where h is Planck's constant and κ is a constant that does not depend on temperature.

---

[4] See Exercise 6.1.

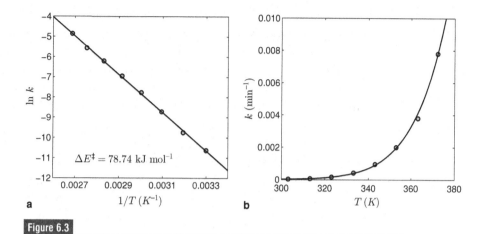

Figure 6.3

Relationship between rate constant (k) for conversion of phosphocreatine to creatinine and temperature (T). (a) $\ln k$ measured at pH = 5 plotted versus T^{-1}, where k is in units of min^{-1}. Solid line is fitted to $k = ae^{-\Delta E^{\ddagger}/RT}$ with $\Delta E^{\ddagger} = 78.74$ kJ mol^{-1}. (b) The same data and model fit on linear axis scales. Data are taken from Diamond [17].

6.1.4 Example: temperature dependence of a chemical reaction

The effectiveness of these transition state theories can be explored by analyzing some real data. Figure 6.3 shows measured data on the effectively irreversible hydrolysis of phosphocreatine to creatinine. Since this reaction proceeds by mass action in solution, it is described by a single rate constant k.

From the Arrhenius equation, we expect a linear relationship between $\ln k$ and T^{-1}, as seen in Figure 6.3. The slope of this relationship provides an estimate of the transition energy barrier $\Delta E^{\ddagger}$, which here is estimated to be 78.74 kJ mol^{-1}. The dramatic relationship between k and T predicted by the Arrhenius equation is shown in Figure 6.3(b).[5]

6.2 Enzymes

In the above example an effectively irreversible uni-unimolecular reaction reaction occurs spontaneously in solution. Therefore the kinetics at a given temperature can effectively be described by mass action, with a single rate constant k. More commonly (particularly in biological systems and for multiple-substrate reactions), reactions proceed via specific catalytic mechanisms. In biology catalysts usually take the form of *enzymes*, and catalytic mechanisms are called *enzyme mechanisms*.

---

[5] See Exercise 6.2.

Enzymes are specialized proteins with structure and function shaped by billions of years of evolution. Many are *specific* in that they catalyze only one or a handful of closely related biochemical reactions. Furthermore, most, if not all, enzymes are regulated in vivo by a complex system of chemical and genetic mechanisms.

Since dealing with enzymes – developing models of how they work, estimating parameter values for these models, and simulating their operation in isolation and in interacting systems of many enzymes – is the bread and butter of analyzing and simulating biochemical reactions, the remainder of this chapter is devoted to enzyme-catalyzed biochemical reaction kinetics. Indeed, comprehensive texts on this subject exist. (The book by Cornish-Bowden [16] is highly recommended.) So, rather than attempt a comprehensive exposition on the subject, we first introduce the canonical Michaelis–Menten enzyme mechanism. Following that, a detailed case study is developed, reviewing a half century of experiments and analysis on a single enzyme, revealing (it is hoped) some practical techniques and lessons. In Section 7.2.3 we apply some of the ideas developed for single reactions to a system of many reactions operating in vivo.

6.2.1 The Michaelis–Menten rate law

The so-called *Michaelis–Menten reaction mechanism* is the starting point for all treatments on the theory of enzyme kinetics. The mature field of enzyme kinetics grew up from the early work of Michaelis, Menten, and contemporaries such as Henri, Briggs, and Haldane [16]. In fact, the canonical Michaelis–Menten mechanism provides a useful vehicle for describing the two most important concepts in the analysis of enzyme-mediated kinetic mechanisms: the quasi-equilibrium model and the quasi-steady state model.

In its simplest form, the Michaelis–Menten mechanism is a model for the irreversible conversion A $\rightarrow$ B. The mechanism is

$$A + E \underset{k_{-1}}{\overset{k_{+1}}{\rightleftharpoons}} EA \overset{k_2}{\rightarrow} E + B, \tag{6.17}$$

where the first step A $+$ E $\rightleftharpoons$ EA involves the reversible association of substrate A with enzyme E to form a complex EA. In the second step, EA $\rightarrow$ E $+$ B, the product B is produced and dissociated from the enzyme complex, yielding unbound enzyme E. The parameters k_{-1}, k_{+1}, and k_2 are the rate constants associated with the individual reaction steps.[6] Thus Eq. (6.17) introduces a convention where the rate constants associated with reaction steps are indicated in the chemical reaction equation. These reaction steps and associated rate constants specify a

---

[6] The individual (elementary) steps in the mechanism are assumed to follow mass-action kinetics. Given this assumption, the overall reaction A $\rightarrow$ B does not obey simple mass action, as shown in this section.

kinetic scheme. Based on the mass-action principles we have already discussed, the kinetics of the species involved in Eq. (6.17) are described by the ordinary differential equations

$$da/dt = -k_{+1}ea + k_{-1}c$$

$$de/dt = +(k_{-1} - k_2)c + k_{+1}ea$$

$$dc/dt = -(k_{-1} + k_2)c + k_{+1}ea$$

$$db/dt = +k_{+2}c, \tag{6.18}$$

where $a = [A]$, $b = [B]$, $e = [E]$, and $c = [EA]$. Here we have assumed that the system is well mixed, and maintained under conditions (such as constant temperature) such that the rate constants k_{-1}, k_{+1}, and k_2 remain constant.

From a mathematical perspective, we might immediately notice that $de/dt = -dc/dt$. Thus the above equations are not independent; either the state variable e or c may be eliminated, yielding a third-order nonlinear system. This mathematical dependence arises as a consequence of physical mass conservation: the sum $e + c$ remains constant in the system defined by these equations. Furthermore, the variable b does not appear on the right-hand side of any of these equations, because the product B does not participate as a substrate of any reaction step. Thus the first two equations may be taken as an independent second-order system.

This system may be simulated by integrating the associated differential equations. More complex systems (with more substrates and products, more intermediate states, and/or networks of coupled reactions) may be treated in a similar manner. However, it is often practical and sometimes necessary to invoke simplifying assumptions in analyzing and simulating enzymatic reactions.

One broadly useful tool is to assume that certain steps in a reaction mechanism occur relatively quickly compared with others and are thus maintained in a *quasi-equilibrium*. For example, Michaelis and Menten [42] assumed that the first step in Eq. (6.17) was maintained in equilibrium: $c/(ea) = 1/K_1$, where K_1 is a constant. Using E_o to denote total enzyme concentration ($E_o = e + c$), we have

$$c = ea/K_1$$

$$= a(E_o - c)/K_1$$

$$= \frac{E_o a}{K_1 + a}. \tag{6.19}$$

Since the rate of production of B is given by $k_2 c$, we may define the quasi-equilibrium flux

$$J = \frac{k_2 E_o a}{a + K_1} = \frac{V_{max} a}{a + K_m}, \tag{6.20}$$

where we have introduced two definitions: the parameter $V_{max} = k_2 E_o$ is the maximum *velocity* or reaction flux, and $K_m = K_1 = k_{-1}/k_{+1}$ is called the *Michaelis–Menten constant*. (The behavior of this equation is explored below.)

The major value of this analysis is that it yields an expression for the enzyme-catalyzed reaction rate as a function of only substrate concentration a. This expression potentially simplifies analysis of kinetic data and facilitates a simplified approach to simulating enzyme kinetics.

An alternative simplifying tool is the *quasi-steady state* model, based on the assumption that the enzyme acts in a steady state. From Eq. (6.18) we have

$$dc/dt = - (k_{-1} + k_2) c + k_{+1} ea = 0, \text{ or}$$

$$c = \frac{k_{+1}}{k_{-1} + k_2} a (E_o - c)$$

$$= \frac{E_o a}{a + K_m}, \tag{6.21}$$

where $K_m = (k_{-1} + k_2)/k_{+1}$. In this case, the flux expression takes the same form as Eq. (6.20):

$$J = \frac{V_{max} a}{a + K_m}. \tag{6.22}$$

But the constant K_m takes a different form under the quasi-steady approximation than under the quasi-equilibrium assumption.

As we shall see in the following section, both the quasi-steady approximation and quasi-equilibrium approximations find application in real-world examples. Yet before considering more complex examples, let us briefly explore the behavior of Eq. (6.22) – known as the *Michaelis–Menten equation* or the *Michaelis–Menten rate law*.

The behavior of Eq. (6.22) is illustrated in Figure 6.4. When substrate concentration is equal to the Michaelis–Menten constant (when $a = K_m$), the flux is one half of V_{max}. As the substrate concentration increases beyond K_m, the predicted flux approaches V_{max} asymptotically. This *saturation* behavior is typical of the kinetics of enzyme-catalyzed reactions. At low substrate concentrations, the reaction proceeds at a rate proportional to substrate concentration. In fact, in the limit $a \to 0$, the Michaelis–Menten equation approaches the limiting behavior

$$\lim_{a \to 0} = \left(\frac{V_{max}}{K_m} \right) a, \tag{6.23}$$

where V_{max}/K_m acts as a quasi-first-order rate constant.

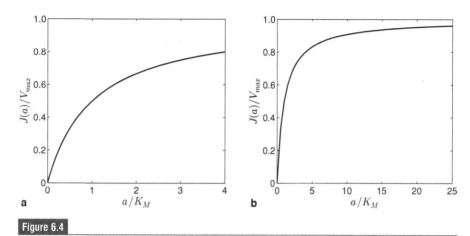

Behavior of the Michaelis–Menten equation. Reaction flux predicted by Eq. (6.22) is plotted as a function of substrate concentration a. The two plots illustrate behavior of the equation over two different ranges of substrate concentration.

Under certain circumstances, the Michaelis–Menten rate law may be used to simplify the governing equations used to model the complete reaction system. Specifically, applying Eq. (6.22) to model the overall reaction turnover of A → B, then the system of Eq. (6.18) is reduced to

$$da/dt = -J(a)$$
$$db/dt = +J(a). \tag{6.24}$$

This system, which ignores the kinetics of the free enzyme and the intermediate complex, accurately captures the behavior of $a(t)$ and $b(t)$ in the full system of Eq. (6.18) under conditions where the rates of change of e and c are small compared with the rates of change of a and b. This happens, for example, when reactant concentrations are large compared with total enzyme concentration ($a/E_o \gg 1$). This also happens when the reaction is maintained near a quasi-steady state with $|dc/dt| \ll |da/dt|$ over the time course of interest.[7]

6.2.2 Case study: mechanism and kinetics of fumarase

The Michaelis–Menten model represents an idealized model for a single-substrate single-product biochemical reaction. Although many of the basic principles remain the same, realistic mechanisms and associated rate laws often take forms that are substantially more complex than illustrated in the previous example. As a

---

[7] The range of validity of this approximation is treated in detail in Chapter 4 of Beard and Qian [8].

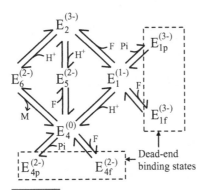

Figure 6.5

Fumarate catalytic mechanism described by Mescam *et al.* [41]. Fumarate and malate are indicated by F and M, respectively. The mechanism includes two distinct pathways for reaction (6.25) as described in the text, as well as four inhibitory dead-end binding steps.

case study, let us consider the fumarate hydration reaction, which is the seventh step of the tricarboxylic acid (TCA) cycle, a critical part of mitochondrial metabolism:

$$C_4O_4H_2^{2-} + H_2O \rightleftharpoons C_4O_5H_4^{2-}$$

$$fumarate^{2-} + H_2O \rightleftharpoons malate^{2-}. \tag{6.25}$$

This reversible hydration/dehydration reaction is catalyzed by the enzyme fumarase (also called fumarate hydratase). The enzyme is believed to be activated though allosteric binding of inorganic phosphate, fumarate, and other anions [41]. Allosteric activation, as well as binding of inhibitors is accounted for in the model developed here.

A mechanism describing the kinetics of the fumarase-catalyzed reaction is illustrated in Figure 6.5. This mechanism[8] includes two pathways for the hydration of fumarate to malate. One pathway involves fumarate binding to enzyme state 4 (in the step $E_4 \rightarrow E_5$), followed by loss of a hydrogen ion (in the step $E_5 \rightarrow E_2$), followed by subsequent addition of an H^+ (in the step $E_2 \rightarrow E_6$), followed by loss of malate from the complex (in the step $E_6 \rightarrow E_4$), regenerating the unbound enzyme E_4. In the other pathway enzyme state E_4 loses a hydrogen ion before fumarate addition, and the full cycle is $E_4 \rightarrow E_1 \rightarrow E_2 \rightarrow E_6 \rightarrow E_4$. In addition to these catalytic cycles the mechanism includes four inhibitory binding steps, where inorganic phosphate or fumarate reversibly binds to form dead-end complexes that cannot participate in catalysis.

---

[8] In Mescam *et al.* [41] a six-state model was reduced to the five-state model of Figure 6.5 by eliminating the state labeled E_3 in the original six-state model. This is why the states in the figure are labeled E_1, E_2, E_4, E_5, and E_6.

Quasi-equilibrium binding

In developing equations to represent the quasi-steady turnover rate for this mechanism, we use the following notations for the concentration variables: $e_1 = [E_1]$, $e_2 = [E_2]$, $e_4 = [E_4]$, $e_5 = [E_5]$, $e_6 = [E_6]$, f = chemical activity of fumarate^{2-}, m = chemical activity of malate^{2-}, and $h = 10^{-pH}$. (Expressions for computing the chemical activities of fumarate and malate ions from the total reactant concentrations are given below.) As determined by Mescam *et al.*, we can assume that the proton association/dissociation steps are relatively rapid, and the reactions are maintained in equilibrium with the following equilibrium constants:

$$K_{26} = \frac{k_{62}}{k_{26}} = h\frac{e_2}{e_6}$$

$$K_{25} = \frac{k_{52}}{k_{25}} = h\frac{e_2}{e_5}$$

$$K_{14} = \frac{k_{41}}{k_{14}} = h\frac{e_1}{e_4}. \tag{6.26}$$

Here lower case k's (e.g., k_{26}) are rate constants, and capital K's (e.g., K_{26}) are equilibrium dissociation constants. The subscripts on each rate constant indicate the reaction that it is associated with. For example, k_{26} is the rate constant for the step $E_2 \rightarrow E_6$, with $J_{2\rightarrow6} = k_{26}he_2$. From these equilibrium expression we can express all enzyme state concentrations in terms of e_2 and e_4:

$$e_6 = \frac{h}{K_{26}}e_2$$

$$e_5 = \frac{h}{K_{25}}e_2$$

$$e_1 = \frac{K_{14}}{h}e_4. \tag{6.27}$$

The equilibrium dissociation constants for the dead-end binding steps in Figure 6.5 are denoted $K_{i1,p}$, $K_{i1,f}$, $K_{i4,p}$, and $K_{i4,f}$, with $K_{i1,f}[E_{1f}^{(3-)}] = f[E_1^{(1-)}]$, $K_{i1,p}[E_{1p}^{(3-)}] = p[E_1^{(1-)}]$, $K_{i4,f}[E_{4f}^{(2-)}] = f[E_4^{(0)}]$, and $K_{i4,p}[E_{4p}^{(2-)}] = p[E_4^{(0)}]$.

Quasi-steady state flux

Next we invoke quasi-steady state for this mechanism by equating the net flux from E_6 to E_4 to the net flux from E_4 to E_5 plus the net flux from E_1 to E_2:

$$J_{6\rightarrow4} = J_{4\rightarrow5} + J_{1\rightarrow2}$$

$$e_6k_{64} - e_4k_{46}m = e_1k_{12}f - e_2k_{21} + e_4k_{45}f - e_5k_{54}$$

$$e_2\frac{h}{K_{26}}k_{64} - e_4k_{46}m = e_4\frac{K_{14}}{h}k_{12}f - e_2k_{21} + e_4k_{45}f - e_2\frac{h}{K_{25}}k_{54}.$$

Rearranging, we have

$$e_2 = \left(\frac{\frac{K_{14}}{h} k_{12} f + k_{45} f + k_{46} m}{\frac{h}{K_{26}} k_{64} + k_{21} + \frac{h}{K_{25}} k_{54}} \right) e_4 = R e_4, \tag{6.28}$$

an expression that we will make use of shortly. Ignoring (for the moment) the dead-end inhibition steps, the total enzyme concentration e_0 is the sum

$$
\begin{aligned}
e_0 &= e_1 + e_2 + e_4 + e_5 + e_6 \\
&= \frac{K_{14}}{h} e_4 + e_2 + e_4 + \frac{h}{K_{25}} e_2 + \frac{h}{K_{26}} e_2 \\
&= \left(\frac{K_{14}}{h} + R + 1 + \frac{h}{K_{25}} R + \frac{h}{K_{26}} R \right) e_4, \tag{6.29}
\end{aligned}
$$

yielding

$$e_4 = \frac{e_0}{\frac{K_{14}}{h} + R + 1 + \frac{h}{K_{25}} R + \frac{h}{K_{26}} R}. \tag{6.30}$$

Here R is the ratio e_2/e_4 defined in Eq. (6.28). These expressions allow us to formulate an equation for the net steady-state flux:

$$
\begin{aligned}
J &= k_{64} e_6 - k_{46} m e_4 \\
&= k_{64} \frac{h}{K_{26}} e_2 - k_{46} m e_4 \\
&= \frac{e_0 \left(k_{64} \frac{h}{K_{26}} R - k_{46} m \right)}{\frac{K_{14}}{h} + R + 1 + \frac{h}{K_{25}} R + \frac{h}{K_{26}} R}. \tag{6.31}
\end{aligned}
$$

This expression is modified to account for allosteric activation of the enzyme by fumarate and inorganic phosphate as follows:

$$J = \frac{p/K_{a,p} + f/K_{a,f}}{1 + p/K_{a,p} + f/K_{a,f}} \cdot \frac{e_0 \left(k_{64} \frac{h}{K_{26}} R - k_{46} m \right)}{\frac{K_{14}}{h} + R + 1 + \frac{h}{K_{25}} R + \frac{h}{K_{26}} R}, \tag{6.32}$$

where p is the chemical activity of the inorganic phosphate ion, $[\mathrm{HPO_4^{2-}}]$. The term $(p/K_{a,p} + f/K_{a,f})/(1 + p/K_{a,p} + f/K_{a,f})$ in Eq. (6.32) is a multiplying factor that is derived from the simplifying assumption that fumarate and phosphate bind competitively to a regulatory site, and if either is bound the enzyme is catalytically active. If neither is bound, the enzyme is inactive. Thus the factor goes to zero if neither f nor p is present. The allosteric activation constants $K_{a,p}$ and $K_{a,f}$ are determined from kinetic data.

Influence of ionic strength on kinetics

Mescam *et al.* analyzed data obtained over a large range in ionic strength, with I ranging from 0.003 to 0.485 M. In order to account for the effects of ionic strength on the kinetics, species concentrations in the kinetic expressions were scaled by the appropriate *activity coefficients*. An activity coefficient for a given chemical species is defined as the ratio between the chemical activity and the chemical concentration of the species. Activity coefficients vary as a function of ionic strength owing to screening of charged species by other dissolved charges, such as salt ions. (Recall that I is defined in Eq. (5.59) in Chapter 5.)

The activity coefficient for a species with valance $\pm z$, which is denoted γ_z, may be calculated using an approximation from the extended Debye–Hückel equation [40] for nondilute solutions:

$$\ln \gamma_z = -\frac{Az^2 I^{1/2}}{1 + BI^{1/2}},\qquad(6.33)$$

where the parameter B is equal to $1.6\,\mathrm{M}^{-1/2}$ [1]. The quantity A varies with temperature; an empirical function that reproduces apparent thermodynamic properties over a temperature range of approximately 273 to 313 K is [15, 2]

$$A = 1.10708 - (1.54508 \times 10^{-3})t + (5.95584 \times 10^{-6})T^2.\qquad(6.34)$$

Using activity coefficients, the concentration variables[9] in the kinetic equations are replaced by chemical activities computed as

$$f = \gamma_2[\text{fumarate}^{2-}] = \gamma_2 \frac{[\Sigma\text{fumarate}]}{1 + h/K_{h,f}}$$

$$m = \gamma_2[\text{malate}^{2-}] = \gamma_2 \frac{[\Sigma\text{malate}]}{1 + h/K_{h,m} + [\text{Na}^+]/K_{Na,m}}$$

$$p = \gamma_2[\text{HPO}_4^{2-}] = \gamma_2 \frac{[\Sigma\text{Pi}]}{1 + h/K_{h,Pi} + [\text{Na}^+]/K_{Na,Pi}},$$

where $K_{h,f}$, $K_{h,m}$, and $K_{h,Pi}$ are the proton dissociation constants for fumarate, malate, and inorganic phosphate, respectively; and $K_{Na,m}$ and $K_{Na,Pi}$ are the sodium ion dissociation constants for malate and inorganic phosphate. The above expressions account for the binding of hydrogen ions and Na^+ cations to fumarate, malate, and inorganic phosphate. The concentrations $[\Sigma\text{fumarate}]$, $[\Sigma\text{malate}]$, and $[\Sigma\text{Pi}]$ are the total concentrations of all species making up fumarate, malate, and inorganic phosphate. Recall that Section 5.7.1 describes how expressions such as those in Eq. (6.35) arise. Here it is assumed that H^+ and Na^+ reversibly associate with malate and inorganic phosphate. (Fumarate does not significantly associate

---

[9] The pH is defined to be $-\log_{10}$ of the *activity* of hydrogen ion. Therefore the variable $h = 10^{-pH}$ already represents a chemical activity.

Table 6.1: Cation dissociation constants used for
the fumarase kinetic model.

Dissociation constant	Value
$K_{h,f}$	8.13×10^{-5} M
$K_{h,m}$	1.93×10^{-5} M
$K_{h,Pi}$	1.66×10^{-7} M
$K_{Na,m}$	0.525 M
$K_{Na,Pi}$	0.245 M

In Mescam *et al.* [41] these values vary as a function
of ionic strength. Here, they are assumed constant.

with Na$^+$ under the conditions employed in the experiments analyzed below.)
Values of the dissociations constants used by Mescam *et al.* are tabulated in
Table 6.1. Furthermore, accounting for the charges assigned to the enzyme states
in Figure 6.5, the effective rate constants depend on I as well. For example, for k_{12}:

$$k_{12} = \gamma_1 k_{12}^0,$$

where k_{12}^0 is an ionic-strength-independent constant. The ionic-strength
dependence of the full set of rate constants is given by

$$\begin{aligned}
k_{12} &= \gamma_1 k_{12}^0, & k_{21} &= \gamma_3 k_{21}^0, \\
k_{45} &= k_{45}^0, & k_{54} &= \gamma_2 k_{54}^0, \\
k_{46} &= k_{46}^0, & k_{64} &= \gamma_2 k_{64}^0.
\end{aligned} \tag{6.35}$$

The dissociation constants may be similarly corrected. For example,

$$K_{26} = \frac{\gamma_2 k_{62}^0}{\gamma_3 k_{26}^0} = \frac{\gamma_2}{\gamma_3} K_{26}^0,$$

where K_{26}^0 is an ionic-strength-independent constant. The ionic-strength
dependence of the full set of dissociation constants is given by

$$\begin{aligned}
K_{14} &= \frac{1}{\gamma_1} K_{14}^0, & K_{25} &= \frac{\gamma_2}{\gamma_3} K_{25}^0, & K_{26} &= \frac{\gamma_2}{\gamma_3} K_{26}^0, \\
K_{i1,p} &= \frac{\gamma_3}{\gamma_1} K_{i1,p}^0, & K_{i1,f} &= \frac{\gamma_3}{\gamma_1} K_{i1,f}^0, \\
K_{i4,p} &= \gamma_2 K_{i4,p}^0, & K_{i4,f} &= \gamma_2 K_{i4,f}^0.
\end{aligned} \tag{6.36}$$

Thermodynamic constraints on kinetic parameters

In addition to ionic screening effects, the values of the rate constants must obey
two independent thermodynamic constraints associated with catalytic cycles in

the mechanism [8]. One of the thermodynamic constraints is associated with the catalytic cycle $E_4 \rightarrow E_1 \rightarrow E_2 \rightarrow E_6 \rightarrow E_4$. Turnover of this cycle is associated with turnover of the reaction with the equilibrium constant K_{eq}^0:

$$K_{eq}^0 = \left(\frac{m}{f}\right)_{eq} = \frac{k_{41}^0 k_{12}^0 k_{26}^0 k_{64}^0}{k_{14}^0 k_{21}^0 k_{62}^0 k_{46}^0} = \frac{K_{14}^0 k_{12}^0 k_{64}^0}{K_{26}^0 k_{21}^0 k_{46}^0}. \tag{6.37}$$

The other is associated with the cycle $E_4 \rightarrow E_1 \rightarrow E_2 \rightarrow E_5 \rightarrow E_4$, which results in no overall reaction:

$$1 = \frac{k_{41}^0 k_{12}^0 k_{25}^0 k_{54}^0}{k_{14}^0 k_{21}^0 k_{52}^0 k_{54}^0} = \frac{K_{14}^0 k_{12}^0 k_{54}^0}{K_{25}^0 k_{21}^0 k_{45}^0}. \tag{6.38}$$

These equations effectively reduce the number of kinetic parameters by two. Here these equations are applied to compute k_{12}^0 and k_{64}^0 in terms of the other kinetic parameters:

$$k_{12}^0 = \frac{K_{25}^0 k_{21}^0 k_{45}^0}{K_{14}^0 k_{54}^0} \tag{6.39}$$

$$k_{64}^0 = \frac{K_{26}^0 k_{21}^0 k_{46}^0}{K_{14}^0 k_{12}^0} K_{eq}^0. \tag{6.40}$$

Flux expression with inhibition

Finally, accounting for inhibition steps (and expressing the flux in terms of ionic-strength dependent rate constants) we have the flux expression for the model:[10]

$$J = \frac{p/K_{a,p} + f/K_{a,f}}{1 + p/K_{a,p} + f/K_{a,f}}$$

$$\times \frac{e_0 \left(k_{64}\frac{h}{K_{26}}R - k_{46}m\right)}{\frac{K_{14}}{h}(1 + \frac{f}{K_{i,f1}} + \frac{p}{K_{i,p1}}) + R + (1 + \frac{f}{K_{i,f4}} + \frac{p}{K_{i,p4}}) + \frac{h}{K_{25}}R + \frac{h}{K_{26}}R}. \tag{6.41}$$

The enzyme mechanism of Figure 6.5 has certainly proved to be more complex than the simple Michaelis–Menten model from the previous section! Although we have applied the same quasi-steady analysis, the resulting flux expression is substantially more complex than Eq. (6.22), with many more kinetic parameters. At this point, one may wonder whether or not a model of this level of complexity is appropriate, and whether it can be identified from available data. Indeed, the questions of identifiability and the appropriate level of complexity are tied together. The model studied here is in fact the least complex (of several alternative models considered) that is able to effectively match a set of kinetic data obtained to probe the operation of the enzyme [41]. In other words, whether or not this is the *right*

---

[10] Exercise 6.3 asks the reader to derive this expression.

model depends on the data. The data (and the model's behavior in comparison to the data) are precisely what we will examine next.

Analysis of kinetic data

To probe this enzyme's catalytic mechanism, Mescam *et al.* [41] measured the kinetics of fumarate consumption by reaction (6.25) under a range of initial fumarate concentrations and buffer conditions. Figure 6.6 plots 20 measurements over the time course of fumarate ([Σfumarate]) under buffer conditions indicated in the figure panels. In these kinetic progress curves fumarate concentrations vary over a range of five orders of magnitude (from approximately 0.02 to 100 mM), and inorganic phosphate concentration ranges from 1 to 100 mM. Therefore relatively broad ranges of substrate and activator concentrations are probed, providing a robust test of the model. To simulate the time courses in the figure, we use the flux expression of Eq. (6.41) in a differential equation for fumarate concentration:

$$\frac{df}{dt} = -J(f, m)$$

$$\frac{dm}{dt} = +J(f, m), \tag{6.42}$$

with initial conditions applied according to the individual experiments.

Model predictions, obtained using parameter values provided in Table 6.2, are plotted in Figure 6.6 for comparison with the experimentally measured time courses. This data set, along with four additional time courses measured at pH = 6 and four additional time courses measured at pH = 8, were used by Mescam *et al.* to identify this model. Thus 28 progress curves were used to estimate the 14 adjustable parameters in Table 6.2 and identify the model. (Recall that although there are 16 parameters listed in the table, because of the two thermodynamic constraints there are only 14 independent values.)

Note that here we have slightly simplified the model developed by Mescam *et al.* by assuming that the cation dissociation constants in Table 6.1 are constant and do not vary with ionic strength. As a result, the model predictions here agree slightly less well with the data in Figure 6.6 than in Mescam *et al.* [41]. Nevertheless, the match between the model and the data in Mescam *et al.* is not exact, illustrating a trade-off between a model's complexity and its ability to match data. In Mescam *et al.* the model predictions are consistently no more than 10% different from the measured data. Given the imperfect nature of the biochemical measurements, not to mention the simplifying assumptions in the Debye–Hückel theory and other aspects of the kinetic model, we can safely judge a ≤ 10% difference between model and data as good enough!

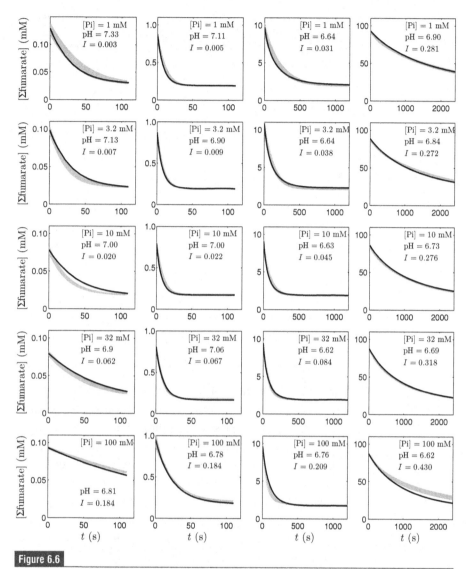

Fumarase kinetics. Total fumarate concentration is plotted versus time for the fumarase-catalyzed hydration of fumarate to malate at 25° C in five different buffer phosphate concentrations, ranging from 1 to 100 mM. Enzyme concentrations are $e_0 = 0.54 \times 10^{-7}$ for the left column and $e_0 = 0.54 \times 10^{-6}$ for the other three columns. The gray shaded bars indicate mean ± standard error of the experimental data; the solid black lines are model fits. Data are re-plotted from Mescam *et al.* [41], with permission.

6.2.3 Systematic approaches to enzyme kinetics

Since there is almost no biochemical reaction in a cell that is not catalyzed by an enzyme, the operation of enzymes is central in biology. While we have devoted a

Table 6.2: Kinetic parameter values for the fumarase model.

Parameter	Value	Units
K_{eq}^0	3.5006	Unitless
k_{12}^0	1.13×10^7	$M^{-1}\,s^{-1}$
k_{21}^0	5.82×10^4	s^{-1}
k_{45}^0	4.46×10^5	$M^{-1}\,s^{-1}$
k_{54}^0	54.12	s^{-1}
k_{46}^0	1.91×10^6	$M^{-1}\,s^{-1}$
k_{64}^0	3.72×10^5	s^{-1}
K_{14}^0	7.69×10^{-9}	M
K_{25}^0	1.81×10^{-10}	M
K_{26}^0	8.82×10^{-8}	M
$K_{a,p}$	0.58×10^{-3}	M
$K_{a,f}$	31.5×10^{-3}	M
$K_{i,p1}^0$	14.3×10^{-3}	M
$K_{i,f1}^0$	4.0×10^{-3}	M
$K_{i,p4}^0$	41.7×10^{-3}	M
$K_{i,f4}^0$	90.7×10^{-3}	M

few pages to the study of the kinetics of enzyme-mediated reactions, it is worth noting that the field of enzyme kinetics is much more vast that this brief treatment. The two examples worked through above do illustrate important concepts in enzyme kinetics, namely how quasi-equilibrium and quasi-steady approximations can be used to determine expressions for saturating steady-state reaction fluxes. It is hoped that the lessons learned here may be easily translated to other enzyme mechanisms. Furthermore, we saw in the fumarase example how a realistic mechanism can yield complex kinetic behavior, requiring relatively large numbers of parameters to represent. As with all models studied in this book, the ultimate utility of the model of fumarase is a function of whether or not it helps to answer questions concerning the operation of biological processes. Thus the reader might be concerned that here we have not used the model for a purpose beyond illustrating the model's ability to match some in vitro data. That concern will hopefully be dispelled in Chapter 7, where validated models of enzyme mechanisms are combined to simulate and probe the integrated function of biochemical systems.

For more comprehensive treatments of enzyme kinetics, the reader is referred to other sources. Beard and Qian [8] devote a chapter to the theoretical study of enzyme mechanisms. Perhaps the most useful text on the subject is Cornish-Bowden's [16].

6.3   Biochemical reaction systems

In Sections 6.2.1 and 6.2.2 we examined models of catalytic mechanisms for single biochemical reactions. In cells, individual reactions participate in a system of many interdependent reactions, called a *biochemical reaction network*. (Specific applications may invoke terms such as *metabolic networks*, *signaling networks*, and *gene regulatory networks* as appropriate.) In Chapter 7 we will examine how to combine models for individual reactions to simulate the integrated function of biochemical networks. Before moving on to such applications, let us consider an example where the function of a complex biochemical system may be simulated by treating a whole network of reactions as a single lumped process with a phenomenological rate law inspired by models of individual enzyme mechanisms.

6.3.1   Example: feedback control of oxidative phosphorylation

Figure 6.7 illustrates the overall process of mitochondrial ATP synthesis: a biochemical process fundamental to many eukaryotic cells, including most of the cells in the human body. Inside the mitochondrion (an organelle located in the cytoplasm) ATP (andenosine triphosphate) is synthesized from ADP (adenosine diphosphate) and Pi (inorganic phosphate). In the cytoplasm, ATP hydrolysis ($ATP \rightarrow ADP + Pi$) is the driving force for a host of critical cellular processes. Since the synthesis process in mitochondria requires oxygen, it depends on oxygen delivery via blood flow. This oxygen requirement is why organs such as the heart and brain suffer irreversible damage when the blood flow is interrupted for a significant time.

Figure 6.7 does not detail the individual reaction steps (including the oxygen-consuming step) in the overall process of ATP synthesis. Indeed, here we are not concerned with the individual steps. Rather, we seek a higher-level model to describe the kinetics of the synthesis reaction $ADPPi \rightarrow ATP$. Can a high-level model that ignores the underlying detail and explains available data be constructed? If so, what insights will it yield?

To construct a model for cardiac muscle (the tissue for which we will analyze relevant data below), we do need to account for a handful of additional biochemical reactions and associated reactants in the cytoplasm. Specifically, we account for adenosine monophosphate (AMP), creatine (Cr), and creatine phosphate (CrP), reactants that participate in the adenylate kinase and creatine kinase reactions. In a cardiomyocyte (heart muscle cell), over the relatively short timescale of

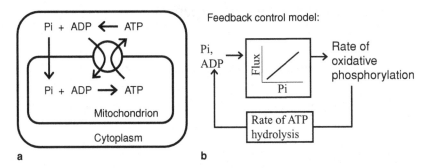

Figure 6.7

Feedback control of oxidative phosphorylation in the heart. (a) An overview of mitochondrial oxidative ATP synthesis – oxidative phosphorylation. ATP is hydrolyzed in the cytoplasm into ADP and inorganic phosphate (Pi). Pi enters the mitochondria through a phosphate carrier, and cytoplasmic ADP is exchanged for mitochondrial ATP via a transporter. The rate of mitochondrial synthesis is physiologically activated by products of ATP hydrolysis, of which Pi is the most important controller, as illustrated in (b). (Panel (b) from Beard [5], with permission.)

minutes, these reactants plus ATP, ADP, and Pi contribute to three conserved (constant) metabolite pools: conserved pools of adenine nucleotides, creatine, and exchangeable phosphate. These pools are defined:

$$a_1 + a_2 + a_3 = A_o$$

$$c_0 + c_1 = C_o$$

$$(1 + \gamma)p + a_2 + 2a_3 + c_1 = P_o, \tag{6.43}$$

where we use the definitions $a_1 = [AMP]$, $a_2 = [ADP]$, $a_3 = [ATP]$, $c_0 = [Cr]$, $c_1 = [CrP]$, $p = [Pi]$, and A_o and C_o are the total cytoplasmic adenine nucleotide and creatine pools, set here to physiologically reasonable values of 10 mM and 40 mM, respectively. The phosphate pool P_o is estimated below. These equations assume that inorganic phosphate, which is transported across the mitochondrial inner membrane, is in approximately equal concentration in the cytosolic and mitochondrial water space. The factor γ is the ratio of mitochondrial water space to cytosolic (non-mitochondrial cytoplasmic) water space, set here to 0.5.

In addition to the conservation equations, we have equilibrium ratios that are maintained by the adenylate kinase and creatine kinase enzymes:

$$\frac{a_1 a_3}{a_2^2} = R_{AK} \approx 2$$

$$\frac{a_3 c_0}{a_2 c_1} = R_{CK} \approx 166. \tag{6.44}$$

These equations arise from the chemical reactions

$$\text{ATP}^{4-} + \text{AMP}^{2-} \rightleftharpoons 2\,\text{ADP}^{3-}$$

$$\text{ATP}^{4-} + \text{creatine}^0 \rightleftharpoons \text{ADP3}{-}\text{H}^+ + \text{creatine-P}^{2-}.$$

Together Eqs (6.43) and (6.44) give five equations for six unknown concentrations in terms of the conserved pool concentrations. Thus there is one degree of freedom, which is set by the interaction between the rate of ATP synthesis (in the heart mainly in mitochondria) and the rate of ATP hydrolysis.

We may analyze this system with a few simplifying assumptions. First, we note that a_3 remains essentially constant in response to physiological changes in work rate and rate of oxygen consumption, and that c_1 decreases from approximately 23.5 mM at resting levels of ATP consumption to approximately 20 mM at near-maximal levels of ATP consumption in the heart [66]. Given this range, we can compute the range of ADP concentration to be approximately 42 μM (at rest) to 60 μM (at maximal ATP consumption). Introducing the superscripts R and M to denote rest and maximal work rate, we compute (holding a_3 constant): $c_1^R = 23.5$ mM, $c_1^M = 20$ mM, $a_1^R = 0.36$ μM, $a_1^M = 0.73$ μM, $a_2^R = 42$ μM, $a_2^M = 60$ μM. These values yield $a_3^R - a_3^M \approx 18$ μM, validating the assumption that a_3 is effectively constant in the equilibrium ratio expression.

From these values we can compute the change in cytosolic inorganic phosphate concentration (from rest to maximal work) to be

$$\Delta p = p^M - p^R = (1 + \gamma)^{-1} \left[2(a_2^R - a_2^M) + a_2^R - a_2^M + c_1^R - c_1^M \right] \approx 2.3 \text{ mM},$$
(6.45)

which is reasonably close to the value of 2.0 mM estimated by analysis based on a detailed kinetics model [66]. Using the value of $p^R = 0.29$ mM from Wu *et al.* [66], we have $p^M = 2.59$ mM.

From the conservation and equilibrium expressions we can find approximate expressions for a_1, a_3, c_1, and p as functions of a_2:

$$a_1(a_2) = R_{AK} \frac{a_2^2}{a_3} \approx R_{AK} \frac{a_2^2}{\text{A}_o}$$

$$a_3(a_2) = \text{A}_o - a_1(a_2) - a_2$$

$$c_1(a_2) = \frac{\text{C}_o}{1 + K_{CK} a_2/a_3} \approx \frac{\text{C}_o}{1 + K_{CK} a_2/\text{A}_o}$$

$$p(a_2) = (1 + \gamma)^{-1} \left[\text{P}_o - a_2 - 2a_3(a_2) - c_1(a_2) \right],$$
(6.46)

where $a_2 \in \left[a_2^R, a_2^M \right]$ defines the observed physiological range. To simulate feedback control, we may use the simple two-parameter model for Pi- and ADP-driven

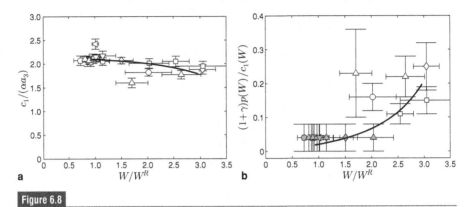

Figure 6.8

Comparison of predictions of Eq. (6.47) with data obtained from ^{31}P-magnetic resonance spectroscopy (^{31}P-MRS). Data are adapted from [66]. Shaded data points indicate the situation when Pi is below the limit of detection. Model predictions use $W_{max} = 5.14 \, W^R$. The factor α (set to 1.12) accounts for the fact that the ATP signal measured from ^{31}P-MRS arises from cytosolic plus mitochondrial ATP, while CrP is confined to the cytosol.

control of the rate of oxidative phosphorylation introduced by Katz *et al.* [33]:

$$W(a_2, p) = W_{max} \frac{a_2 p/(K_{a2}k_p)}{1 + a_2/K_{a2} + p/K_p + a_2 p/(K_{a2}K_p)}, \qquad (6.47)$$

where K_{a2} and K_p are effective half-activation constants, and W_{max} is the maximal rate. Using the values $K_{a2} = 20 \, \mu M$ and $K_p = 0.8$ mM obtained from in vitro experiments, varying a_2 over the estimated physiological range for heart, applying Eq. (6.46) to estimate $p(a_2)$, Eq. (6.47) predicts the curves plotted in Figure 6.8.

Because Pi concentration varies in vivo from well below (at rest) to well above (at maximal work) the apparent K_p value, it is a more important physiological controller of the rate of oxidative ATP synthesis in the heart than ADP.

6.4   Recapitulation and lessons learned

This and the previous chapter have laid out concepts in thermodynamics and kinetics important in the study of chemical reaction systems. While the emphasis here is on biochemical thermodynamics and biochemical kinetics, there is fundamentally nothing special about biochemical systems in terms of the physical principles in play. While biochemical thermodynamics requires a detailed accounting of the multiple rapidly converting species that make up a biochemical reactant, unique laws of solution chemical thermodynamics do not exist for living versus nonliving systems. Thus we started our study of biochemical thermodynamics by working

though the basic principles of dilute solution thermodynamics to see how fundamental physical concepts such as entropy and free energy help us to understand
biological processes. Similarly, in this chapter we saw how basic concepts of
chemical kinetics find application in the study of enzyme-mediated reactions in
biochemistry.

With the fundamental physical chemical concepts established, and with an
understanding of biochemical thermodynamics and enzyme-mediated kinetics,
we are now prepared to study systems of biochemical reactions in earnest.

Problems

6.1   Transition state theory 1. Show that Eqs (6.14), (6.15), and (6.16) are con-
      sistent with the expected form of the equilibrium constants for the NVT and
      NPT ensembles:

$$K_{NVT} = e^{-\Delta_r F^o/RT} = e^{-(\Delta_r E^o - T\Delta_r S^o)/RT}$$

$$K_{NPT} = e^{-\Delta_r G^o/RT} = e^{-(\Delta_r H^o - T\Delta_r S^o)/RT}.$$

6.2   Transition state theory 2. In biological systems the empirical factor Q_{10}
      is sometimes defined as a measure of the amount by which a given rate
      changes with temperature. Specifically, for a chemical reaction with rate k,
      Q_{10} is defined as the amount by which k changes when the temperature is
      increased by $10°$ C. Assuming that, over a given temperature range, rates
      vary exponentially with temperature, we have

$$Q_{10} = \left(\frac{k(T_2)}{k(T_1)} \right)^{\frac{10}{T_2-T_1}}.$$

      Determine an approximate relationship between the transition energy barrier
      height $\Delta E^{\ddagger}$ and Q_{10} that is valid over a small temperature range. What is the
      Q_{10} for the data set shown in Figure 6.3? How much would the factor $k_B T$
      from Eq. (6.16) influence the data fit and theoretical prediction in Figure 6.3?
6.3   Enzyme kinetics. Show that accounting for the dead-end binding steps in
      Figure 6.5 leads to Eq. (6.41). Show that when product concentration goes
      to zero ($m = 0$) and when inhibition and activation are not included, the flux
      expression for fumarase

$$J = \frac{e_0 \left(k_{64} \frac{h}{K_{26}} R - k_{46} m \right)}{\frac{K_{14}}{h} + R + 1 + \frac{h}{K_{25}} R + \frac{h}{K_{26}} R}$$

from Section 6.2.2 is equivalent to the irreversible Michaelis–Menten flux expression. What is the effective Michaelis–Menten constant K_m for this case?

6.4 Biochemical systems simulation. (Adapted from Exercise 5.4 of [8].) Consider the system of three irreversible reactions

$$A \xrightarrow{k_1} X, \quad X + Y \xrightarrow{k_2} 2Y, \quad Y \xrightarrow{k_3} B.$$

Construct a computer code to simulate the concentrations [X] and [Y] with fixed concentrations $[A] = a$ and $[B] = b$. Show that the concentrations of X and Y oscillate. (Alternatively, this can be shown mathematically without using computer simulation.) Next consider the reversible system

$$A \underset{k_{-1}}{\overset{k_{+1}}{\rightleftharpoons}} X, \quad X + Y \underset{k_{-2}}{\overset{k_{+2}}{\rightleftharpoons}} 2Y, \quad Y \underset{k_{-3}}{\overset{k_{+3}}{\rightleftharpoons}} B.$$

Show that the equilibrium constant ratio b/a is $K_{AB} = \frac{k_1 k_2 k_3}{k_{-1} k_{-2} k_{-3}}$. Show that when $b/a = K_{AB}$ there is no oscillation.

6.5 Control of oxidative phosphorylation. Consider the model of Section 6.3.1 for oxidative ATP synthesis. How does the predicted relationship between work rate and metabolite levels change as the phosphate pool concentration P_o is varied? How does the conclusion that Pi concentration represents the crucial feedback control signal depend on P_o?

7 Chemical reaction systems: large-scale systems simulation

Overview

So far in the examples of biochemical systems that we have studied, we have looked at systems of one or a handful of reactions. Yet the ultimate quest to understand (and simulate and predict) cellular function calls for the capability to synthesize many thousands of simultaneous chemical reaction and transport processes. Can we realistically expect to simulate cellular biochemistry with many important reactions accounted for at the level of rigor and detail of the examples in the previous two chapters? The answer is a qualified "yes," because in principle there is no reason why realistic simulations of systems of hundreds or thousands of reactions cannot be constructed. One approach to doing to is to adopt a systematic approach based on the fundamentals developed in the previous two chapters. This approach is practical when data exist for building the necessary thermodynamic and kinetic models of the individual enzymes and transporters at the level of detail of the model of fumarase in Chapter 6. However, when that level of detail is either not possible or not practical, model-based simulations may still be developed.

This chapter describes examples of two contrasting approaches to simulation and analysis of large-scale systems – a simulation of the kinetics of a metabolic pathway based on detailed biochemical thermodynamics and kinetics, and a less biochemically rigorous analysis of gene expression data to identify regulatory network structure. Model development in the first example is based on substantial prior knowledge of the enzymes and the associated reactions in the system. In the second example a model is built from genetic regulatory motifs to attempt to find a network structure that can explain a data set. These contrasting examples highlight the different sorts of questions we can ask in the different domains. For the metabolic system, we will ask our simulation to reveal and quantify the mechanistic operation of the reaction network. For the gene expression system, we will be asking the more basic question: what is the network? Or, more precisely, what hypothetical regulatory network structures can we find that at least have the capacity to explain the available data?

7.1 Biochemical systems in living cells

Is it possible to build a practical, manageable, valid simulation of the biochemical processes occurring in a living cell? Recall Eduard Buchner's Nobel lecture quoted on the first pages of this book: "We ... must have every confidence that the day will dawn when even those processes of life which are still a puzzle today will cease to be inaccessible to us natural scientists." One hundred years later, so many complex "processes of life" have been determined that the idea of synthesizing them into a realistic simulation of even a living cell seems daunting. Yet today, with modern computing and a vast knowledge of biochemistry, we can certainly make a reasonable attempt to integrate the key processes necessary to represent particular cellular functions in realistic simulations. The key is to develop the model simulation that is appropriate for the particular job it will be asked to do. There is no model of *everything* encapsulating all biochemical function of any cell; perhaps there never will be. But we can build models of pathways or components that are extremely useful in interpreting and understanding function. For example, in Section 2.3.1 we built a model representing six reactions in yeast. Those six reactions by no means represent each and every thing going on in the strains of yeast studied. But the model developed was complete enough to yield insight into the operation of these reactions in these organisms under the conditions assayed in the experiments analyzed.

Like the example of Section 2.3.1, here we are going to deal with biochemical reactions that may be simulated using differential equations. That is, reactions that tend to proceed in a deterministic manner and do not require the introduction of stochastic processes to represent them. Furthermore, in contrast to Section 2.3.1, here we will see how the physical chemical principles developed in Chapters 5 and 6 may be explicitly adopted as a foundation of a simulation. When we have solid information on the thermodynamics and biochemical kinetics of the reactions involved in a particular system, these principles allow us to build models with a good deal of confidence in their underlying structure.

7.2 General approach to metabolic kinetics

In Section 6.2 we saw how a *rate law* for an enzyme can be constructed from a given mechanism, how a rate law can be parameterized from fitting data, and how the rate law an be integrated into a biochemical system simulation. While the system simulated in Section 6.2.2 was a single reaction proceeding *in vitro*, the general principle of mass-action kinetics invoked in Eq. (6.42) applies equally well to larger-scale systems. One critical thing that is needed to simulate larger systems is a set of validated rate laws for the enzymes catalyzing the reactions to be studied.

7.2.1   Enzyme rate laws and biochemical networks

Useful formulae and associated parameter estimates are available for many (but certainly not all) reactions. And thus it is possible to construct reasonable models for many systems without having to do *de novo* kinetic studies of the sort illustrated in Section 6.2.2 for all enzymes in a system.[1] Often, data for a particular reaction can be obtained from published kinetic models, while a consensus model and associated parameter values (particularly for a given enzyme isoform isolated from a given tissue of a given species) are not likely to exist. Thus, a good deal of searching through the literature and perhaps some independent analysis are required to assemble a set of rate laws to use to represent the kinetics of the enzymes catalyzing the reactions in a given system. As discussed in Chapter 6, several excellent sources on the topic of enzyme kinetic mechanisms exist.

Assuming that we have validated models, associated rate laws, and parameter values to compute reaction fluxes as functions of the dynamically changing concentrations in a system, constructing a model is a matter of expressing the rates of change of dynamic variables in terms of the reaction fluxes through mass conservation principles. For example, let us consider the set of reactions of glycogenolysis (glycolytic breakdown of glycogen):[2]

$$\text{GP}: \quad \text{GLY}_n + \text{HPO}_3^{2-} \rightleftharpoons \text{GLY}_{n-1} + \text{G1P}^{2-}$$

$$\text{PGLM}: \quad \text{G1P}^{2-} \rightleftharpoons \text{G6P}^{2-}$$

$$\text{PGI}: \quad \text{G6P}^{2-} \rightleftharpoons \text{F6P}^{2-}$$

$$\text{PFK}: \quad \text{F6P}^{2-} + \text{ATP}^{4-} \rightleftharpoons \text{FBP}^{4-} + \text{ADP}^{3-} + \text{H}^+$$

---

[1] The word "reasonable" here is not precisely defined. What is reasonable for one application may not be reasonable for another. Perhaps it is better to say that it is sometimes possible to build an initial working model for a system based on established enzyme rate laws.

[2] Abbreviations are GP: glycogen phosphorylase; PGLM: phosphoglucose mutase; PGI: phosphoglucose isomerase; PFK: phosphofructose kinase; ALD: aldolase; TPI: triose phosphate isomerase; G3PDH: glycerol-3-phosphate dehydrogenase; GAPDH: glyceraldehyde-3-phosphate dehydrogenase; PGK: phosphoglycerate kinase; PGM: phosphoglyceromutase; EN: enolase; PK: pyruvate kinase; LDH: lactate dehydrogenase; MCT: monocarboxylate transporter; CK: creatine kinase; AK: adenylate kinase; ATPase: ATP hydrolysis; OxPhos: oxidative phosphorylation; GLY$_n$ glycogen (*n* residues); GLY$_{n-1}$ glycogen (*n*−1 residues); Pi: inorganic phosphate; G1P: glucose-1-phosphate; G6P: glucose-6-phosphate; F6P: fructose-6-phospate; ATP: adenosine triphosphate; ADP: adenosine diphosphate; AMP: adenosine monophosphate; FBP: fructose-1,6-bisphosphate; DHAP: dihydroxyacetone phosphate; GAP: glyceraldehyde-3-phosphate; NAD: oxidized nicotinamide adenine dinucleotide; NADH: reduced nicotinamide adenine dinucleotide; GL3P: glycerol-3-phosphate; BPG: 1,3-bisphosphoglycerate; PG3: 3-phosphoglycerate; PG2: 2-phosphoglycerate; PEP: phosphoenolpyruvate; PYR: pyruvate; LAC: lactate; PCr: phosphcreatine; Cr: creatine. The subscript "out" in Eq. (7.1) denotes reactants outside of the cell.

$$ALD: \quad FBP^{4-} \rightleftharpoons DHAP^{2-} + GAP^{2-}$$

$$TPI: \quad GAP^{2-} \rightleftharpoons DHAP^{2-}$$

$$G3PDH: \quad GL3P^{2-} + NAD^{-} \rightleftharpoons DHAP^{2-} + NADH^{2-} + H^{+}$$

$$GAPDH: \quad GAP^{2-} + NAD^{-} + HPO_3^{2-} \rightleftharpoons BPG^{4-} + NADH^{2-} + H^{+}$$

$$PGK: \quad BPG^{4-} + ADP^{3-} \rightleftharpoons PG3^{3-} + ATP^{4-}$$

$$PGM: \quad PG3^{3-} \rightleftharpoons PG2^{3-}$$

$$EN: \quad PG2^{3-} \rightleftharpoons PEP^{3-} + H_2O$$

$$PK: \quad PEP^{3-} + ADP^{3-} + H^{+} \rightleftharpoons PYR^{-} + ATP^{4-}$$

$$LDH: \quad PYR^{-} + NADH^{2-} + H^{+} \rightleftharpoons LAC^{-} + NAD^{-}$$

$$MCT: \quad LAC^{-} + H^{+} \rightleftharpoons LAC_{out}^{-} + H_{out}^{+}$$

$$CK: \quad Cr + ATP^{4-} \rightleftharpoons PCr^{2-} + ADP^{3-} + H^{+}$$

$$AK: \quad ATP^{4-} + AMP^{2-} \rightleftharpoons 2\,ADP^{3-}$$

$$ATPase: \quad ATP^{4-} + H_2O \rightarrow ADP^{3-} + HPO_3^{2-} + H^{+}$$

$$OxPhos: \quad ADP^{3-} + HPO_3^{2-} + H^{+} \rightarrow ATP^{4-} + H_2O. \tag{7.1}$$

This metabolic pathway is illustrated in Figure 7.1. Here oxidative synthesis of ATP (OxPhos) is expressed simply as the reverse of the ATP hydrolysis reaction. This reaction, which will be used to simulate recovery after transient anoxia in the example of Section 7.2.3, ignores any carbon substrate coupling between glycolysis and oxidative phosphorylation. This is a model simplification that will prove reasonable to analyze transient data from resting muscle in the applications explored here.

In Eq. (7.1) reaction formulae are for chemical reactions written in terms of chemical species with defined elemental makeup and charge. The governing equations for the reactant concentrations in this system are obtained from mass balance:

$$\frac{d[\text{ATP}]}{dt} = -J_{CK} - J_{AK} - J_{PFK} + J_{PGK} + J_{PK} - J_{ATPase} + J_{OxPhos}$$

$$\frac{d[\text{ADP}]}{dt} = +J_{CK} + 2J_{AK} + J_{PFK} - J_{PGK} - J_{PK} + J_{ATPase} - J_{OxPhos}$$

$$\frac{d[\text{AMP}]}{dt} = -J_{AK}$$

$$\frac{d[\text{NADH}]}{dt} = +J_{GAPDH} + J_{G3PDH} - J_{LDH}$$

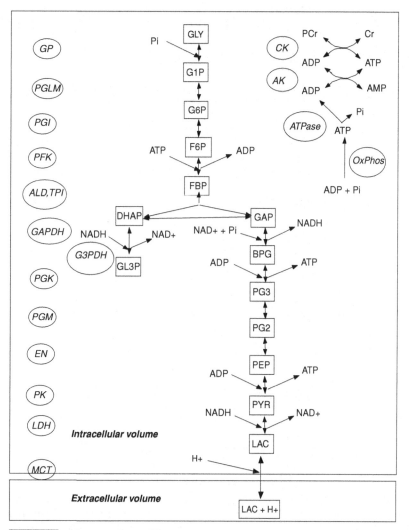

Figure 7.1

Schematic diagram of glycogenolysis. Enzymes and transporters are indicated in ovals; biochemical reactants in boxes. Abbreviations and reactions are defined in text. Figure reproduced from Vinnakota *et al.* [65] with permission.

$$\frac{d[\text{NAD}]}{dt} = -J_{GAPDH} - J_{G3PDH} + J_{LDH}$$

$$\frac{d[\text{Pi}]}{dt} = -J_{GP} - J_{GAPDH} + J_{ATPase} - J_{OxPhos}$$

$$\frac{d[\text{GLY}]}{dt} = -J_{GP}$$

$$\frac{d[\text{G1P}]}{dt} = +J_{GP} - J_{PGLM}$$

$$\frac{d[\text{G6P}]}{dt} = +J_{PGLM} - J_{PGI}$$

$$\frac{d[\text{F6P}]}{dt} = +J_{PGI} - J_{PFK}$$

$$\frac{d[\text{FBP}]}{dt} = +J_{PFK} - J_{ALD}$$

$$\frac{d[\text{DHAP}]}{dt} = +J_{ALD} + J_{G3PDH} + J_{TPI}$$

$$\frac{d[\text{GAP}]}{dt} = +J_{ALD} - J_{TPI} - J_{GAPDH}$$

$$\frac{d[\text{GL3P}]}{dt} = -J_{G3PDH}$$

$$\frac{d[\text{BPG}]}{dt} = +J_{GAPDH} - J_{PGK}$$

$$\frac{d[\text{PG3}]}{dt} = +J_{PGK} - J_{PGM}$$

$$\frac{d[\text{PG2}]}{dt} = +J_{PGM} - J_{EN}$$

$$\frac{d[\text{PEP}]}{dt} = +J_{EN} - J_{PK}$$

$$\frac{d[\text{PYR}]}{dt} = +J_{PK} - J_{LDH}$$

$$\frac{d[\text{LAC}]}{dt} = +J_{LDH} - J_{MCT}, \tag{7.2}$$

where the J's are reaction fluxes and subscripts identify the reactions. For example, J_{CK} is the creatine kinase (CK) flux. Here the concentration [GLY] denotes concentration of glycogen monomer available to the GP reaction. All concentrations in Eq. (7.2) represent biochemical reactant (sums of species) concentrations.

The stoichiometry of the reactions in Eq. (7.1) is reflected on the right-hand side of Eq. (7.2). For example, since the CK, AK, PFK, and ATPase reactions are defined to consume ATP, the reaction fluxes associated with these reactions appear with stoichiometry -1 in the equation for $d[\text{ATP}]/dt$. Similarly, since the PGK and PK reactions generate ATP, the associated fluxes appear with positive stoichiometry. Integration of this system of equations, governing the intracellular kinetics of these biochemical reactants, requires means of computing the fluxes on the right-hand side of this system of equations in terms of the reactant concentrations. (Such expressions are provided in the model analysis presented below.) Moreover, we wish to use this model to analyze experimental data collected under conditions where the intracellular pH changes as a result of the operation of these reactions during transient anoxia (under conditions where oxygen supply is temporarily cut off.) Thus it is necessary to simulate both the pH dynamics and the influence of pH on the individual reactions of glycolysis.

7.2.2 Simulating pH kinetics

When ATP hydrolysis is coupled to anaerobic glycogenolysis yielding lactate, there is a net generation of one hydrogen ion for every lactate molecule produced. We can see this by summing the reactions from Eq. (7.1) in the following stoichiometry:

$$GP + PGLM + PGI + PFK + ALD + 2TIP + 2GAPDH$$

$$+ \, 2\,PGK + 2\,PGM + 2\,EN + 2\,PK + 2\,LDH + 3\,\text{ATPase}$$

yielding

$$\mathrm{GLY}_n \rightleftharpoons \mathrm{GLY}_{n-1} + 2\,\mathrm{LAC}^- + 2\,\mathrm{H}^+. \tag{7.3}$$

Thus for each glycogen monomer consumed two LAC^- and two H^+ are generated, along with three ATP molecules. If the pathways defined above were to operate in steady state, then the ATP hydrolysis rate would match the ATP synthesis rate (with three ATP molecules hydrolyzed for every glycogen monomer) and the generated lactate and hydrogen ions would be removed from the system at a steady rate to match the production rate. In a dynamic non-steady state, dynamic binding of hydrogen ions to metabolites and other buffers means that the pH dynamics can be more complex than a simple steady-state stoichiometric balance.

Appropriate equations for pH kinetics in non-steady-state systems are obtained from a mass balance of hydrogens, developed by defining the quantity H_o, the concentration of hydrogen atoms in the system:

$$H_o = [\mathrm{H}^+] + [\mathrm{H}^+_{\text{bound}}] + [\mathrm{H}^+_{\text{reference}}], \tag{7.4}$$

where $[\mathrm{H}^+_{\text{bound}}]$ and $[\mathrm{H}^+_{\text{reference}}]$ represent reversibly bound hydrogen ions and hydrogen atoms incorporated into reference chemical species, respectively.[3] Differentiating Eq. (7.4) with respect to time,

$$\frac{dH_o}{dt} = \frac{d[\mathrm{H}^+]}{dt} + \frac{d[\mathrm{H}^+_{\text{bound}}]}{dt} + \frac{d[\mathrm{H}^+_{\text{reference}}]}{dt}, \tag{7.5}$$

---

[3] Equation (7.4) does not include the $W - [\mathrm{OH}^-]$ term that was included in Eq. (5.49). Strictly speaking this term should be included. See Exercise 7.1.

which is equal to the net flux of hydrogen ion into the system:

$$\frac{d[\mathrm{H}^+]}{dt} + \frac{d[\mathrm{H}^+_{\mathrm{bound}}]}{dt} + \frac{d[\mathrm{H}^+_{\mathrm{reference}}]}{dt} = J^H_{transport}. \tag{7.6}$$

The individual terms in Eq. (7.6) are obtained as follows. First, the rate of change of bound hydrogen ion is obtained by accounting for all of the buffers in the system:

$$\begin{aligned}
\frac{d[\mathrm{H}^+_{\mathrm{bound}}]}{dt} &= \frac{d}{dt}\left(\sum_i c_i \frac{[\mathrm{H}^+]/K_i}{1+[\mathrm{H}^+]/K_i}\right) + \frac{d}{dt}\left(A_o \frac{[\mathrm{H}^+]/K_o}{1+[\mathrm{H}^+]/K_o}\right) \\
&= \sum_i \frac{dc_i}{dt}\frac{[\mathrm{H}^+]/K_i}{1+[\mathrm{H}^+]/K_i} + \frac{d[\mathrm{H}^+]}{dt}\left(\sum_i c_i \frac{1/K_i}{(1+[\mathrm{H}^+]/K_i)^2}\right) \\
&\quad + A_o \frac{d[\mathrm{H}^+]}{dt}\frac{1/K_o}{(1+[\mathrm{H}^+]/K_o)^2}.
\end{aligned} \tag{7.7}$$

The sum $\sum_i$ in Eq. (7.7) is over all reactants in the system, c_i represents the total concentration of reactant i, and K_i is the hydrogen ion dissociation constant for reactant i. Thus $c_i([\mathrm{H}^+]/K_i)/(1+[\mathrm{H}^+]/K_i)$ is the concentration of the hydrogen ion bound form of the reactant. (Here we are accounting for two species of each reactant: hydrogen ion-bound and nonbound, as in Section 5.6.) The second term in Eq. (7.7) accounts for intrinsic buffers (in addition to the reactants that are explicitly included in the kinetic model.) The intrinsic buffer is assumed to be described by an overall buffering capacity A_o and an effective dissociation constant K_o.

The rate of change of concentration of hydrogen ion incorporated into reference species is given by

$$\frac{d[\mathrm{H}^+_{\mathrm{reference}}]}{dt} = -\sum_j \mu_j^H J_j, \tag{7.8}$$

where the summation is over all chemical reactions, μ_j^H is the stoichiometric coefficient for hydrogen ion for reaction j, and J_j is the flux of reaction j. For example, for PFK, $\mu_{PFK}^H = +1$. Thus a net positive flux through the PFK reaction results in a negative contribution to $d[\mathrm{H}^+_{\mathrm{reference}}]/dt$.

Combining Eqs (7.6)–(7.8), we obtain:

$$\frac{d[\mathrm{H}^+]}{dt} = \frac{J^H_{transport} + \sum_j \mu_j^H J_j + \sum_i \frac{dc_i}{dt}\frac{[\mathrm{H}^+]/K_i}{1+[\mathrm{H}^+]/K_i}}{1 + \sum_i c_i \frac{1/K_i}{(1+[\mathrm{H}^+]/K_i)^2} + A_o \frac{1/K_o}{(1+[\mathrm{H}^+]/K_o)^2}}, \tag{7.9}$$

which is a governing equation for free hydrogen ion concentration (or, more formally, hydrogen ion activity) accounting for dynamic binding to biochemical reactants with time-varying concentrations, $c_i(t)$. This equation obviously takes a much more complex form than the simple stoichiometric balance of Eq. (7.2).

In fact, here we are accounting only for hydrogen ion binding. More complex equations arise when one wishes to track dynamic binding of multiple cations (such as Mg^{2+} and K^+) to biochemical reactants. In many applications, such as the example below, this simplification is reasonable when the free concentrations of magnesium and potassium are not expected to vary significantly.

7.2.3 Example: glycogenolysis in skeletal muscle

In a 2010 study Vinnakota *et al.* [65] measured metabolic kinetics in mouse muscles *in vitro*. The muscles were isolated from the animal and placed in a flow chamber where oxygen and other substrate concentrations could be controlled and cellular pH and phosphate metabolites measured as functions of time. In their experiments on two different muscle types, oxygen was transiently removed from the inflow to the chamber, then reintroduced in order to measure how the systems responded to the perturbation.

Figure 7.2 illustrates measured data on intracellular PCr, Pi, and pH in extensor digitorum longus (EDL) and soleus (SOL) muscle under resting conditions (0 to 20 minutes), anoxia (20 to 80 minutes in the EDL and 20 to 66 minutes in the SOL), and recovery. (ATP concentration remains essentially constant during these experiments.) The EDL and SOL are composed primarily of glycolytic and oxidative muscle fibers, respectively. Because the muscles were not stimulated in these experiments, one of the goals of this study was to identify differences in resting state energy metabolic rate and control between these two muscle types.

To simulate glycogenolysis in this experimental system, Vinnakota *et al.* constructed a model that accounts for the reactions of Eq. (7.1) in addition to several other processes that were part of the experiments. First, the model accounts for the hydration of CO_2 via the reaction

$$CO_2 + H_2O \rightleftharpoons HCO_3^- + H + .$$

This reaction proceeds spontaneously in aqueous solution, and is catalyzed by carbonic anhydrase in cells. Thus flux through this reaction is denoted J_{CA} in cells and J_{CO2hyd} in the buffer in the flow chamber. Furthermore, CO_2 is generated via the oxidative phosphorylation process, with a flux that is here denoted by J_{CO2}. Finally, there is transport of CO_2 between the cell and the surrounding buffer (with flux denoted J_{trCO2}) and transport into the chamber from inflowing buffer. Combining these processes together, we have the following governing equations

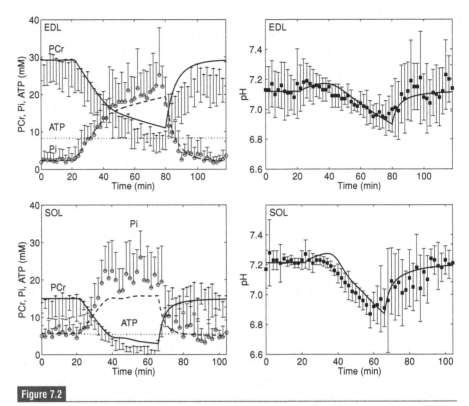

Figure 7.2

Muscle cellular energetics during baseline, anoxia, and recovery in extensor digitorum longus (EDL) and soleus (SOL) muscle. The experimental protocol is described in the text. Model simulations are compared to observed data on [PCr], [Pi], and pH. Figure reproduced from [65] with permission.

for CO_2 and HCO_3^- in the cell and flow chamber:

$$\frac{d[CO_2]}{dt} = +J_{CO2} - J_{CA} - J_{trCO2}$$

$$\frac{d[HCO_3]}{dt} = +J_{CA}$$

$$\frac{d[CO_2]_{cham}}{dt} = +J_{trCO2}V_{cell}/V_{cham} - J_{CO2hyd}$$

$$+ F([CO_2]_{input} - [CO_2]_{cham})/V_{cham}$$

$$\frac{d[HCO_{3cham}]}{dt} = +J_{CO2hyd} + F([HCO_3]_{input} - [HCO_3]_{cham})/V_{cham}. \quad (7.10)$$

Here the transport flux is assumed to be expressed in units of mass per unit time per unit cell volume. The chamber is treated as a well-mixed system where concentrations $[CO_2]_{input}$ and $[HCO_3]_{input}$ are the inflow concentrations and F is the flow.

Addition components of the buffer transported into and out of the flow chamber that must be accounted for are lactate and a buffering compound, 3-(N-morpholino) propanesulfonic acid, commonly referred to as MOPS:

$$\frac{d[\text{LAC}]_{cham}}{dt} = +J_{MCT} V_{cell}/V_{cham} - F[\text{LAC}]_{cham}/V_{cham}$$

$$\frac{d[\text{H}^+]_{cham}}{dt} = \frac{+J_{MCT} V_{cell}/V_{cham} + J_{CO2hyd} + F([\text{H}^+]_{input} - [\text{H}^+]_{cham})/V_{cham}}{1 + A_{MOPS}\frac{1/K_{MOPS}}{(1+[\text{H}^+]/K_{MOPS})^2}},$$

$$(7.11)$$

where the equation for $d[\text{H}^+]_{cham}/dt$ is derived from Eq. (7.9), accounting for hydrogen ions generated from the MCT flux, hydration of CO_2 in the buffer, and advective transport. The concentration and hydrogen ion dissociation constant for MOPS are denoted A_{MOPS} and K_{MOPS}, respectively. Lactate is transported into the buffer via the MCT equation and is washed out via flow through the well-mixed chamber.

Equations (7.2), (7.9), (7.10), and (7.11) provide the governing equations for the Vinnakota *et al.* model used to simulate their experiments. The hydrogen ion fluxes in the numerator of Eq. (7.9) are computed from the stoichiometric balance of hydrogen ions generated, consumed, transported in, and transported out of cells:

$$J^H_{transport} + \sum_j \mu^H_j J_j = +J_{PFK} + J_{G3PDH} + J_{GAPDH} - J_{PK} - J_{LDH}$$

$$- J_{MCT} + J_{CK} + J_{ATPase} + J_{OxPhos} + J_{CA}. \quad (7.12)$$

These equations involve a large number of fluxes, which are expressed using appropriate rate laws in Vinnakota *et al.* [65]. The rate laws for all reactions and transport processes are listed in Section 9.8 of the Appendices, along with all associated parameter values.

For most reaction flux expressions, a pH-dependent factor multiplies the rate law. These factors account for the pH-dependence of the apparent catalytic activity of the enzymes. Some of the expressions used (such as for glycogen phosphorylase) are based on assumptions of protonation/deprotonation of the active site [16]; others (such as for aldolase) are simply phenomenological functions. In all cases, parameter values are obtained by fitting kinetic data.

Simulated time courses of ATP, PCr, Pi, pH, and lactate obtained from the model are plotted in Figures 7.2 and 7.3. For both muscle types the model predictions fall largely within the error bars of the experimental data, although the model does not match the data perfectly. Particularly in the SOS, the model underpredicts the rise in [Pi] and overpredicts the rise in [LAC] observed during anoxia. Thus the rate of glycogenolysis predicted by the model is perhaps slightly too high. Yet,

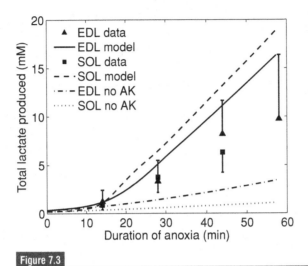

Figure 7.3

Lactate accumulation during experimental protocols of Figure 7.2. Figure reproduced from Vinnakota *et al.* [65] with permission.

given the biological variability and the complexity of the system, the agreement between the model and data is satisfactory, particularly when noting that of the eight parameters adjusted to match the data, six can be compared to prior reported estimates and in all six cases the the optimal values are close to the independent estimates.[4]

One of the primary findings from this analysis is the determination that the resting rates of ATP hydrolysis in both muscles are equivalent, at 0.7 mM min^{-1}. In addition the model can be analyzed to probe its integrated behavior to yield insight into the biological system. For example, recall from Section 6.3.1 that the major controller of oxidative ATP synthesis in the heart was determined to be buildup of inorganic phosphate concentration, stimulating ATP synthesis in response to an increasing rate of ATP hydrolysis. Is a similar mechanism at work here? We can test the hypothesis that inorganic phosphate is a controller of glycogenolysis flux under resting conditions by artificially clamping the Pi concentration in the model for the protocol of Figure 7.2 and determining the predicted response to the anoxia perturbation. This computational experiment is illustrated in Figure 7.4, where [Pi] is set to its initial resting value for both muscle types. Thus these simulations do not match the order of magnitude increase in [Pi] observed following anoxia. Yet for both muscle types the simulations are still able to match the observed phosphocreatine dynamics, and do predict a stable ATP concentration. Therefore we can conclude that feedback via inorganic phosphate does not impart a significant influence on glycogenolytic ATP synthesis in these muscles, at least under

---

[4] See Table 2 of Vinnakota *et al.* [65].

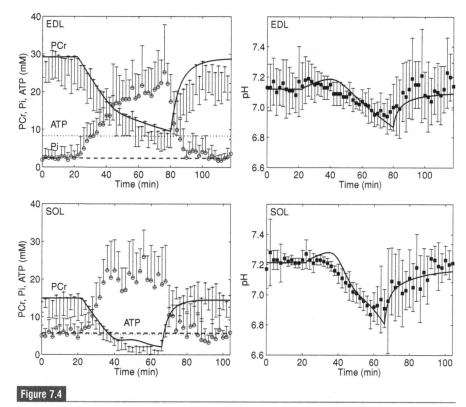

Computed cellular energetics with inorganic phosphate concentration clamped. Model simulations are compared to observed data on [PCr], [Pi], and pH in extensor digitorum longus (EDL) and soleus (SOL) muscle. Figure derived from Vinnakota *et al.* [65].

resting conditions and during acute anoxia at rest. This is because by clamping the phosphate concentration we have effectively removed from the model inorganic phosphate's potential to act as a signal affecting flux through the pathway. Since doing this effects no substantial changes in model predictions (except on [Pi] itself), it is apparent that phosphate concentration is not acting as a primary controller of flux in the model.

Vinnakota *et al.* conclude that [AMP] is playing a significant role in controlling flux. Indeed the model accounts for two forms of glycogen phosphorylase: an active form (GPa) and a less active form (GPb). (See Eqs (9.39) and (9.40) in Section 9.8.) The enzyme is primarily in the inactive GPb form at rest, with only 1% present as GPa. Transitions between GPb to GPa are physiologically regulated by cellular signaling processes that are modulated by the endocrine system. In addition the GPb form can be allosterically activated by AMP, as modeled by Eq. (9.40). AMP concentrations rise during the anoxia protocol of Vinnakota

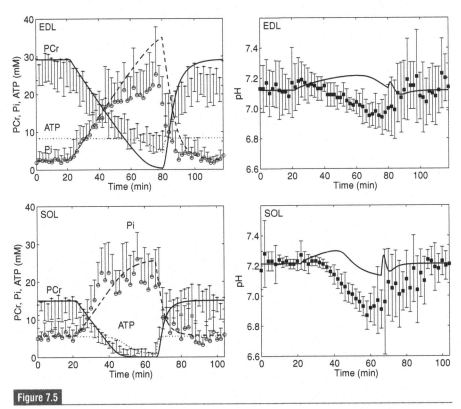

Computed cellular energetics with adenylate kinase reaction disabled. Model simulations are compared with observed data on [PCr], [Pi], and pH in extensor digitorum longus (EDL) and soleus (SOL) muscle. Figure derived from Vinnakota *et al.* [65].

et al.; as ATP concentration is maintained at an effectively constant level, [ADP] increases, and the adenylate kinase reaction proceeds in the direction $2\,ADP \rightarrow ATP + AMP$. Thus, buildup of ADP leads to buildup of AMP. The AMP buildup can be taken out of the model by setting he adenylate kinase flux to zero.

Computational experiments with $J_{AK} = 0$ were performed by Vinnakota *et al.* to investigate the role of AMP as a modulator of pathway flux in their model. Predicted lactate generation from these simulations is shown in Figure 7.3, illustrating that, with no AK flux, the glycogenolysis pathway is severely inhibited. The ATP, PCr, and Pi concentrations associated with the $J_{AK} = 0$ simulations for both muscle types are shown in Figure 7.5. Here we see that without this reaction, and without the associated stimulation of GPb, the model cannot match the observed energetics. Specifically, in anoxia the ATP concentration cannot be maintained and predicted PCr drops below its observed concentrations because the glycogenolysis pathway cannot effectively simulate ATP. Thus it is concluded that this feedback pathway is necessary for the model to be able to predict realistic rates of glycogenolysis under the conditions imposed in these experiments.

The analyses illustrated in Figure 7.4 and 7.5 show how a model of complex system may be integrated to reveal mechanistic insight into system behavior. Yet it is important to keep in mind that the model remains an imperfect representation of reality. The conclusion that AMP concentration is a much more significant controller of non-oxidative ATP synthesis in these muscles is based on the predictions of the model. Because the model was constructed based on detailed knowledge of the kinetics and thermodynamics of the individual reactions and transport processes involved, and because the kinetic models of the enzymes and transporters were parameterized mostly independently of the experiments analyzed here, we should take its predictions seriously. However, it should be noted that key parameters related to the AMP-mediated activation of GPb were estimated to obtain the fits of Figure 7.2. (Specifically, K_{AMP} and n_H from Eq. (9.40) for the AMP activation were treated as adjustable parameters by Vinnakota *et al.*) Therefore the model-based predictions may be accepted as a viable hypothesis, one that cannot be disproved by these experiments, yet not necessarily the only viable hypothesis,[5] nor necessarily the complete elucidation of the mechanisms of controlling this pathway *in vivo*.

7.3   Reverse engineering and model discovery

Often in biological research we lack the prior knowledge necessary to build the sort of detailed model analyzed in the previous section. In such cases we might have the capacity to collect data on certain variables without knowing in advance how they are mechanistically connected. In these cases a primary goal may be to *discover* putative mechanisms that may explain the data. This paradigm is illustrated in Figure 7.6, in which trial kinetic models are constructed to compare to kinetic data on a handful of variables.

In this figure we imagine that there are four variables of interest. Trial models are illustrated as graphs where the nodes represent variables and the edges interactions between the variables. For example, the connection $1 \rightarrow^+ 2$ would indicate that variable x_1 stimulates the production or activation of variable x_2. Similarly, the connection $1 \rightarrow^- 2$ indicates an inhibitory process. The kinetics of activation

---

[5] Vinnakota *et al.* confidently conclude that the only way to explain the data that is consistent with what is known about the kinetics of the enzymes in the pathway is through AMP-mediated activation of GPb. The likelihood of this hypothesis being basically correct is greatly strengthened, viewing this study in the context of the field. It is certainly reasonable to conclude that alternative explanations do not exist.

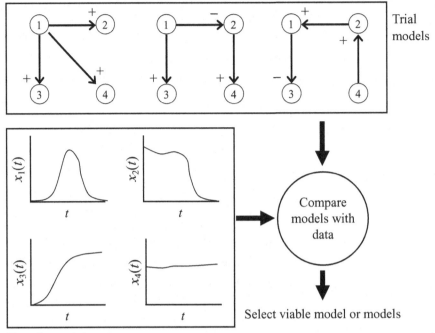

Experimental data

Figure 7.6

System discovery paradigm. Data on variables of interest are analyzed using trial models representing the stimulatory and inhibitory interactions among the variables. Models capable of explaining are selected as putative regulatory networks.

and inhibition may be represented by linear or nonlinear mathematical models, depending on the particular application.

The challenge underlying the sort of model discovery process illustrated in Figure 7.6 is that, even for relatively small systems, there exist large numbers of models to test. For real applications, exhaustive searches are not feasible. For example, for a network of N nodes, there are $N(N-1)$ single-edge directed graphs connecting the nodes. Accounting for inhibiting and activating edges separately, there are $2N(N-1)$ single-edge models, or 24 for the 4-node example. There are over 1000 2-edge 4-node networks. In general, the number of models grows in proportion to $2^E N^{2E}$, where E is the number of edges.[6] Thus even if we know the number of edges to assign to a model (and we typically do not), the number of possible models grows so large for modest N that exhaustive searches are not possible. Therefore in practice what is done is a directed (nonexhaustive) search. One key to designing a useful algorithm for reverse engineering of biological

---

[6] See Exercise 7.2.

networks is to employ a search strategy that works for the problem at hand. Furthermore, even an appropriate search strategy will not yield meaningful results unless the underlying biophysics is somehow captured by the mathematical model structure underlying the interactions represented by the graph edges. Nor can the approach work without data on the kinetic behavior of the interacting nodes obtained under an experimental protocol (or protocols) for which the mechanistic interactions to be revealed are active. Of course, one cannot know whether one has informative data until one determines (discovers) a working model. Therefore experimental design and model determination must be mutually informative in an iterative manner.

7.3.1   Example: gene interaction in *Dictyostelium*

As a practical example, let us consider the data set shown in Figure 7.7, which is obtained from a study by Iranfar *et al.* [31]. These data are measures of messenger RNA (mRNA) levels for 14 genes in a specific subset of cells (called prestalk cells) in a growing population of *Dictyostelium* amoebae. As the population grows, the cells change morphologically as the homogeneous collection of single-celled organisms aggregate and differentiate into different components of heterogeneous multicellular entity. The 14 genes tracked in Figure 7.7 are those that are found in the experiments of Iranfar *et al.* to be expressed at levels more than 2.5-fold higher than in the initial measured mRNA level in the pre-stalk cells. (These data from this study were obtained online from the URL http://www.biology.ucsd.edu/loomis-cgi/microarray/TC2.html.)

In one strategy for determining putative gene regulatory interactions that may explain mRNA data such as these, it is assumed that a protein product of each gene has the potential to act as a transcription factor influencing the transcription of other genes. As a specific example, imagine the regulatory structure illustrated in Figure 7.8, where the gene product of tpsA acts as an activator and the product of ampA acts as an inhibitor of expression of cprA.

The simulations shown in Figure 7.8 for cprA are based on the following governing equation:

$$\frac{dx_1(t)}{dt} = r_1 \frac{\left(\frac{x_9(t-\tau)}{K_{A9,1}}\right)^n}{1 + \left(\frac{x_9(t-\tau)}{K_{A9,1}}\right)^n + \left(\frac{x_5(t-\tau)}{K_{I5,1}}\right)^n} - d_1 x_1, \qquad (7.13)$$

where $x_j(t)$ is the (normalized) mRNA level of gene j. The index $j = 1$ is associated with cprA, the first gene in the data set illustrated in Figure 7.7. Here it is

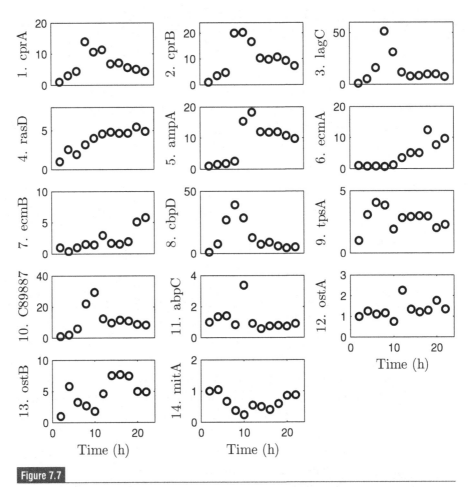

Time courses of 14 genes in the prestalk complex of a growing Dictyostelium colony. Measured mRNA levels are plotted relative to initial level, as reported by Iranfar *et al.* [31].

assumed that gene products of tpsA and ampA bind competitively to a regulatory site associated with cprA. The time delay τ is interpreted to account for a finite delay between mRNA expression and protein availability. The parameters r_1 and d_1 represent the maximal rate of transcription and a degradation rate constant. The constant $K_{A9,1}$ represents a binding constant (measured relative to normalized mRNA level) for the x_9 (tpsA) gene product; $K_{I5,1}$ represents a binding constant for the x_5 (ampA) gene product. Activation terms appear in the numerator and denominator of the first term in Eq. (7.13). Inhibition terms appear only in the denominator. The competitive binding interaction of all activation and inhibition factors determines whether or not synthesis of cprA occurs at a given time. (The Hill coefficient n is fixed at a constant value 4 in these simulations.)

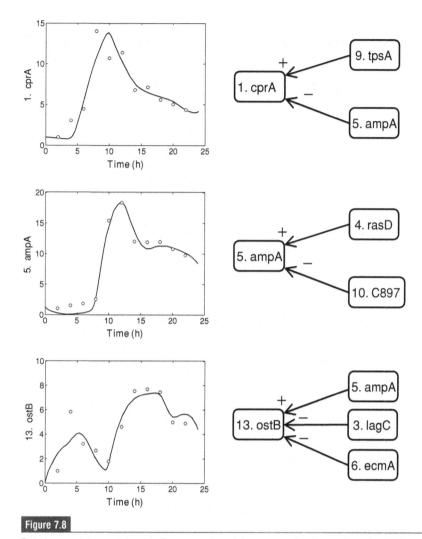

Figure 7.8

Putative regulatory mechanism for expression of three genes (cprA, ampA, and ostB) from the Iranfar *et al.* [31] data set. The plots on the left show measured data and model predictions (solid lines) associated with the regulatory structures illustrated on the right. Equation (7.13) is the model for cprA. Parameter values for this case are $r_1 = 5.09$, $d_1 = 0.167$, $K_{A9,1} = 2.81$, $K_{I5,1} = 7.58$, and $x_1(0) = 0.988$.

To obtain the simulations of $x_1(t)$ shown in the top panel of Figure 7.8, Eq. (7.13) is integrated with $x_9(t)$ and $x_5(t)$ determined by fitting piecewise continuous cubic polynomials through the data for these genes. Thus the model for the dynamics of a single gene has only a single dynamic state variable. (A similar assumption is used to obtain the dynamic model predictions illustrated for ampA and ostB.[7]) This

---
[7] See Exercise 7.4.

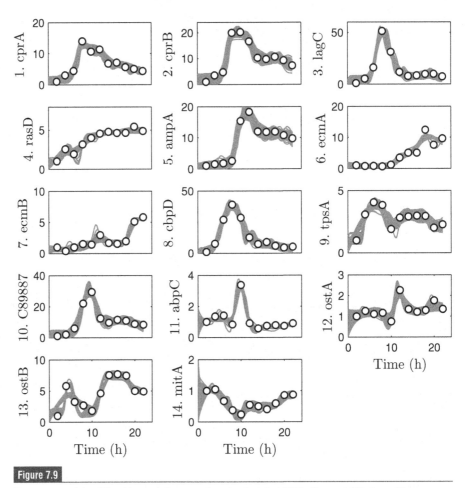

Figure 7.9

Time-course data of the 14 pre-stalk genes from Iranfar *et al.* [31] with model fits from 25 independently determined regulatory networks for each gene.

computational approach allows us to rapidly search for candidate subnetworks associated with single genes and to easily distribute the single-gene calculations to independent computer processes, making search algorithms easily scalable in distributed computing environments.

While the results in Figure 7.8 appear reasonable, these candidate networks that fit the data for cprA, ampA, and ostB do not represent the only possible models. For each gene there are many possible sets of inhibitors and activators that can produce essentially equally good matches to the data. This phenomenon is illustrated in Figure 7.9, in which model predictions from ensembles of >50 candidate models for each of the 14 genes in the Iranfar data set are shown. This

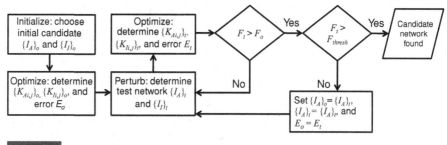

Algorithm for searching for candidate networks to explain biological network data.

degree of nonuniqueness seriously confounds our ability to determine what real interactions might be revealed by the mRNA data set.

One strategy for trying to determine which of the many candidate network interactions in the many trial models are important and biologically meaningful is simply to construct an ensemble of candidate models and to determine which network interactions appear most frequently in the ensemble. The assumption here is that if a particular interaction appears in many or most candidate models, its likelihood of representing a real biological relationship is high. In order to apply this assumption it is necessary to systematically generate many independent model realizations. One may generate an ensemble of candidate networks for this data set using a generic model of the form of Eq. (7.13). Specifically, the rate of translation of gene j is represented by

$$\frac{dx_j(t)}{dt} = r_j \frac{\sum_{i \in I_{Aj}} \left(\frac{x_i(t-\tau)}{K_{Ai,j}}\right)^n + e_j}{1 + \sum_{i \in I_{Aj}} \left(\frac{x_i(t-\tau)}{K_{Ai,j}}\right)^n + \sum_{i \in I_{Ij}} \left(\frac{x_i(t-\tau)}{K_{Ii,j}}\right)^n + e_j} - d_j x_j. \quad (7.14)$$

Here I_{Aj} and I_{Ij} are the sets of indices of variables that act as activators and inhibitors of x_j production. The constant e_j accounts for potential externally stimulated or constitutive transcription that is not brought about directly through the explicit model variables.

Candidate networks are determined through the algorithm diagrammed in Figure 7.10. For a given variable, the algorithm is initialized by choosing an initial network structure, defined by the sets of inhibitors and activators $\{I_A\}_o$ and $\{I_I\}_o$. (For example, one approach is to set $\{I_A\}_o = \emptyset$ and $\{I_I\}_o = \emptyset$.) At each iteration, the candidate network is randomly perturbed by adding or removing connections from other variables and a new optimal to the data is determined.

The variable F_t represents a measure of the likelihood of a given network explaining the data. (For example, the likelihood function employed by Bazil

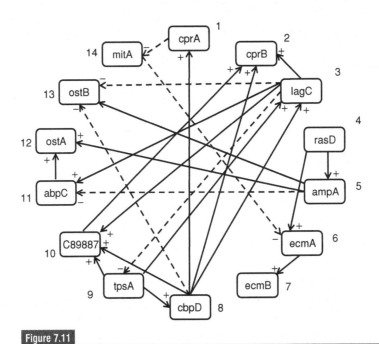

Figure 7.11

Illustration of 23 most frequently appearing network interactions in 14-gene network of Iranfar *et al.* [31]. Activator interactions are illustrated as solid lines, inhibitor interactions as dashed lines.

et al. is proportional to the negative of the mean squared difference between the optimal model prediction and the measured data. Additional factors penalize candidate networks for increasing complexity and associated number of adjustable parameters.) Changes to trial networks are accepted if the likelihood increases. Finally, a candidate network is accepted if the likelihood is less than some threshold F_{thresh}. For details see Bazil *et al.* [4].

Based on this algorithm one can obtain many independent putative network interactions and obtain multiple model fits, as illustrated in Figure 7.9. Here, a consensus 14-node network for the 14-gene data set is assembled by selecting the network interactions that appear with the greatest frequency in independent trials of the search algorithm. The 23 most highly selected edges for the 14-node network are illustrated in Figure 7.11.

This illustrated network represents nothing more (or less) than the most commonly invoked network connections determined by the algorithm described above. It does not represent a unique way to explain the data. This nonuniqueness of solutions to this problem (and reverse engineering problems in biological networks, in general) represents a major confounding weakness. When there is a very large (possibly effectively infinite) number of possible solutions to a problem, it is difficult to have much confidence in one.[8] Thus the standard scientific approach of systematically disproving all alternative explanation is not feasible here. This is why tools like this one are best applied to generate hypotheses from data.[9] The network illustrated in Figure 7.11 represents a systematically generated set of hypothetical gene interactions that may be tested in follow-up experiments.

7.4 Recapitulation and lessons learned

The two distinctly different examples studied in this chapter illustrate two important types of applications in modern biological research. In the glycolysis example, a model was constructed based on a tremendous amount of prior knowledge. The model used to analyze the kinetic data was built from knowledge of the existence and stoichiometry of the biochemical reactions in the pathway. Information on the mechanisms of the enzyme-catalyzed reactions and values for kinetic and thermodynamic constants were obtained from the prior literature. Moreover, the integrated model was formulated on the basis of an existing physical theory of the kinetics of metabolic systems.

The gene regulatory network analysis was undertaken with a far less complete foundation of knowledge. While the underlying model employed arose as a statement of mass conservation, and the production term in Eq. (7.14) is inspired by a model of nonlinear competitive binding of multiple transcription factors, the model cannot be taken as representing biochemical mechanisms at the level of detail of the enzyme models of Sections 6.2.2 and 7.2.3. In essence, it is a generic phenomenological model, inspired by knowledge of transcription regulation. The utility of the model is in its ability to generate credible hypotheses, not in its mechanistic underpinnings.

The two approaches to large-scale systems analysis reflect the state of knowledge in the fields. While the detailed analysis of a metabolic pathway in Section 7.2.3 allows us to draw conclusions regarding the control of the biochemical systems *in vivo*, the gene regulation example of Section 7.3.1 distills a data set into a set of untested hypotheses related to possible interactions among the measured variables. Yet if the goal set out at the beginning of this chapter – simulating the integrated biochemical functions of a cell – is to be realized, then molecular genetic

---

[8] On the other hand, the existence of a large number of solutions that are able to fit the data may represent a key to the success of this algorithm. If there were only one unique solution, perhaps this algorithm would fail to find it! Thus the problem of nonuniqueness encountered here does afford the practical advantage that we can easily find lots of solutions.

[9] Hypothesis-generating studies are a common feature of *discovery science*. The idea is to start with large-scale data collection, such as from gene expression profiling, and to determine hypotheses from the data.

processes must be functionally linked with metabolism. The enzymes catalyzing metabolic processes are dynamically synthesized and degraded; the metabolic state of a cell influences gene transcription. As our knowledge of gene transcription regulation and cellular signaling approaches our level of understanding of cellular metabolism, and as increasingly complete and validated models of metabolism are created, we will move closer to a realization of Buchner's vision of a synthesized understanding of "the processes of life."

Problems

7.1 Dynamic buffering. How must Eq. (7.9) be modified to account for the water self-buffering captured in Eq. (5.49)? How important is this additional term for biochemical systems?

7.2 System identification. Show that an N-node network, the number of graphs made up of R directed edges, where each edge can represent either an inhibitor or activator, grows as $2^E N^{2E}$. How many such graphs exist for 10 nodes and 10 edges?

7.3 Gene regulation and feedback. (Adapted from [51].) Consider a reaction system that involves three processes, binding of a transcription factor (TF) to DNA, synthesis of the transcription factor, and degradation of the transcription factor:

$$\text{DNA} + n\text{TF} \underset{k_{-1}}{\overset{k_{+1}}{\rightleftharpoons}} \text{DNA} \cdot (\text{TF})_n,$$

$$\text{DNA} \xrightarrow{g_0} \text{TF}, \quad \text{DNA} \cdot (\text{TF})_n \xrightarrow{g_1} \text{TF},$$

$$\text{TF} \xrightarrow{k_D} .$$

The DNA is assumed to bind n copies of TF; the transcription factor is assumed to be synthesized at two different rates, depending on whether or not the DNA is bound to the TF. Write down the kinetic equations for dx/dt and dy/dt, where $x = [\text{DNA} \cdot (\text{TF})_n]$ and $y = [\text{TF}]$. Show that in the case of negative feedback ($g_0 > g_1$) and $n = 2$, only one unique steady state is possible. Show that in the case of positive feedback ($g_0 < g_1$) multiple stable steady states are possible. (Hint: Start by finding unitless versions of the governing equations and plot the *null clines* – plots of x versus y obtained by applying the steady-state conditions $dx/dt = 0$ and $dy/dt = 0$.)

7.4 Computer exercise. Using the data of Iranfar *et al.* [31] (http://www.biology.ucsd.edu/loomis-cgi/microarray/TC2.html) determine the activation and inhibition parameter values that are associated with the model fits in Figure 7.8. (The underlying model for the upper panel is given in Eq. (7.13);

$K_{A9,1}$, $K_{I5,1}$, r_1, d_1, and $x_1(0)$ are the parameters to be determined. The equations for the other examples are defined by the illustrated network structures.) How are the model fits to the data sensitive to τ and n, which are set to $\tau = 1$ h and $n = 4$ in the figure? Can the data for these genes be fitted with $n = 1$?

8 Cellular electrophysiology

Overview

This chapter is devoted to the study and simulation of the electrical potential across cell membranes. Practically every type of animal cell actively maintains an electrostatic potential difference between its cytoplasm and external milieu. Organelles in the cell, such as mitochondria for example, maintain electrical gradients across their membranes as well. The electrostatic gradient across cell membranes changes dynamically in response to chemical, mechanical, and/or electrical stimuli. In turn, dynamic changes in cell membrane potential are tied to all manner of downstream processes in various cell types. For example, muscle cells contract when calcium ions enter the cytoplasm in response to depolarization of the cell membrane potential. Depolarization, caused by the influx of sodium and other positively charged ions, may be initiated by the combined action of several processes, including stimulation by neurotransmitters and transmission of the electrical signal from neighboring cells.

The focus here is on simulating the dynamics of the cell membrane potential difference, and the ion currents that influence changes in the potential difference. A substantial fraction of the chapter is devoted to working through the details of the celebrated Hodgkin–Huxley model of the nerve impulse. This exposition allows us to introduce the basic formulation of that classic model, which is still widely used in the field. Reviewing the work of Hodgkin and Huxley allows us to see how their model was developed based on data from experiments carefully designed to reveal a quantitative description of the biophysical processes at work.

8.1 Basic concepts of cell electrophysiology

The cell membrane provides a barrier separating the aqueous solutions inside and outside a cell. Most of the time – through the action of ion pumps and ion selective channels – the ionic compositions on either side of the membrane are substantially different. Because ions moving across the membrane carry electrical charges, their movements influence and are influenced by electrical forces. Thus the influence of electrical forces must be accounted for in a study of the biophysics of cellular electrophysiology.

8.1.1 Thermodynamics of ion fluxes

To work out some basic concepts, let us imagine that the potassium ion concentrations inside and outside a cell are drastically different. In fact, this is the usual circumstance, with inside concentration, $[K^+]_i$, exceeding outside $[K^+]_o$. This concentration difference is associated with an entropy-driven thermodynamic driving force. In a typical cell (under resting conditions[1]) there is an electrical force driving K^+ ions into the cell that balances the entropic force associated with the concentration gradient. Recall that in the definition of Gibbs free energy

$$\Delta_r G = \Delta_r G^o - RT \ln\left([K^+]_o/[K^+]_i\right)$$

the first term accounts for the free energy of formation in the solution of the participating species under the conditions imposed. Here we have assumed that the forward process is a movement of K^+ from the outside to the inside of the cell. Since the underlying chemical process

$$K_o^+ \rightleftharpoons K_i^+ \tag{8.1}$$

does not involve the synthesis or degradation of any chemical entity (and assuming that both sides of the membrane have equal temperature, pressure, and overall ionic strength, and that movement of K^+ ions does not incur a meaningful change in volume), the reference free energy change associated with the process is the electrical energy needed to move a charge from one side of the membrane to the other. If we adopt the sign convention that electrostatic potential (voltage) is measured relative to the outside of the cell, then a positive potential difference $(\Delta\Psi)$ will be associated with passive energy dissipation when positive ions move from the inside to the outside. Thus

$$\Delta_r G = F\Delta\Psi - RT \ln\left([K^+]_o/[K^+]_i\right) \tag{8.2}$$

for the process of Eq. (8.1). (The factor $F \approx 96\,485$ coulomb per mole is called Faraday's constant, used to convert voltage measured in joules per coulomb to units of joules per mole.) According to Eq. (8.2), there is no thermodynamic driving force when

$$\Delta\Psi = \Delta\Psi_K = \frac{RT}{F} \ln\left(\frac{[K^+]_o}{[K^+]_i}\right).$$

[1] We will see what we mean by "resting conditions" shortly.

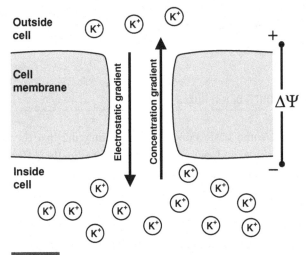

Outside cell

Cell membrane

Inside cell

Electrostatic gradient

Concentration gradient

$\Delta\Psi$

$+$

$-$

K^+ ... (ions labeled throughout figure)

Figure 8.1

Illustration of a potassium ion concentration gradient across a cell membrane. The concentration gradient and the electrostatic potential oppose and balance one another in thermodynamic equilibrium. Figure reprinted from Beard and Qian [8], with permission.

Taking typical values of $[K^+]_i = 140$ mM and $[K^+]_o = 5$ mM, we have $\Delta\Psi_K \approx -70$ mV, meaning that under these conditions there would be no net passive K^+ flux when the intracellular environment is at an electrostatic potential 70 mV less than that outside the cell. This situation is illustrated in Figure 8.1.

The potential $\Delta\Psi_K = (RT/F)\ln([K^+]_o/[K^+]_i)$ is called the *Nernst equilibrium potential* (or simply the Nernst potential) for potassium ion. If there existed channels by which K^+ could passively permeate across the cell membrane, and if no other ions traveled across the membrane, and if the concentration gradient were maintained, then $\Delta\Psi$, the "membrane potential," would attain the value of the Nernst potential for K^+. In fact, this is the approximate situation in an unstimulated nerve or muscle cell, for which the resting membrane potential is around -70 mV. When other currents are involved it is possible to attain membrane potentials anywhere within the range of the Nernst potentials of the participating ions.

For example, imagine that potassium flux were governed by the linear difference between the membrane potential and the potassium Nernst potential:

$$I_K = -g_K(V - V_K). \tag{8.3}$$

Here we have adopted the convention to denote ion fluxes using the variable I, indicating a current. We are also using a shorthand V for membrane potential difference and $V_K = (RT/F)\ln([K^+]_o/[K^+]_i)$ for the Nernst potential for K^+. The minus sign appears in Eq. (8.3) because the convention is that current is inward

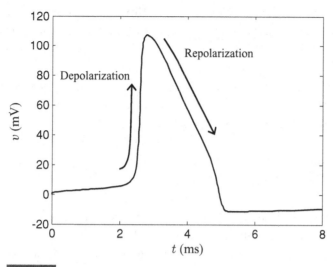

Figure 8.2

Action potential predicted by the Hodgkin–Huxley model. Parameter values are given in the text. Applied current is set to a constant, $I_{app} = 6.2 \, \mu A \, cm^{-2}$.

while V is measured as inside minus outside potential. If membrane channels facilitated the flux of sodium ions governed by the same sort of relation,

$$I_{Na} = -g_{Na}(V - V_{Na}), \qquad (8.4)$$

then net current would be zero and the membrane potential would remain constant when

$$V = \frac{g_K V_K + g_{Na} V_{Na}}{g_K + g_{Na}}, \qquad (8.5)$$

where $V_{Na} = (RT/F)\ln([Na^+]_o/[Na^+]_i)$.

As we will see in the next section, cases exist where the basic model of Eqs (8.3) and (8.4) reasonably represents observed behavior. The key to the classic Hodgkin–Huxley model of nerve activation is to determine an effective way to account for dynamic changes in the effective conductances g_K and g_{Na}.

8.2 The Hodgkin–Huxley model of the squid giant axon

In a tour de force of experimental observations and theory development in the late 1940s and early 1950s Alan Hodgkin and Andrew Huxley elucidated the biophysical underpinnings of nerve excitation. One of the most important contributions of their work was the determination of the mechanisms that produce the *action potential* illustrated in Figure 8.2. This curve illustrates membrane voltage as a function of time predicted by the Hodgkin–Huxley model of 1952 [28]. Here voltage is measured relative to the resting unstimulated membrane potential: $v = V - V_o$, where V_o is the potential observed when no external current is applied.

The action potential occurs as a sharp (positive) departure from the resting potential, termed the *depolarization* phase. Following peak depolarization, the potential recovers to near the resting potential (the *repolarization* phase). Following repolarization there is a period of latency, termed the *refractory period*, during which the neuron is temporarily not excitable. After the refractory period, another action potential may be elicited. (The refractory phenomenon will be explored at the end of this section.)

Hodgkin and Huxley worked extensively with a nerve axon from a squid. Specifically, inserting an electrode into the giant axon of the Atlantic squid, it is possible to impose prescribed voltages and measure the current across the membrane. At the time, this work required state-of-the-art electronic equipment to clamp voltages precisely and measure small currents. Indeed, the investigations of Hodgkin and Huxley, which were on the leading edge of engineering, biophysics, and computing, were ultimately recognized with the award of the 1963 Nobel Prize in Physiology or Medicine. This work established a new paradigm for precise predictive simulation of biological systems, and set a standard for clarity and insight that we should try to live up to. The basic Hodgkin–Huxley model formulation still serves as the foundation for electrophysiology modeling today. For all these reasons, it will be valuable to step systematically through some of their work.

Like all scientific stories, an analysis of the Hodgkin–Huxley model starts somewhere in the middle of the story. Before they worked through their seminal study "A quantitative description of membrane current and its application to conduction and excitation in nerve" [28], Hodgkin and Huxley understood a few things about the action potential illustrated in Figure 8.2. They know that the resting potential was close to the Nernst potential for K^+. They also knew that the depolarization phase was apparently due to a rapid influx of sodium ions. (One source of evidence for this was the fact that the depolarization was abolished when extracellular Na^+ was replaced by choline, a substitute monovalent cation that apparently cannot travel through the pathways available to sodium ions.) They also had a good idea that a "slow" potassium current was responsible for the repolarization back to resting potential, based on experiments we will review below.

Based on these understandings, Hodgkin and Huxley developed the conceptual representation of the membrane illustrated in Figure 8.3. One important concept here is that each ion (sodium or potassium) is assumed to travel through its own conductance pathway. The I_{app} current illustrated in the figure represents an externally applied current. Although they had no knowledge of the existence of ion-selective channels, Hodgkin and Huxley postulated that these conductance pathways were associated with "carrier molecules" in the membrane. Another

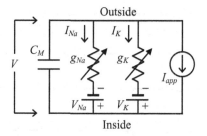

Figure 8.3

The Hodgkin–Huxley model of the nerve cell membrane. Sodium and potassium currents pass through variable conductors and depend on cell membrane potential V and on concentration-dependent Nernst potentials V_K and V_{Na}.

important concept illustrated in the figure is that, given the sign convention applied, each ion current is driven by the difference between the membrane potential V and the Nernst potential for a given ion. Since V is measured as inside potential minus outside potential, for a typical cell with a value of V_K of approximately -70 mV the membrane potential and the Nernst potential work in opposition when $V \approx -70$ mV. The sodium concentration, however, is typically higher on the outside of the cell, and V_{Na} may be in the range of $+50$ mV. Thus when $V = -70$ mV, the thermodynamic driving force for Na$^+$ influx is $-(V - V_{Na}) \approx +120$ mV.

The governing equation associated with this circuit model is

$$C_m \frac{dv}{dt} = -g_{Na}(v - v_{Na}) - g_K(v - v_K) + I_{app}, \qquad (8.6)$$

which is expressed in terms of membrane voltages relative to resting: $v = V - V_o$, $v_{Na} = V_{Na} - V_o$, and $v_K = V_K - V_o$. The average resting potential measured by Hodgkin and Huxley in squid axons in seawater at $8\,°C$ was $V_o = -56$ mV. The capacitance of the membrane takes on a fixed value of 1×10^{-6} μF cm^{-2}. (See Section 9.7.) The action potential emerges from this model as a consequence of voltage-dependent changes in the conductivities g_{Na} and g_K. These conductivities are nowadays known to represent the open versus closed probability of *voltage-gated* ion channels. The Hodgkin–Huxley model represents two such channels: a (relatively) fast-opening and slow-closing sodium channel and a slow potassium channel.

8.2.1 The potassium conductance

To characterize the relatively slowly changing potassium current, Hodgkin and Huxley made use of the experimental preparation referenced above, in which sodium in the extracellular bath was replaced by choline. Therefore measured

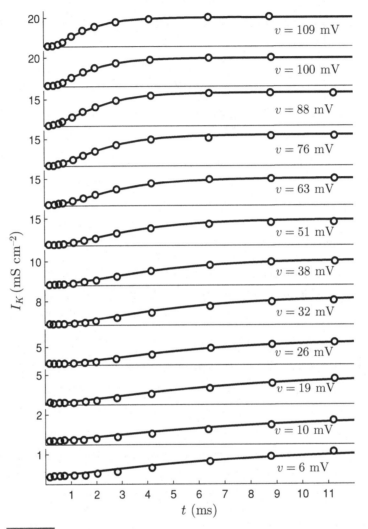

Figure 8.4

Plot of measured potassium conductance as a function of time for various voltage clamp protocols. The different data sets correspond to experiments in which the voltage was suddenly changed from resting $v = 0$ to the value indicated (between 6 and 109 mV). Figure adapted from Hodgkin and Huxley [28], Figure 3.

currents could be reasonably assumed to reflect current carried by the movement of K^+ across the membrane. By applying a step change in membrane voltage, the measured current revealed the apparent potassium ion conductance as a function of time, as indicated in Figure 8.4.

Based on these data, Hodgkin and Huxley postulated a model where potassium conductivity varies between zero and some maximal value, $\bar{g}_K$. Specifically, they used the following kinetic equations:

$$g_K = n^4 \bar{g}_K$$

$$\frac{dn}{dt} = \alpha_n(1 - n) - \beta_n n, \qquad (8.7)$$

where α_n and β_n are rate constants, and n is a unitless number that varies between 0 and 1. The number n can be interpreted as the probability of a potassium channel *gate* being open. Equation (8.7) assumes that an individual channel behaves as if it is made up of four independent identical gates, and thus n^4 is the probability that all four gates are open for an individual channel. (In practice the power of 4 used in this expression is justified by the fact that it gives a satisfactory fit to the data. Hodgkin and Huxley concluded that "better agreement might have been obtained with a fifth or sixth power, but the improvement was not considered to be worth the additional complication.")

While it is assumed that the rate constants α_n and β_n are functions of voltage for voltage-gated channels, the experiments of Figure 8.4 apply a constant voltage, thus reducing Eq. (8.7) to a first-order linear constant-coefficient equation, with solution

$$n(t) = n_\infty - (n_\infty - n_o)e^{-t/\tau_n}$$

$$n_\infty = \alpha_n/(\alpha_n + \beta_n)$$

$$\tau_n = 1/(\alpha_n + \beta_n), \qquad (8.8)$$

where n_∞ is the constant steady-state solution.

Hodgkin and Huxley fitted the above expression to the data in Figure 8.4 to quantify the dependence of the rate constants on v. Specifically, $g_K(t)$ predicted by Eq. (8.7), given $n(t)$ from Eq. (8.8), was fitted to these data, estimating n_∞ and τ_n for each value of the applied voltage in their experiments. Then, from expressions for n_∞ and τ_n in Eq. (8.8), the expressions

$$\alpha_n = \frac{n_\infty}{\tau_n}$$

$$\beta_n = \frac{1 - n_\infty}{\tau_n} \qquad (8.9)$$

were used to transform the estimated parameters to rate constants to be used in Eq. (8.7). The resulting transformed data are plotted in Figure 8.5 for a range of applied voltages and several independent experimental preparations. Note that in their figure (Figure 4 of [28]), Hodgkin and Huxley switched the sign convention used in Figure 8.4. Here we are using the same convention throughout, where a positive voltage v corresponds to an increase in intracellular potential compared with resting potential.

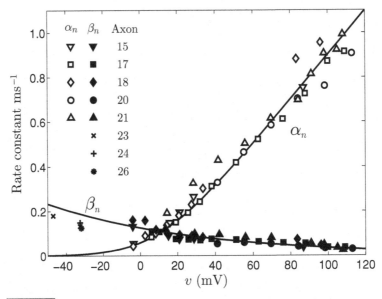

Estimated opening and closing rate constants for potassium conductance as functions of membrane voltage. The ordinate represents estimates of α_n and β_n from Eq. (8.9). The abscissa represents the applied membrane potential difference measured as inside potential minus outside potential, measured relative to the resting potential. Figure adapted from Hodgkin and Huxley [28], Figure 4. (The sign convention for voltage in this figure is the same as that used in the text and in Figure 8.4. In the original publication the opposite sign convention was used for this figure.)

Figure 8.5 demonstrates several key concepts. First, the potassium channel opening rate (α_n) increases with applied voltage, while the closing rate (β_n) increases with decreasing voltage. Second, these data illustrate the reproducibility and robustness of the result over many experiments. Finally, this plot provides the basis for determining phenomenological functions to represent α_n and β_n as functions of voltage. Hodgkin and Huxley determined the following functions:

$$\alpha_n = 0.01 \, \frac{10 - v}{\exp\left(\frac{10-v}{10}\right) - 1}$$

$$\beta_n = 0.125 \, \exp\left(\frac{-v}{80}\right), \tag{8.10}$$

which were deemed to "give a satisfactory formula for the steady state potassium conductance ($g_K(\infty)$) at any membrane potential (V)." Here we have expressed the rate constants in units of ms^{-1} as functions of v, measured in units of mV. Combining these expressions using Eq. (8.8) we can obtain an expression for the steady-state value of n (that is, n_∞) as a function of voltage, which is plotted in Figure 8.6. Again, in the original publication the opposite sign convention was

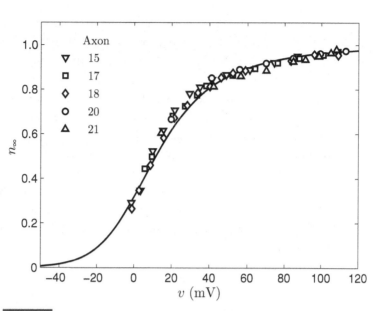

Figure 8.6

Steady-state values of n_∞ as a function of membrane voltage. The abscissa represents the applied membrane potential difference measured as outside potential minus inside potential, relative to the resting potential. Figure adapted from Hodgkin and Huxley [28], Figure 5. (The sign convention for voltage in this figure is the same as that used in the text and in Figure 8.4. In the original publication the opposite sign convention was used for this figure.)

used for this figure (Figure 5 of Hodgkin and Huxley [28]). The ordinate in this text's Figure 8.6 is the cell membrane potential measured as inside potential minus outside potential, and measured relative to resting potential.

8.2.2 The sodium conductance

The sodium current was more difficult to characterize, for a number of reasons. First, because the electrochemical gradient of K^+ is always outward, it is not possible to remove the potassium current by replacing the potassium in the bath with a substitute cation. As a result, Hodgkin and Huxley were forced to design a clever protocol to estimate sodium current in their voltage-clamp experiments. The basis of the approach was to measure two current recordings for the same voltage clamp – one with the choline solution and one with seawater bathing the neuron. The seawater experiment was assumed to include both Na^+ and K^+ currents, while the Na^+ current was assumed to be abolished in the choline bath experiments. Thus the Na^+ current was taken by subtracting the current from the choline bath experiments from that of the seawater experiment. (This approach was complicated by the fact that the resting potential in choline is approximately -60 mV compared with -56 mV in seawater. This difference, which is due to the presence of a small depolarizing Na^+ under resting conditions, is accounted for in the detailed procedure described in Hodgkin and Huxley [29].)

Estimated sodium conductances for protocols analogous to those of Figure 8.4 are illustrated in Figure 8.7. These data display a qualitatively different character from the potassium conductance in that, in response to a depolarization, the sodium conductance increases, reaches a maximum within one or two milliseconds, and then decays toward zero. As Hodgkin and Huxley describe [28], this sort of biphasic behavior might be "determined by a single variable that obeys a second-order differential equation" or might be "determined by two variables, each of which obeys a first-order equation." They chose to use two gating variables for the sodium conductivity:

$$g_{Na} = m^3 h \bar{g}_{Na}$$

$$\frac{dm}{dt} = \alpha_m(1 - m) - \beta_m m$$

$$\frac{dh}{dt} = \alpha_h(1 - h) - \beta_h h, \tag{8.11}$$

where m and h are gating variables analogous to n in Eq. (8.7).

As in the analysis of the potassium conductance, the equations for $m(t)$ and $h(t)$ are first-order linear constant-coefficient equations for the constant voltage-clamp experiment. Their solutions are

$$m(t) = m_\infty - (m_\infty - m_o)e^{-t/\tau_m}$$

$$m_\infty = \alpha_m/(\alpha_m + \beta_m)$$

$$\tau_m = 1/(\alpha_m + \beta_m) \tag{8.12}$$

and

$$h(t) = h_\infty - (h_\infty - h_o)e^{-t/\tau_h}$$

$$h_\infty = \alpha_h/(\alpha_h + \beta_h)$$

$$\tau_h = 1/(\alpha_h + \beta_h), \tag{8.13}$$

where m_∞ and h_∞ are the steady-state solutions for a given applied voltage. Model predictions of $g_{Na}(t)$ were fitted to the experimental data of Figure 8.7, yielding estimates for α_m, β_m, α_h, and β_h as functions of applied voltage.

Fitting these data to the theoretical model is somewhat more challenging than for the potassium conductance, because the model for the potassium conductance

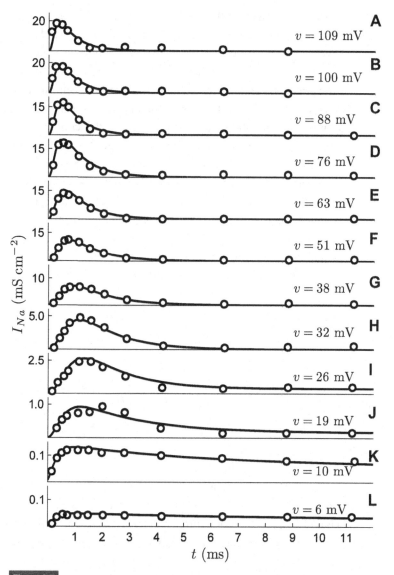

Figure 8.7

Plot of measured sodium conductance as a function of time for various voltage clamp protocols. The different data sets correspond to experiments in which the voltage was suddenly changed from resting $v = 0$ to the value indicated (between 6 and 109 mV). Figure adapted from Hodgkin and Huxley [28], Figure 6.

data – Eq. (8.8) – involves three parameters: n_o, the resting value; n_∞, the steady-state value; and τ_n, the time constant. (The estimates of α_n and β_n depend only on n_∞ and τ_n.) The sodium conductance curves are more complex, and are fitted by a six-parameter model – Eqs (8.12) and (8.13). Nowadays we could use a desktop computer to easily fit the model predictions to these data, perhaps using a least-squares optimization, as illustrated in Section 9.4. Hodgkin and Huxley did not have such powerful computational resources at their fingertips.

For depolarizations greater than $+30$ mV (curves A through H in Figure 8.7), parameters m_o and h_∞ were assumed to be zero, assuming that the sodium conductance in the resting state "is very small compared with the value attained during a large depolarization," and, similarly, that the conductance after long times in Figure 8.7 is small compared with the transient peak. For these cases, the theoretical model "was fit to an experimental curve by plotting the latter on double log paper and comparing it with a similar plot of a family of theoretical curves drawn with different ratios of τ_m to τ_h." For the other cases (curves I through L), fits assumed "that h_∞ and τ_h had values calculated from experiments described in a previous paper." In essence, the parameter values were estimated based on visually matching model predictions to the data. Given estimates of m_∞, τ_m, h_∞, and τ_h, the rates constants α_m, β_m, α_h, and β_h may be computed:

$$\alpha_m = \frac{m_\infty}{\tau_m}$$

$$\beta_m = \frac{1 - m_\infty}{\tau_m}$$

$$\alpha_h = \frac{h_\infty}{\tau_h}$$

$$\beta_h = \frac{1 - h_\infty}{\tau_h}, \tag{8.14}$$

at each value of applied voltage.

The associated estimates of α_m and β_m are plotted in Figure 8.8. Positive depolarizations are associated with positive values of v, as indicated in the legend. Thus, similar to the potassium channel "n" gate opening rate, the sodium channel "m" gate opening rate increases with increasing depolarization. The α_m and β_m rate constants are ten times greater than α_n and β_n, indicating that the sodium conductance responds much more rapidly than the potassium conductance. The predicted steady state $m(t)$ (m_∞) is plotted in Figure 8.9.

Estimates of $\alpha_m(v)$ and $\beta_m(v)$ are fit by the phenomenological functions

$$\alpha_m = 0.1 \frac{25 - v}{\exp\left(\frac{25-v}{10}\right) - 1}$$

$$\beta_m = 4 \exp\left(\frac{-v}{18}\right), \tag{8.15}$$

where the rates constants are in units of ms^{-1} and v is in mV.

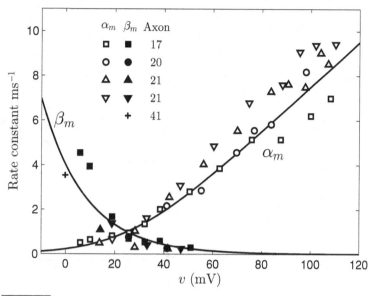

Figure 8.8

Estimated opening and closing rate constants for sodium conductance as functions of membrane voltage. The ordinate represents estimates of α_m and β_m from Eq. (8.14). Figure adapted from Hodgkin and Huxley [28], Figure 7. (The sign convention for voltage in this figure is the same as that used in the text and in Figure 8.7. In the original publication the opposite sign convention was used for this figure.)

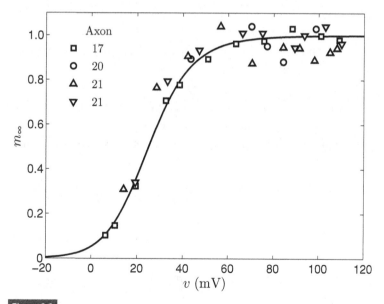

Figure 8.9

Steady-state values of m_∞ as a function of membrane voltage. Figure adapted from Hodgkin and Huxley [28], Figure 8. (The sign convention for voltage in this figure is the same as that used in the text and in Figure 8.7. In the original publication the opposite sign convention was used for this figure.)

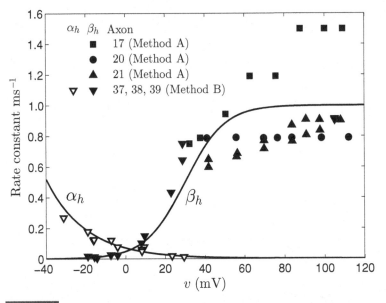

Figure 8.10

Estimated opening and closing rate constants for sodium conductance as functions of membrane voltage. The ordinate represents estimates of α_h and β_h from Eq. (8.14). Figure adapted from Hodgkin and Huxley [28], Figure 9. (The sign convention for voltage in this figure is the same as that used in the text and in Figure 8.7. In the original publication the opposite sign convention was used for this figure.)

Estimates of α_h and β_h, and associated h_∞, are plotted in Figures 8.10 and 8.11. In contrast to the behavior of the "m" gate, the "h" gate closing rate (β_h) increases with increasing depolarization. The α_h and β_h rate constants are substantially smaller than α_m and β_m. The relatively fast opening and slow closing in response to a depolarization are responsible for the transients observed in Figure 8.7. There is a good deal more variability in the reported estimates of β_h at large values of depolarization compared with estimates of β_n and β_m. As indicated in the original figure, estimates are provided by two different data-fitting methods, detailed in the original publications [30, 28].

These estimates of $\alpha_h(v)$ and $\beta_h(v)$ are fit by the phenomenological functions

$$\alpha_h = 0.07 \ \exp\left(\frac{-v}{20}\right)$$

$$\beta_h = \frac{1}{\exp\left(\frac{30-v}{10}\right)+1}, \tag{8.16}$$

where, as above, rates constants are in units of ms^{-1} and v is in mV.

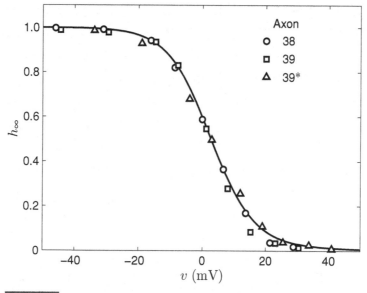

Figure 8.11

Steady-state values of h_∞ as a function of membrane voltage. Figure adapted from Hodgkin and Huxley [28], Figure 10. (The sign convention for voltage in this figure is the same as that used in the text and in Figure 8.7. In the original publication the opposite sign convention was used for this figure.)

8.2.3   Summary of model equations

To synthesize the full system of equations of the Hodgkin–Huxley model of membrane potential kinetics in the giant axon of the Atlantic squid, we have a system of four state variables governing membrane potential, and three gating variables:

$$C_m \frac{dv}{dt} = -\bar{g}_K n^4 (v - v_K) - \bar{g}_{Na} m^3 h (v - v_{Na}) - \bar{g}_L (v - v_L) + I_{app}$$

$$\frac{dm}{dt} = \alpha_m (1 - m) - \beta_m m$$

$$\frac{dn}{dt} = \alpha_n (1 - n) - \beta_n n$$

$$\frac{dh}{dt} = \alpha_h (1 - h) - \beta_h h. \tag{8.17}$$

The term $-\bar{g}_L (v - v_L)$ represents a "leak" current [28], and I_{app} represents any externally applied current. The rate constants are determined as functions of membrane potential:

$$\alpha_m = 0.1 \, \frac{25 - v}{\exp\left(\frac{25-v}{10}\right) - 1}$$

$$\beta_m = 4 \, \exp\left(\frac{-v}{18}\right)$$

$$\alpha_n = 0.01 \, \frac{10 - v}{\exp\left(\frac{10-v}{10}\right) - 1}$$

$$\beta_n = 0.125 \, \exp\left(\frac{-v}{80}\right)$$

$$\alpha_h = 0.07 \, \exp\left(\frac{-v}{20}\right)$$

$$\beta_h = \frac{1}{\exp\left(\frac{30-v}{10}\right) + 1}. \tag{8.18}$$

Experimentally determined equilibrium potentials (which depend on the ion gradients across the membrane) for the model are

$$v_K = -12 \text{ mV}, \quad v_{Na} = 115 \text{ mV}, \quad v_L = 10.6 \text{ mV}.$$

(Recall that, like v, these Nernst potentials are measured relative to the resting potential of approximately -56 mV.) The conductivity constants $\bar{g}_K$, $\bar{g}_{Na}$, and $\bar{g}_L$ are

$$\bar{g}_K = 36 \text{ mS cm}^2, \quad \bar{g}_{Na} = 120 \text{ mS cm}^2, \quad \bar{g}_L = 0.3 \text{ mS cm}^2.$$

Finally, the capacitance has an experimentally estimated value $C_m = 1 \times 10^{-6}$ μF cm^{-2}. Thus the currents and the capacitance are all measured per cell membrane area.

One way to appreciate the simple beauty and power of this model is to consider a computer code that simulates the governing differential equations. A code (taken from Beard and Qian [8]) to compute the time derivatives of the state variables (the right-hand side of Eq. (8.17)) is provided in Box 8.1.

One cannot help but be struck by the brevity of this code (it contains 24 non-comment commands!), given the years of work that went into the development of the model and the fields of study that it launched. Despite the code's simplicity, the reader should not fail to ponder the extraordinary effort required by Andrew Huxley to integrate this system of nonlinear differential equations using a hand-crank mechanical calculator! For users of modern desktop computers, integration of these equations to practically arbitrary precision is a trivial matter. For example, using MATLAB, one could execute the commands in Box 8.2 to obtain the simulation reported in Figure 8.12.

Box 8.1   MATLAB code for Hodgkin–Huxley model.

```
function [f] = dXdT_HH(t,x,I_app);
% FUNCTION dXdT_HH
%  Inputs: t - time (milliseconds)
%          x - vector of state variables {v,m,n,h}
%          I_app - applied current (microA cm^{-2})
%
%  Outputs: f - vector of time derivatives
%              {dv/dt,dm/dt,dn/dt,dh/dt}
% Resting potentials, conductivities, and capacitance:
V_Na = 115;
V_K  = -12;
V_L  = 10.6;
g_Na = 120;
g_K  = 36;
g_L  = 0.3;
C_m  = 1e-6;
% State Variables:
v = x(1);
m = x(2);
n = x(3);
h = x(4);
% alphas and betas:
a_m = 0.1*(25-v)/(exp((25-v)/10)-1);
b_m = 4*exp(-v/18);
a_h = 0.07*exp(-v/20);
b_h = 1 ./ (exp((30-v)/10) + 1);
a_n = 0.01*(10-v)./(exp((10-v)/10)-1);
b_n = 0.125*exp(-v/80);
% Computing currents:
I_Na = (m^3)*h*g_Na*(v-V_Na);
I_K  = (n^4)*g_K*(v-V_K);
I_L  = g_L*(v-V_L);
% Computing derivatives:
f(1) = (-I_Na - I_K - I_L + I_app)/C_m;
f(2,:) = a_m*(1-m) - b_m*m;
f(3) = a_n*(1-n) - b_n*n;
f(4) = a_h*(1-h) - b_h*h;
```

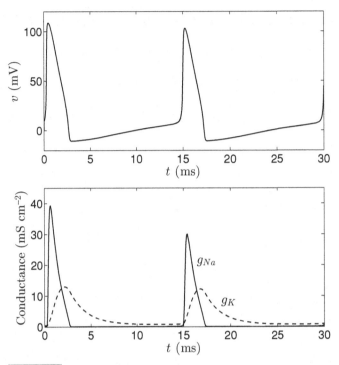

Figure 8.12

Simulated action potential from the Hodgkin–Huxley model. The upper panel plots action potential; the lower panel plots the predicted conductances of the sodium and potassium channels as functions of time. The applied current for these simulations is set to $I_{app} = 6.2 \; \mu A \, cm^{-2}$, for which sustained period firing of the nerve cell is predicted.

Box 8.2   MATLAB executable script for simulation in Figure 8.12.

```
% Initial equilibration with I_app = 0 to
% Generate initial condition xo for simulation:
I_app = 0;
[t,x] = ode15s(@dXdT_HH,[0 30],[0 0 0 0],[],I_app);
xo = x(end,:);
% Add nonzero applied current:
I_app = 6.2;
[t,x] = ode15s(@dXdT_HH,[0 30],xo,[],I_app);
% Plot computed action potential
plot(t,x(:,1));
```

The dynamic behavior of the model is illustrated in Figure 8.12, which shows the predicted membrane potential transients for a value of applied current that results in a periodic train of action potentials. Starting from resting conditions, the applied current causes the membrane to depolarize enough to trigger opening of the sodium conductance. Following activation of the sodium conductance, the membrane hyperpolarizes owing to the associated sodium current. Following hyperpolarization, the sodium channel closes and the potassium channel opens, returning the cell to its resting potential near v_K, which is -12 mV. The next action potential occurs when the membrane is depolarized enough to again trigger the sodium conductivity.

8.2.4 Refractory period

As illustrated in the conductance measurements in Figure 8.7, the sodium conductance in the neuron does not stay open following a steady-state voltage clamp depolarization. Depending on the applied voltage, the conductivity reaches its peak within one or a few milliseconds and decays to zero. In the model, this phenomenon is due to a transient closing of the "h" gate. As illustrated by the steady-state plot (Figure 8.11), the value of h will not be driven to nonzero values until the membrane is repolarized. Because this resetting of the value of h to a nonzero value following an action potential is not instantaneous, there is a finite period, called the *refractory period*, following an action potential, during which it is not possible to elicit the sodium conductance necessary to generate another action potential.

This phenomenon can be explored through model simulations. Figures 8.13 and 8.14 show the predicted membrane potential generated from period trains of input current spikes. In Figure 8.13 the time between current injections is 12 ms, allowing for recovery of $h(t)$ between action potentials. When the period is reduced to 6 ms in Figure 8.14, action potentials occur with alternating input spikes. In between elicited action potentials, the applied current causes a small depolarization, but the sodium conductance is not able to respond to generate a full action potential. (For both simulated cases the input current spikes are modeled as square waves with amplitude 10 mA cm^{-2} and duration 0.5 s.)

8.2.5 The legacy of the Hodgkin–Huxley model

The work of Alan Hodgkin and Andrew Huxley elucidating the biophysical basis of the nerve action potential established and guided the field of cellular biophysics. In

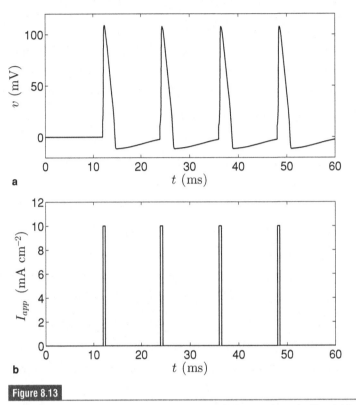

Figure 8.13

Simulated action potential associated with periodic input pulse train. Panel (a) plots action potential associated with the period injections of current illustrated in panel (b). For the given period ($T = 12$ ms) an action potential is elicited at each input spike.

his 1963 Nobel lecture, Andrew Huxley suggested that their "equations should be regarded as a first approximation which needs to be refined and extended in many ways in the search for the actual mechanism of the permeability changes on the molecular scale." Indeed, over the past half century, sophisticated models of cardiac cell electrophysiology and ion handling have been developed, accounting for many more ion currents than the two channels invoked in the Hodgkin–Huxley model, for distinct intracellular compartments, and in many different cardiac cell types and individual animal species. Yet, as cell membrane potential is linked to a wide variety of cellular processes, the basic formulation of Eq. (8.6) is applied in simulating numerous physiological processes, ranging from metabolism to endocrine signaling.

One important component of current research is identifying molecular mechanisms underlying channel kinetics. While the phenomenological representations of Eqs (8.17) and (8.18) for the sodium and potassium conductance kinetics simulate observed behavior with reasonable accuracy, they do not necessarily reveal the

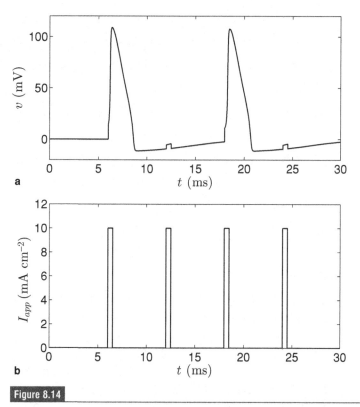

Figure 8.14

Simulated action potential associated with periodic input pulse train. Panel (a) plots action potential associated with the period injections of current illustrated in panel (b). For the given period ($T = 6$ ms) an action potential is elicited for alternating input spikes.

biophysical basis of voltage-dependent conductance gating. In the remainder of this chapter we will explore some simple models developed to explain ion channel kinetics. As in Section 6.2 on enzyme kinetics, the intent is not an exhaustive treatment, but an introduction to basic approaches employed to characterize and simulate mechanisms.

8.3 Models of ion channel gating

Although Eqs (8.12)–(8.14) for sodium channel gating were not developed to represent a specific multistate channel kinetic model (in fact the existence of ion channel membrane proteins was not revealed until the 1970s), these model equations can be thought of as representing the eight-state model illustrated in Figure 8.15(a). In this figure the eight states are labeled using a notation where the first three digits denote the value of three independent m variables, and the last

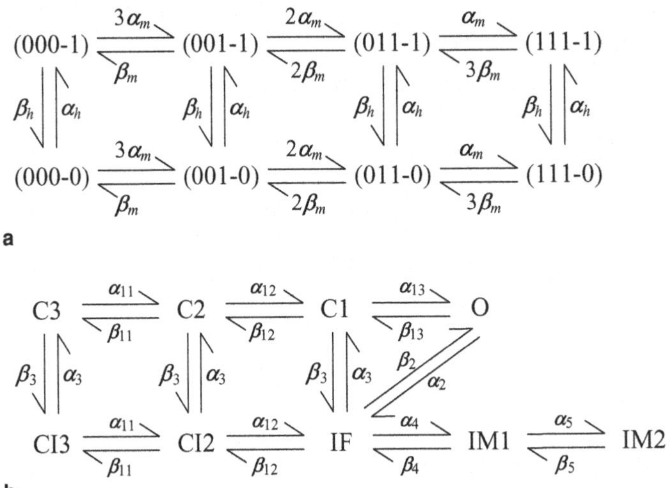

$$(000\text{-}1) \xrightleftharpoons[\beta_m]{3\alpha_m} (001\text{-}1) \xrightleftharpoons[2\beta_m]{2\alpha_m} (011\text{-}1) \xrightleftharpoons[3\beta_m]{\alpha_m} (111\text{-}1)$$

$$(000\text{-}0) \xrightleftharpoons[\beta_m]{3\alpha_m} (001\text{-}0) \xrightleftharpoons[2\beta_m]{2\alpha_m} (011\text{-}0) \xrightleftharpoons[3\beta_m]{\alpha_m} (111\text{-}0)$$

a

$$C3 \xrightleftharpoons[\beta_{11}]{\alpha_{11}} C2 \xrightleftharpoons[\beta_{12}]{\alpha_{12}} C1 \xrightleftharpoons[\beta_{13}]{\alpha_{13}} O$$

$$CI3 \xrightleftharpoons[\beta_{11}]{\alpha_{11}} CI2 \xrightleftharpoons[\beta_{12}]{\alpha_{12}} IF \xrightleftharpoons[\beta_4]{\alpha_4} IM1 \xrightleftharpoons[\beta_5]{\alpha_5} IM2$$

b

Figure 8.15

Models of sodium channel kinetics: (a) the Hodgkin–Huxley model of the sodium channel; (b) the Bébarová *et al.* [9] model of the cardiac fast sodium channel.

denotes the value of the h variable. For example, state 011-1 has two of three m variables in the "1" (open) state, and the h variable in the open "1" state. The channel is open when all gate variables are in the "1" state: that is, state 111-1.

As illustrated in the diagram, the transitions from $h = 0$ to $h = 1$ occur with rate α_h; transitions from $h = 1$ to $h = 0$ occur with rate β_h. Similarly, rate constants α_m and β_m are associated with transitions between $m = 0$ and $m = 1$ states. Because the "m" gates are independent, the rate of transition from 000-0 to 001-0 (and from 000-1 to 001-1) is $3\alpha_m$. The statistical factor 3 appears because in transitioning from a state with three closed "m" gates to a state with two closed "m" gates, there are three closed gates that can open. Analogous statistical factors (1, 2, or 3) multiply all α_m and β_m rate constants in Figure 8.15(a). The assignment of these factors comes from the assumption of independence: because each gate operates independently, the transition from 000-0 to 001-0 is three times as likely (and thus occurs at three times the effective rate) as the transition from 011-0 to 111-0.

With rate constants α_m, β_m, α_h, and β_h defined as functions of voltage as in Eqs (8.15) and (8.16), this model is identical to the sodium channel model of Hodgkin and Huxley.[2]

Bébarová *et al.* [9] developed a nine-state model of the cardiac fast sodium channel (Figure 8.15(b)) to help understand the effects of a loss-of-function mutation resulting in inherited cardiac electrophysiological dysfunction. In the

---

[2] See Exercise 8.6.

model the "O" state is the open conducting state, and all other states are nonconducting.

The rate constants in the Bébarová *et al.* model are functions of membrane voltage given by the following empirical formulae:

$$\alpha_{11} = \frac{1}{0.02701e^{-(20+V)/17.0} + 0.3586e^{-(20+V)/100.0}}$$

$$\alpha_{12} = \frac{1}{0.02701e^{-(20+V)/22.75} + 0.006049e^{-(20+V)/100.0}}$$

$$\alpha_{13} = \frac{1}{0.02701e^{-(20+V)/12.0} + 0.3190e^{-(20+V)/100.0}}$$

$$\beta_{11} = 1.333 \left(e^{-(13.438+V)/20.3} + e^{-(13.438+V)/15.0} \right)$$

$$\beta_{12} = 0.1208 \left(e^{-(19.5+V)/20.3} + e^{-(13.438+V)/15.0} \right)$$

$$\beta_{13} = 0.5276 \left(e^{-(19.0+V)/20.3} + e^{-(13.438+V)/100.0} \right)$$

$$\alpha_3 = \left(1.0254 \times 10^{-6} \right) e^{-V/7.7}$$

$$\beta_3 = 0.0084 + \left(1.209 \times 10^{-5} \right) V$$

$$\alpha_2 = 18.356 \left(e^{(V-10.0)/45.0} + \frac{1}{0.001 + e^{(V+30)/10} + e^{-(V+30)/10}} \right)$$

$$\beta_2 = \frac{\alpha_{13}\alpha_2\alpha_3}{\beta_{13}\beta_3}$$

$$\alpha_4 = \alpha_2$$

$$\beta_4 = \frac{1}{3.523e^{V/182.73} + 2.583e^{(V+64.31)/18.324}}$$

$$\alpha_5 = \frac{\alpha_2}{65610.0}$$

$$\beta_5 = \frac{\beta_4}{14.11}. \tag{8.19}$$

Here the αs and βs are determined in units of ms^{-1} and membrane voltage V is given in units of mV. (Recall that the Hodgkin–Huxley model was expressed in terms of v, the voltage measured relative to the resting potential. Here V represents the membrane potential difference: inside potential minus outside potential.)

The behavior of the Bébarová *et al.* model is compared with experimental data in Figure 8.16. Specifically, the current is assumed to be determined by

$$I_{Na} = f_o(t)\bar{g}_{Na} \left(V - V_{Na} \right), \tag{8.20}$$

where f_o is the fraction in the "O" state, $\bar{g}_{Na} = 70$ mS μF^{-1}, and $V_{Na} = +55$ mV is the Nernst potential for Na$^+$.

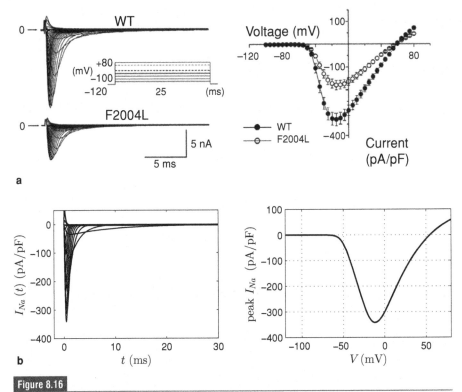

Figure 8.16

Kinetics of Bébarová *et al.* model of the SCN5A cardiac fast sodium channel. (a) Experimental measurements of whole-cell sodium current following voltage steps from initial potential of $V = -120$ mV to clamped values from -100 to $+80$ mV. Also shown is peak current as a function of applied voltage obtained from this protocol. The "WT" data correspond to cells expressing the wild-type channel; the "F2004L" correspond to cells expressing a mutant channel, as described in the text. (b) Simulations of the protocol illustrated above based on the Bébarová *et al.* WT model. Sodium current following voltage steps from initial polarization of $V = -120$ mV to clamped voltages of -100 to $+80$ mV is shown. Figure 8.16(a) reproduced from Bébarová *et al.* [9] with permission.

Model predictions should be compared with data from cells expressing the wild-type (WT) channel. The plots on the left illustrate transient sodium current following voltage clamp protocols described in the figure legend. The plots on the right show peak current as a function of membrane potential in these experiments. We can see that the model behavior effectively matches the observations, particularly on peak I_{Na} versus V. Simulations show that for relatively small depolarizations (such as from -120 to -50) the inactivation time is much longer than observed.[3]

---

[3] See Exercise 8.7.

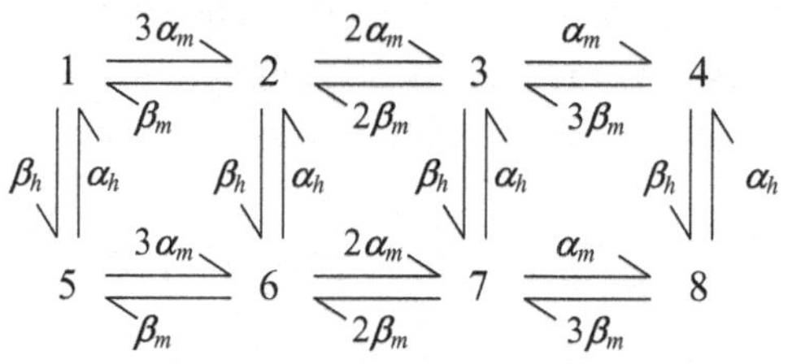

Figure 8.17

Hodgkin–Huxley model of the sodium channel with states labeled 1 through 8.

8.4 Stochastic simulations

The simulations in the preceding sections of this chapter – and, for that matter, in most of this book – are deterministic in nature. However, often in the examples studied the underlying processes simulated, such as chemical reactions or molecular diffusion, emerge as the average behavior of an ensemble of individual components, each of which behaves effectively randomly. The whole-cell currents illustrated, for example, in Figure 8.4 or Figure 8.16, are the sum of currents through thousands of individual channels in the cell membrane. Yet since each individual channel exists at any given instant in either an open conducting state or a closed state, the current through an individual channel varies stochastically with time.

The stochastic operation of individual channels can be simulated based on their kinetic mechanisms, such as those illustrated in Figure 8.15. Specifically, we can simulate the random movement from state to state on a kinetic diagram as a biased random walk in the state space.

A stochastic simulation of a system like this makes use of the *transition rate matrix* $\mathbb{Q}$, which stores the rate constants associated with the state transitions in the kinetic mechanism. For an N-state system, $\mathbb{Q} \in \Re^{N \times N}$ and Q_{ij} is defined as the first-order rate constant for the transition from state i to state j. For example, if we index the states in the Hodgkin–Huxley sodium channel model as 1 through 8, as in Figure 8.17, then

$$
\mathbb{Q} =
\begin{bmatrix}
0 & 3\alpha_m & 0 & 0 & \beta_h & 0 & 0 & 0 \\
\beta_m & 0 & 2\alpha_m & 0 & 0 & \beta_h & 0 & 0 \\
0 & 2\beta_m & 0 & \alpha_m & 0 & 0 & \beta_h & 0 \\
0 & 0 & 3\beta_m & 0 & 0 & 0 & 0 & \beta_h \\
\alpha_h & 0 & 0 & 0 & 0 & 3\alpha_m & 0 & 0 \\
0 & \alpha_h & 0 & 0 & \beta_m & 0 & 2\alpha_m & 0 \\
0 & 0 & \alpha_h & 0 & 0 & 2\beta_m & 0 & \alpha_m \\
0 & 0 & 0 & \alpha_h & 0 & 0 & 3\beta_m & 0
\end{bmatrix}.
\tag{8.21}
$$

Since in the model the transition probabilities depend only on the current state (where the Q_{ij} are first-order rate constants) this system is what is known as a *continuous Markov process* in mathematics. The state probabilities evolve according to

$$\frac{dp_j(t)}{dt} = \sum_{i\neq j} Q_{ij} p_i - \sum_{i\neq j} Q_{ji} p_j, \qquad (8.22)$$

where $p_i(t)$ is the probability that the system is in state i at time t. The first term sums the rates of transition into state j from states $i \neq j$. The second term sums the rates of transition out of state j into other states. Equation (8.22) is known as the *master equation* in the chemical physics literature.

An algorithm for simulating the stochastic movement through the state space is constructed by first recognizing that the transition out of a particular state is a first-order decay process, with a rate constant given by the sum of rate constants for all transitions out of the state. In other words, the dwell time in state i is exponentially distributed

$$f_{T_i}(t) = q_i e^{-q_i t}, \qquad (8.23)$$

where

$$q_i = \sum_{j\neq i} Q_{ij}.$$

For example, for state 3 in Figure 8.17,

$$q_3 = Q_{32} + Q_{34} + Q_{37}$$

$$= 2\beta_m + \alpha_m + \beta_h.$$

Thus the time it takes for the system to transition out of state 3 (into either state 2, state 4, or state 7) is exponentially distributed with mean value

$$\langle f_{T_3}(t)\rangle = \langle q_3 e^{-q_3 t}\rangle = 1/q_3.$$

The relative probability of transitioning from state i to state j is proportional to Q_{ij}. For example, for the starting state of state 3 in Figure 8.17, the probabilities of jumping to state 2, 4, and 7 are given by

$$P_{3\to 2} = Q_{32}/q_3 = 2\beta_m/q_3$$

$$P_{3\to 4} = Q_{34}/q_3 = \alpha_m/q_3$$

$$P_{3\to 7} = Q_{37}/q_3 = \beta_h/q_3.$$

In general, the probability of transition from state i to state j is computed

$$P_{i\to j} = \frac{Q_{ij}}{\sum_{j\neq i} Q_{ij}} = \frac{Q_{ij}}{q_i}. \qquad (8.24)$$

Combining these ideas, an algorithm can be constructed for conducting an appropriately biased random walk around the N-dimensional state space. The basic idea is to, first, determine the amount of time the system stays in a given state, and second, determine to which state the system moves after the dwell time expires. Repeating these steps, the system moves about the state space in a way that mimics the stochastic operation of a single player obeying the kinetic scheme. Given starting state i, the algorithm proceeds as:

1. Draw a random value of the dwell time from the distribution defined in Eq. 8.22. This can be done by drawing r_1 from a uniform distribution on the interval (0, 1) and computing $T_i = -\frac{1}{q_i} \ln r_1$.

2. The next state j ($\neq i$) is chosen from the distribution defined by Eq. (8.24) by drawing a second random number r_2, also from a uniform distribution on (0, 1), and choosing j according to

$$
j = \begin{cases}
1 & \text{if } r_2 \leq P_{i \to 1} \\
2 & \text{if } P_{i \to 1} \leq r_2 < P_{i \to 1} + P_{i \to 2} \\
\vdots & \\
l & \text{if } \sum_{k < l} P_{i \to k} \leq r_2 < \sum_{k \leq l} P_{i \to k} \\
\vdots & \\
N & \text{if } \sum_{k < N} P_{i \to k} \leq r_2 .
\end{cases}
$$

Applying this algorithm to the model of Figure 8.17, with αs and βs defined by Eqs (8.15) and (8.16), we obtain stochastic trajectories, such as those illustrated in Figure 8.18 for two different values of v. The trajectories plotted here represent individual realizations of the stochastic algorithm. If we were to run the algorithm again, with a different initial state and/or different random seed, we would, in general, obtain different results.

These individual realizations reveal some limited insight into how the stochastic model behaves in comparison with the deterministic model. At resting potential ($v = 0$ mV) the channel stays closed. (Recall, state 4 is the open conducting state.) Starting in the initial state 1, when a depolarization of $v = +50$ mV is applied, the channel briefly samples the open state, then remains in closed (inactivated) states 7 and 8 for most of the rest of the simulation. Thus the stationary behavior of the fast sodium channel is not particularly interesting. At a constant applied voltage, the channel remains closed or inactivated.

To illustrate some interesting dynamic behavior, we can simulate a rapid depolarization, analogous to the experiments reported in Figure 8.7. Figure 8.19 illustrates the transient behavior for several individual channels following a depolarization at time $t = 0$ to $v = +50$ mV. The individual channels show transient (and apparently

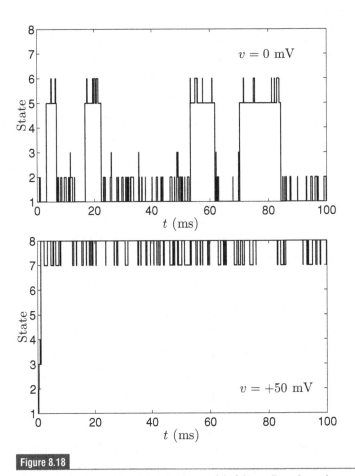

Figure 8.18

Stochastic simulation of Hodgkin–Huxley model of the sodium channel.

purely random) opening and closing. However, a trend emerges when the behavior of a collection of 100 channels is summed. Here we can see that the fraction of channels in state 4, denoted $f_4(t)$, reaches a peak around 1 ms, before inactivating. This behavior can be compared with that observed in Figure 8.7.

8.5 Recapitulation and lessons learned

In this chapter we have been introduced to the basic concepts necessary to understand the biophysics of cell membrane potential dynamics, illustrated how these concepts may be applied to synthesize experimental observations into a simulation of the nerve impulse, and finally explored how Markov models of channel kinetics are constructed and simulated. In a departure from most of the rest of this book, we

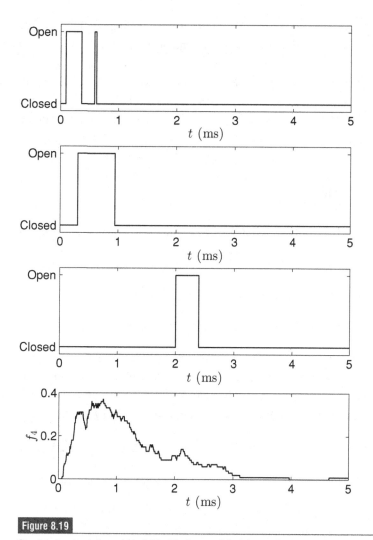

Figure 8.19

Stochastic simulation of Hodgkin–Huxley model of the sodium channel. Upper three panels show single-channel simulations corresponding to a depolarization of $v = +50$ mV. The bottom panel shows the average behavior from 100 individual simulations.

performed a stochastic simulation of a kinetic model, based on drawing random numbers that determine how the state space of a mechanism is sampled kinetically. The application of this kind of algorithm to other multistate mechanisms is straightforward, as the reader is asked to do in Exercise 8.8.

Mathematical modeling and simulation of the cell membrane potential dynamics, and of biophysical phenomena tied to cell membrane potential, is ubiquitous in current applications in computational biology. Membrane potential is fundamentally linked to neurobiology, muscle function, substrate transport, mitochondrial

ATP synthesis, renal solute transport, endocrine signaling, and indeed just about every physiological process. One of the most deeply developed application areas in electrophysiology modeling is cardiac electrophysiology [43], an area that we have not touched on here. In fact, our treatment is rather brief in comparison with others, such as the excellent books by Hille [27] and Keener and Sneyd [35]. Yet, since an important part of the legacy of Hodgkin and Huxley is a close tie between experimental biology and the field of electrophysiology, it is hoped that the exposition here gives the reader what he or she needs to get started in the field.

Problems

8.1   Goldman–Hodgkin–Katz equation. Consider the transport of ions in a homogeneous membrane, governed by the transport equation

$$\frac{\partial c}{\partial t} = -\nabla \cdot \left(\vec{\Gamma}^{drift} + \vec{\Gamma}^{diffusion} \right)$$

$$= -\nabla \cdot \left(-D_m \nabla c - \frac{z}{|z|} u c \nabla \phi \right),$$

where c is the concentration, D_m is the molecular diffusion coefficient, $\nabla \phi$ is the electric field, u is the electrokinetic mobility, and z is the valence of the ions. In steady state, and assuming a uniform electric field across a thin membrane, we have a one-dimensional system:

$$D_m \frac{d^2 c}{dx^2} + \frac{zu}{|z|} \left(\frac{d\phi}{dx} \right) \left(\frac{dc}{dx} \right) = 0$$

$$D_m \frac{d^2 c}{dx^2} - \frac{zu}{|z|} \left(\frac{\Delta \Phi}{d} \right) \left(\frac{dc}{dx} \right) = 0.$$

Using the Einstein relation

$$\frac{u_i}{D_i} = \frac{|z_i| F}{RT},$$

show that flux through the membrane is given by

$$\Gamma = \frac{zDFV}{dRT} \left(\frac{c_0 e^{\frac{zF\Delta\Phi}{RT}} - c_1}{e^{\frac{zF\Delta\Phi}{RT}} - 1} \right), \tag{8.25}$$

where d is the width of the membrane and V is the membrane potential difference. This equation is known as the the *Goldman–Hodgkin–Katz (GHK) equation* for passive flux of an ion through a membrane.

8.2   Goldman–Hodgkin–Katz equation versus linear ohmic behavior. Assuming a value of $V_{Na} = 56$ mV, plot the current–voltage relationships predicted by the GKH equation (see above) and Eq. (8.4). Over what range of V does the linear model approximate the GHK model?

8.3 Show that the kinetic constants in Eqs (8.19) obey the catalytic-cycle thermodynamic constraint described in Section 6.2.2.

8.4 Computer exercise. Use computer simulation of the Hodgkin–Huxley model to determine the relationship between period and applied (constant) current. What happens to the magnitude of the action potential as applied current is increased? Why?

8.5 Computer exercise. What is the length of the refractory period predicted by the Hodgkin–Huxley model? That is, given a current spike that elicits an action potential, how long does it take $h(t)$ to recover enough that the model can generate another action potential? Can the refractory period be estimated from the model equations without performing a simulation?

8.6 Computer exercise. Perform a computer simulation of the eight-state model shown in Figure 8.15(a). Show that the model behavior is identical to that of Eqs (8.12)–(8.14).

8.7 Computer exercise. Show that for relatively small depolarizations (e.g., from $V = -120$ to -50 mV), the Bébarová *et al.* model described in Section 8.3 predicts that inactivation following depolarization is slower than that observed experimentally. Specifically, simulate the behavior of the system for the experimental protocol described in Figure 8.16. How long does it take for the current to inactivate for an applied voltage of -50 mV? What other experiments of Bébarová *et al.* [9] can you simulate? Can you find a parameterization of Eqs (8.19) that improves the match between model simulation and experiments?

8.8 Computer exercise. Use the algorithm of Section 8.4 to simulate the stochastic fluctuations of a single Michaelis–Menten enzyme with mechanism

$$E \underset{k_{-1}}{\overset{k_{+1}[S]}{\rightleftharpoons}} ES \overset{k_{+2}}{\longrightarrow} E + P.$$

Assume the substrate concentration [S] remains constant, yielding a two-state ($N = 2$) system. How does the behavior depend on the values of the rate constants and substrate concentrations? How does the behavior compare with the associated deterministic model?

9 Appendices: mathematical and computational techniques

Overview

This set of appendices describes mathematical and computation tools used throughout this text. This material is appended to this book as a reference for derivations and details of computer implementation that are not required to follow the scientific content in the previous chapters.

9.1 Finite-difference approximations for continuous processes

In Section 1.2.1 of the introductory chapter we used a *finite difference* approximation to estimate the rates of change of concentration variables from discretely sampled data. Later in this chapter we will see how approximate solutions to differential equations are obtained numerically using finite differences. In both applications a finite (or discrete) difference is used to approximate the derivative of a continuous variable.

Recall the definition of the continuous differentiation operator:

$$\frac{df}{dx} = \lim_{\Delta x \to 0} \frac{f(x + \Delta x) - f(x)}{\Delta x}, \qquad (9.1)$$

where $f(x)$ is a smooth function of the continuous variable x. By this we mean that $f(x)$ does not make any instantaneous changes as x is changed infinitesimally. If $f(x)$ is a smooth function, then the limit in Eq. (9.1) exists and is well behaved.

The continuous derivative is defined in defined in Eq. (9.1) as the limiting behavior of a finite difference. From this definition we can formulate a finite approximation simply by taking the ratio in this equation using a finite value for Δx:

$$\frac{df}{dx} \approx \frac{f(x + \Delta x) - f(x)}{\Delta x}, \qquad (9.2)$$

an approximation that is illustrated in Figure 9.1. One can see from this figure that the closer Δx is to 0, the better the finite difference approximation matches the continuous derivative.

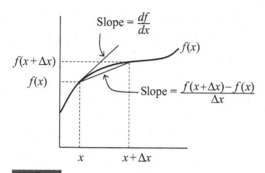

Figure 9.1

Finite-difference approximation to the derivative of a function $f(x)$ with respect to x.

The approximation of the time derivative that is used in Section 1.2.1 is a central-difference approximation

$$\frac{df}{dx} \approx \frac{f(x + \Delta x) - f(x - \Delta x)}{2\Delta x},$$

which is equivalent to Eq. (9.1) for smooth functions. In Chapter 1 the independent variable is t rather than x and the discrete time step is Δt. Later in this chapter (Section 9.6) we use a similar approximation for a second derivative (to simulate the model developed in Section 2.4.2):

$$\frac{d^2 f(x)}{dx^2} = \frac{d}{dx}\left(\frac{df(x)}{dx}\right) \approx \frac{d}{dx}\left[\frac{f(x + \Delta x) - f(x)}{\Delta x}\right]$$

$$\approx \frac{f(x - \Delta x) - 2f(x) + f(x + \Delta x)}{(\Delta x)^2}. \qquad (9.3)$$

In Section 1.2.2 we demonstrated that applying finite differences to noisy data can be messy. To see why, consider a series of data measured at discrete intervals in time:

$$\{f_1, f_2, \ldots, f_N\}.$$

Further consider that the measured data are made up of a nonrandom variable plus random noise:

$$\{f_1 + n_1, f_2 + n_2, \ldots, f_N + n_N\},$$

where f_i is the value of the nonrandom variable and n_i is the added noise at that ith time point. We assume that the noise is a stationary process with

$$\langle n_i \rangle = 0$$

$$\langle n_i^2 \rangle = \sigma^2,$$

where σ^2 is the variance of the noise. If Δt is the sampling time step, then a finite-difference approximation for the derivative is

$$\frac{f_i}{dt} \approx \frac{f_{i+1} + n_{i+1} - f_i - n_i}{\Delta t} = \frac{f_{i+1} - f_i}{\Delta t} + \frac{n_{i+1} - n_i}{\Delta t}. \tag{9.4}$$

The second term in Eq. (9.4) has the properties

$$\left\langle \left(\frac{n_{i+1} - n_i}{\Delta t} \right) \right\rangle = 0$$

$$\left\langle \left(\frac{n_{i+1} - n_i}{\Delta t} \right)^2 \right\rangle = \frac{2\sigma^2}{(\Delta t)^2}.$$

Thus in the limit $\Delta t \to 0$, where the first term in Eq. (9.4) approximates df/dt, the noise term become infinite. This is why the random noise is greatly magnified by the operation of differentiation; this is why the rates approximated from noisy data in Figure 1.3 yield unintelligible results. (The reader can verify that the central-difference approximation used in Section 1.2.2 is actually slightly less sensitive to the added noise than the difference approximation used in Eq. (9.4).)

9.2 Least-squares solution to $\mathbb{A}x = b$

The general algebra problem $\mathbb{A}x = b$, where $\mathbb{A} \in \mathfrak{R}^{m \times n}$, $x \in \mathfrak{R}^{n \times 1}$, and $b \in \mathfrak{R}^{m \times 1}$, is ubiquitous in computational mathematics. When $n > m$ we have more equations (m) than unknowns (n), and it is possible (and likely) that no solution x exists to satisfy $\mathbb{A}x = b$. When no solution exists, only approximate solutions may be sought.

One approach is to seek the x that minimizes the 2-norm of the difference between $\mathbb{A}x$ and b:

$$\|\mathbb{A}x - b\|^2 = \sum_i \left(\sum_j A_{ij} x_j - b_i \right)^2$$

$$= \sum_i \left(\sum_j A_{ij} x_j - b_i \right) \left(\sum_k A_{ik} x_k - b_i \right)$$

$$= \sum_i \left(\sum_j \sum_k A_{ij} A_{ik} x_j x_k - 2 b_i \sum_j A_{ij} x_j + b_i^2 \right).$$

(The operation $\|u\|^2$ is defined as the sum of squares of the entries of the vector: $\|u\|^2 = u^T u = \sum_i u_i^2$.) Therefore the x that minimizes $\|\mathbb{A}x - b\|^2$ is called the *least-squares solution*.

The optimal value of $\mathbf{x}$ is obtained by setting the gradient $\frac{\partial}{\partial x_l}\|\mathbb{A}\mathbf{x}-\mathbf{b}\|^2$ to zero:

$$\sum_i \frac{\partial}{\partial x_l}\left(\sum_j\sum_k A_{ij}A_{ik}x_jx_k\right) - \sum_i \frac{\partial}{\partial x_l}\left(2b_i\sum_j A_{ij}x_j + b_i^2\right) = 0.$$

The first term (inside the sum $\sum_i$) is

$$\frac{\partial}{\partial x_l}\left(\sum_j\sum_k A_{ij}A_{ik}x_jx_k\right) = \sum_{j\neq l} A_{ij}A_{ik}x_j + \sum_{k\neq l} A_{ij}A_{ik}x_k + 2A_{il}A_{il}x_l$$

$$= 2\sum_j A_{ij}A_{il}x_j$$

and the second term is

$$-\frac{\partial}{\partial x_l}\left(2b_i\sum_j A_{ij}x_j + b_i^2\right) = -2b_iA_{il}.$$

Putting it all together, we have

$$\frac{\partial\|\mathbb{A}\mathbf{x}-\mathbf{b}\|^2}{\partial x_l} = \sum_i\left(2\sum_j A_{ij}A_{il}x_j - 2b_iA_{il}\right) = 0,$$

or

$$\sum_j\sum_i A_{ij}A_{il}x_j = \sum_i A_{il}b_i,$$

or, in compact notation,

$$\mathbb{A}^T\mathbb{A}\mathbf{x} = \mathbb{A}^T\mathbf{b}. \tag{9.5}$$

For example applications of this formula, see Section 1.2.2 and Exercises 1.1 and 1.2 of Chapter 1.

9.3   Using computers to integrate ordinary differential equations

Many models of biological systems take the form of systems of ordinary differential equations. (Ordinary differential equations have one independent variable, such as time.) In addition, a common strategy for using computers to solve partial differential equations (differential equations with more than one independent variable, such as time and space variables) is to approximate their behavior with ordinary differential equations that can be integrated computationally. This is because there exist a number of powerful and reliable packages for integrating ordinary differential equations.

Here we show how to integrate the bioreactor example of Section 2.3.1 using the MATLAB computing environment.[1]

The differential equations for the model of Section 2.3.1 are

$$\frac{d[\text{xylose}]}{dt} = -J_1$$

$$\frac{d[\text{xylitol}]}{dt} = J_1 - J_2$$

$$\frac{d[\text{xylulose}]}{dt} = J_2 - 2J_3 - 2J_6$$

$$\frac{d[\text{acetaldehyde}]}{dt} = 3J_3 - J_4 - J_5$$

$$\frac{d[\text{ethanol}]}{dt} = J_4$$

$$\frac{d[\text{acetate}]}{dt} = J_5$$

$$\frac{d[\text{glycerol}]}{dt} = 3J_6, \tag{9.6}$$

where the flux terms J_1 through J_6 appearing on the right-hand side are functions of the concentration state variable and a set of kinetic parameters, as detailed in Section 2.3.1.

To integrate these equations, we construct a MATLAB function that computes the time derivatives of the state variables at any given time. For this application we use the function in Box 9.1.

This code accepts three inputs (t, x, and par) and computes one output, f. The inputs are

t: This is a scalar number representing time. Although this variable is not used in computing the time derivatives, the formal syntax of the solver package requires that the time variable appear as the first input of the derivative function.

x: This is a vector that lists the state variables of the model; the values of the seven concentration variables are input in the vector x. To make the code easy for a

---

[1] While MATLAB is a convenient framework both for real applications and for illustrating basic concepts, there are many other software packages that have the ability to integrate ordinary differential equations; the reader is encouraged to compute in whatever environment he or she chooses.

Box 9.1   MATLAB code for time derivatives of Eq. (9.6).

```
function [f] = dXdT(t,x,par);

k1  = par(1);
k2  = par(2);
k3  = par(3);
k4  = par(4);
k5  = par(5);
k6  = par(6);
km2 = par(7);
km3 = par(8);

xylose        = x(1);
xylitol       = x(2);
xylulose      = x(3);
acetaldehyde  = x(4);
ethanol       = x(5);
acetate       = x(6);
glycerol      = x(7);

J1 = k1*xylose;
J2 = k2*xylitol - km2*xylulose*ethanol;
J3 = k3*xylulose - km3*acetaldehyde*ethanol;
J4 = k4*acetaldehyde;
J5 = k5*acetaldehyde;
J6 = k6*xylulose;

f(1) = -J1;
f(2) = J1 - J2;
f(3) = J2 - 2*J3 - 2*J6;
f(4) = 3*J3 - J4 - J5;
f(5) = J4;
f(6) = J5;
f(7) = 3*J6;
f = f';
```

---

human to read, variables named `xylose`, `xylitol`, etc. are assigned the values of `x(1)` through `x(7)`.

`par`: The eight parameter values are input in the vector `par`. The parameters k_1, k_2, k_3, k_4, k_5, k_6, k_{-2}, and k_{-3} are assigned names k1, k2, k3, k4, k5, k6, km2, and km3 in the code.

Box 9.2   MATLAB code to integrate model of Eq. (9.6).

```
% 1. Parameter values:
par = [7.67e-3 3.60 0.065 0.867 0.045 1.15e-3 88 99];
% 2. Initial conditions:
xo = [0.10724 0 0 0 0 0 0];
% 3. Integrate equations:
[t,x] = ode15s(@dXdT,[0 250],xo,[],par);
```

The function dXdT outputs the vector f, which lists the values of the derivatives in Eq. (9.6) computed for the values of x and par input to the function. See how Eqs (2.40) and (2.41) appear in the code.

To obtain the time courses in Figure 2.7 requires calling a function that integrates the equations coded in dXdT. One convenient function in MATLAB is ode15s, which can be called using the syntax in Box 9.2.

The first command in Box 9.2 assigns values to the parameters. (These values are listed in Table 2.2.) The second command assigns initial values to the seven state variables. The third command calls the function ode15s to integrate the differential equations. Calling ode15s requires passing the name of the function (dXdT) that we have written to compute the time derivatives, the time window ([0 250]) to simulate, in initial conditions (xo), and the parameters (par). The empty vector ([]) in the function call is a placeholder for optional arguments that are not used here. (A full description of ode15s and other integrators can be found in the MATLAB documentation. Try typing help ode15s or doc ode15s at the MATLAB command prompt.)

Simulated time courses of the state variables are output in the array x and can be plotted versus time using the plot command. For example, the time course in Figure 2.7 can be plotted with the commands in Box 9.3, which generate the plot in Figure 9.2 to illustrate the simulate time course.

9.4   Optimization for parameter estimation

Fitting model simulations to experimental data is the standard approach to estimating the values of model parameters. By "fitting" we mean that the model is adjusted to match experimental data as closely as possible. A useful approach is to minimize the sum of squared differences between the model and the data. For the special case when model predictions to be compared with data can be expressed as linear combinations of the unknown parameters, then the linear least-squares

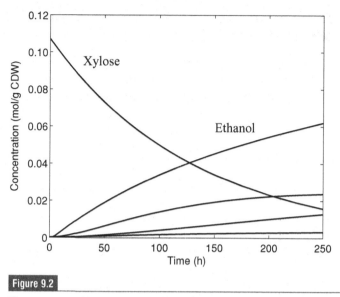

Figure 9.2

Figure generated by simulation in Section 9.3.

Box 9.3   MATLAB code to generate plots of time course in Figure 2.7.

```
plot(t,x(:,1),'k-','linewidth',1.5); hold on;
plot(t,x(:,2),'k-','linewidth',1.5);
plot(t,x(:,5),'k-','linewidth',1.5);
plot(t,x(:,6),'k-','linewidth',1.5);
plot(t,x(:,7),'k-','linewidth',1.5); hold off;
set(gca,'Fontsize',14);
xlabel('Time (hours)');
ylabel('Concentration (mol/g CDW)');
text(30,0.095,'xylose','interpreter','latex',
 'Fontsize',18);
text(150,0.06,'ethanol','interpreter','latex',
 'Fontsize',18);
```

algorithm of Section 9.2 can be applied to estimate the parameter values. More typically, as in the example from Section 2.3, the model is constructed as a system of nonlinear differential equations. These equations are integrated using a computer package in Section 9.3. Similarly, we can use a computer package to estimate the model parameters. Just as the main task to integrate the equations was to construct a program to compute the derivatives of the state variables, the major task here will be to construct a program to compute the function to be optimized (minimized).

Box 9.4 MATLAB code to computer mean-squared error for model of Section 2.3.

```
function [E] = mse(par,t,d1,d2,d5,d6,d7);
% FUNCTION MSE(par,t,d1,d2,d5,d6,d7)
%   Inputs: par - list of parameter values
%           t - time vector for data
%           d1 - xylose data
%           d2 - xylitol data
%           d5 - ethanol data
%           d6 - acetate data
%           d7 - glycerol data
% Step 1. Integrate ODE model
xo = [d1(1) 0 0 0 0 0 0];
[t,x] = ode15s(@dXdT,t,xo,[],par);
% Step 2. Compute error
E = sum( (x(:,1)-d1').^2  ) / sum(d1.^2) + ...
    sum( (x(:,2)-d2').^2  ) / sum(d1.^2) + ...
    sum( (x(:,5)-d5').^2  ) / sum(d1.^2) + ...
    sum( (x(:,6)-d6').^2  ) / sum(d1.^2) + ...
    sum( (x(:,7)-d7').^2  ) / sum(d1.^2) ;
```

The code in Box 9.4 provides a MATLAB function to serve as the *objective function* for the data and model of Section 2.3. This function calls the ode15s integrator to generate simulations to compare with the measured data. The error is computed as the sum of square differences between the model and the data.

The first input to the function mse are par, the vector of parameter values to be estimated. The output is the scalar E. The function is made up of two steps, enumerated by the comments in the code. (Comment lines are preceded by the % symbol.) The first step is to integrate the ordinary differential equations to obtain model predictions at the time points provided in the input vector t. (This ODE integration uses the dXdT function that was built in Section 9.3.) Once the model equations are integrated, the second step is to compute the sum of square differences between the input data (d1, d2, d5, d6, and d7) and the model predictions.

Given the function mse that computes the objective function to be minimized, what is left is to apply an optimization routine to determine optimal parameter values. The built-in MATLAB function fmincon is one possible optimizer to use. This function is called to minimize the output of mse using the script in Box 9.5.

> **Box 9.5   MATLAB code to estimate parameters for model of Section 2.3.**
>
> ```
> % Input data
> t  = [     0    34.0    82.5     154     202     250];
> d1 = [0.1072  0.0846  0.0589  0.0332  0.0211  0.0139];
> d2 = [     0  0.0060  0.0127  0.0197  0.0215  0.0230];
> d5 = [     0  0.0135  0.0276  0.0475  0.0532  0.0621];
> d6 = [     0  0.0010  0.0017  0.0025  0.0026  0.0028];
> d7 = [     0  0.0028  0.0055  0.0090  0.0099  0.0110];
> % Initial parameter guesses
> par0 = [0.01 2 0.1 1.0 0.1 0.001 90 90]% Optimization
> opts = OPTIMSET('display','iter');
> ep = 1e-6;
> LB = ([ep ep ep ep ep ep ep ep]);
> UB = (100*[1 1 1 1 1 1 1 1]);
> par = fmincon(@mse,par0,[],[],[],[],LB,UB,[],opts,t,d1,
>   d2,d5,d6,d7)
> ```

The function `fmincon` accepts a number of linear and nonlinear constraints on the possible values of the estimated parameters. In the syntax used here we have applied sets of upper and lower bounds, defined by the vectors `LB` and `UB`. Many more options, as well as additional optimization routines, can be found in the MATLAB documentation.

The above script returns the following values for the vector `par`: {0.0077, 3.8463, 0.0512, 1.0773, 0.0563, 0.0012, 94.9608, 94.9986}, values that are close to those tabulated in Table 2.2 for the BP000 strain. One can verify that while the parameter values are not identical, the fit to the data obtained with these values is indistinguishable from that of Figure 2.7, indicating that the parameter estimates are not entirely unique.

9.5 The method of lines for the one-dimensional advection equation

The method of lines (also called the *method of characteristics*) is a scheme for solving one-dimensional (that is, with one space dimension) hyperbolic equations, such as the one-dimensional advection equation. In one dimension Eq. (2.51) becomes

$$\frac{\partial}{\partial t}c_i = -v\frac{\partial c_i}{\partial x} + r_i(\mathbf{c}(x,t),x,t)$$

or

$$\frac{\partial}{\partial t}\mathbf{c} = -v\frac{\partial \mathbf{c}}{\partial x} + \mathbf{r}(\mathbf{c}(x,t), x, t), \qquad (9.7)$$

where v is the (assumed constant) velocity in the x-direction, and in general a chemical rate r_i may depend on all the concentrations, expressed as a vector $\mathbf{c}$. This partial differential equation can be transformed to an ordinary differential equation[2] using the variable substitution $\zeta = x + vt + a$, where a is an arbitrary constant. This substitution yields

$$\frac{dc_i}{d\zeta} = \frac{\partial c_i}{\partial x}\frac{dx}{d\zeta} + \frac{\partial c_i}{\partial t}\frac{dt}{d\zeta}$$

$$= \frac{\partial c_i}{\partial x} + \frac{1}{v}\frac{\partial c_i}{\partial t} = \frac{r_i}{v}. \qquad (9.8)$$

What this all means is that the variable substitution $\zeta = x + vt + a$ defines a direction $(dx/d\zeta, dt/d\zeta) = (1, 1/v)$ in the (x, t) plane along which $\mathbf{c}$ varies according to the ordinary differential equation $d\mathbf{c}/d\zeta = \mathbf{r}/v$. These characteristic lines are parallel to $x = vt$, as illustrated in Figure 9.3. In the figure, the limits of integration are defined to be $\zeta = \zeta_0$ to ζ_1. The concentration $\mathbf{c}$ at any point in (x, t) can be found by integrating the ordinary differential equation in ζ along the appropriate characteristic line, and starting with the appropriate $\mathbf{c}(\zeta_0)$ condition.

The characteristic lines illustrated in Figure 9.3 show how the appropriate initial condition for the equation $d\mathbf{c}/d\zeta = \mathbf{r}/v$ is determined for each position in the (x, t) plane. Here, it is assumed that the original problem based on Eq. (9.7) has an initial condition $\mathbf{c}(x, t = 0)$ specified for positions $x \geq 0$, and a boundary condition $\mathbf{c}(x = 0, t)$ for times $t > 0$. Thus at time t_1 and position x_1, the initial condition (in ζ) along a characteristic depends on the initial condition for $x_1 \geq vt_1$ and on the boundary condition for $x_1 < vt_1$.

This makes sense if one realizes that information travels at a finite speed in this advection system. The solution at a position $x_1 > vt_1$ cannot depend on a condition at $(x = 0, t > 0)$, because no fluid from position $x = 0$ will have traveled to positions $x_1 > vt_1$ at time t_1. Thus solutions $\mathbf{c}(\zeta)$ at $(x_1 \geq vt_1, t > 0)$ trace back to the initial condition along $(x, t = 0)$. Solutions at $(x_1 < vt_1, t > 0)$ trace back to the initial condition at $(x = 0, t)$.

Figure 9.3 illustrates how ζ_0 and ζ_1 may be defined for the two regimes $x_1 \geq vt_1$ and $x_1 < vt_1$. It also illustrates an intermediate point $\zeta_0 < \zeta' < \zeta_1$, and shows how an intermediate time t' depends on ζ' for each case. Formally the solution $\mathbf{c}$ at any

---

[2] The partial differential equation (2.51) has two independent variables, x and t. An ordinary differential equation system such as Eq. (9.7) is much more straightforward to integrate computationally than a partial differential equation system.

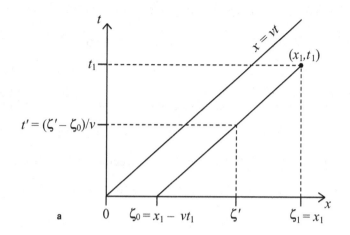

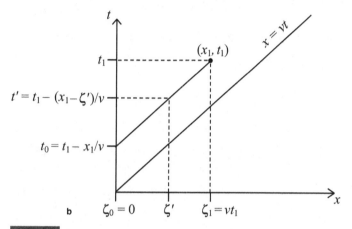

Figure 9.3

Illustration of the method of lines for the equation $\frac{\partial \mathbf{c}}{\partial t} = -v\frac{\partial \mathbf{c}}{\partial x} + \mathbf{r}$, which is equivalent to $\frac{\partial \mathbf{c}}{\partial \zeta} = \mathbf{r}/v$ with $\zeta = x + vt$: (a) $x_1 \geq vt_1$; (b) $x_1 < vt_1$. The solution $\mathbf{c}(x_1, t_1)$ is found by integrating from $\zeta = \zeta_0$ to $\zeta = \zeta_1$ along the appropriate line in the (x, t) plane.

point (x_1, t_1) is obtained by integrating

$$v\frac{d\mathbf{c}}{d\zeta} = \mathbf{r}(\mathbf{c}(\zeta), \zeta, t(\zeta)), \quad \zeta_0 \leq \zeta \leq \zeta_1, \tag{9.9}$$

where the ζ-range is defined

$$\zeta_0 = \begin{cases} x_1 - vt_1, & x_1 \geq vt_1 \\ 0, & x_1 < vt_1 \end{cases}$$

$$\zeta_1 = \begin{cases} x_1, & x_1 \geq vt_1 \\ vt_1, & x_1 < vt_1 \end{cases} \tag{9.10}$$

and the function $t(\zeta)$ is defined

$$t(\zeta) = \begin{cases} (\zeta - \zeta_0)/v, & x_1 \geq vt_1 \\ t_1 - (x_1 - \zeta)/v, & x_1 < vt_1 \end{cases}.$$

Solutions $\mathbf{c}(\zeta_1)$ (or $\mathbf{c}(x_1, t_1)$) are obtained by integrating Eq. (9.9) from $\zeta = \zeta_0$ to ζ_1. For $x_1 \geq vt_1$ the initial condition is specified by the initial condition $\mathbf{c}(x, 0)$:

$$\mathbf{c}(\zeta_0) = \mathbf{c}(x_1 - vt_1, 0).$$

For $x_1 < vt_1$, the initial condition is specified by the boundary condition $\mathbf{c}(0, t)$:

$$\mathbf{c}(\zeta_0) = \mathbf{c}(0, t_1 - x_1/v).$$

9.6 Finite-difference approximation for simulating a FRAP experiment

This section shows how to simulate the model of the fluorescence recovery after photobleaching (FRAP) experiments described in Section 2.4.2 of Chapter 2.

9.6.1 Simulating two-dimensional diffusion

As a first step (not considering bleaching or exchange), we wish to simulate diffusion in a two-dimensional system governed by the diffusion equation

$$\frac{\partial c_m(x, y, t)}{\partial t} = D\nabla^2 c_m(x, y, t).$$

One method to approximate partial differential equations such as this one using computer simulation is to represent a spatial field, such as $c_m(x, y, t)$, on a discrete grid. For the FRAP simulations of Section 2.4.2 we define $C_{i,j}(t)$ on a Cartesian grid where $i, j \in 1, 2, \ldots, N$, as illustrated in Figure 9.4. The grid is defined to have spacings h_x and h_y in the x- and y-directions, respectively. With an equal number of grid steps (N) in both directions, the total number of grid points is $N \times N$. The total area simulated is chosen to approximate the total cell membrane area, which is set to $A_m = 78.5$ μm^2, according to Vinnakota et $al.$ [64]. We set $h_x = h_y$, and thus $N = \sqrt{A_m}/h_x$. In our simulations we use a grid size of $N = 80$ and $h_x = h_y \approx 0.11$ μm.

The Laplacian $\nabla^2 c_m = \left(\partial^2/\partial x^2 + \partial^2/\partial y^2\right)c_m$ may be approximated on the two-dimensional Cartesian grid as [37]

$$(\nabla^2 c_m)_{i,j} \approx \frac{1}{h_x^2}\left[C_{i-1,j} - 2C_{i,j} + C_{i+1,j}\right] + \frac{1}{h_y^2}\left[C_{i,j-1} - 2C_{i,j} + C_{i,j+1}\right],$$

$$(9.11)$$

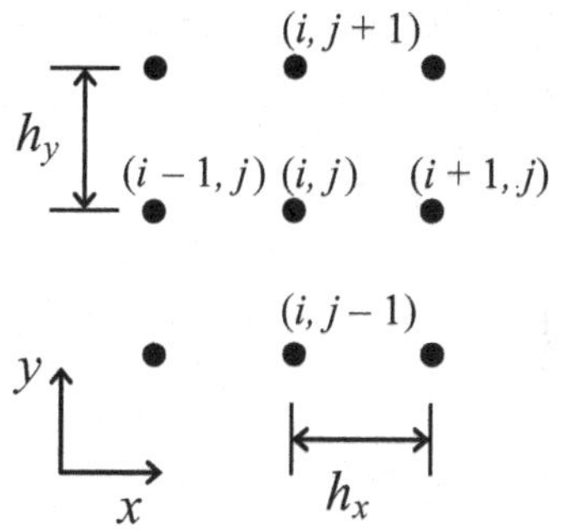

Figure 9.4

Two-dimensional finite difference grid with grid spacing h_x and h_y in the x- and y-directions, respectively.

for an interior grid point – that is, a grid point not on the boundary of the numerical domain.

For simulations of the FRAP experiment of Figure 2.15 we apply periodic boundary conditions on the numerical domain. This boundary condition ensures that the diffusion operator (defined below) is mass-conserving, and this is a reasonable assumption as long as the FRAP simulations are not significantly influenced by boundary effects. To apply period boundary conditions, we construct the numerical Laplace operator so that grid points on the boundary interact with the virtual image of the numerical domain translated across each boundary. For example, for a grid point on the $i = 1$ boundary (with $j \neq 1, N$), the numerical Laplace operator with periodic boundary conditions is

$$(\nabla^2 c_m)_{1,j} \approx \frac{1}{h_x^2}\left[C_{i+N-1,j} - 2C_{1,j} + C_{i+1,j}\right] + \frac{1}{h_y^2}\left[C_{1,j-1} - 2C_{1,j} + C_{1,j+1}\right].$$

(9.12)

Similarly, for $i = N$ ($j \neq 1, N$):

$$(\nabla^2 c_m)_{N,j} \approx \frac{1}{h_x^2}\left[C_{N-1,j} - 2C_{N,j} + C_{1,j}\right] + \frac{1}{h_y^2}\left[C_{N,j-1} - 2C_{N,j} + C_{N,j+1}\right].$$

(9.13)

We define the general Laplace operator (on all grid points) accounting for periodic boundary conditions:

$$(\nabla^2 c_m)_{i,j} \approx \frac{1}{h_x^2}\left[C_{i+i^-,j} - 2C_{i,j} + C_{i+i^+,j}\right] + \frac{1}{h_y^2}\left[C_{i,j+j^-} - 2C_{i,j} + C_{i,j+j^+}\right],$$

(9.14)

where the offsets i^-, i^+, j^-, and j^+ are defined at each grid point (i, j):

$$i^- = \begin{cases} -1, & i \in 2 \dots N \\ N-1, & i = 1 \end{cases}$$

$$i^+ = \begin{cases} +1, & i \in 1 \dots N-1 \\ -(N-1), & i = N \end{cases}$$

$$j^- = \begin{cases} -1, & j \in 2 \dots N \\ N-1, & j = 1 \end{cases}$$

$$j^+ = \begin{cases} +1, & j \in 1 \dots N-1 \\ -(N-1), & j = N \end{cases}. \tag{9.15}$$

To make our notation more compact, let us store all of the $C_{i,j}(t)$ in a vector $\mathbf{u}(t)$, such that

$$u_k(t) = C_{i,j}(t), \ k \in 1, \dots N \times N$$

for

$$k(i, j) = i + N(j-1), \ i, j \in 1, \dots, N.$$

To compute the two-dimensional numerical Laplace operator on $\mathbf{u}$, we define the operator $\mathbb{L}$ so that $\mathbb{L}\mathbf{u}$ approximates the Laplacian operation on $\mathbf{u}$. For each grid point (i, j) the corresponding index k of the $\mathbf{u}$ vector is computed according to the above formula for $k(i, j)$. Applying Eq. (9.14), we have (for $i, j \in 1, \dots, N$)

$$L_{k(i,j),k(i,j)} = -\left(2/h_x^2 + 2/h_y^2\right)$$
$$L_{k(i,j),k(i+i^-,j)} = +1/h_x^2$$
$$L_{k(i,j),k(i+i^+,j)} = +1/h_x^2$$
$$L_{k(i,j),k(i,j+j^-)} = +1/h_y^2$$
$$L_{k(i,j),k(i,j+j^+)} = +1/h_y^2, \tag{9.16}$$

where the offsets i^-, i^+, j^-, and j^+ are defined above. All other entries not defined above are $L_{k,l} = 0$.

With the numerical Laplace operator $\mathbb{L}$ the partial differential equation

$$\frac{\partial c_m}{\partial t} = D\nabla c_m$$

is approximated on a discrete grid by the ordinary differential equation

$$\frac{d\mathbf{u}}{dt} = D\mathbb{L}\mathbf{u}. \tag{9.17}$$

To solve this equation approximately, we can use an implicit time-stepping scheme:

$$\frac{\mathbf{u}(t + \Delta t) - \mathbf{u}(t)}{\Delta t} \approx \frac{D}{2}\left[\mathbb{L}\mathbf{u}(t) + \mathbb{L}\mathbf{u}(t + \Delta t)\right], \tag{9.18}$$

which is a second-order accurate approximation to Eq. (9.17) [37]. Rearranging yields

$$\left(\mathbb{I} - \frac{D}{2\Delta t}\mathbb{L}\right)\mathbf{u}(t + \Delta t) = \left(\mathbb{I} + \frac{D}{2\Delta t}\mathbb{L}\right)\mathbf{u}(t), \tag{9.19}$$

where $\mathbb{I}$ is the identity operator. This equation is a linear system that is solved at each time point to compute $\mathbf{u}(t + \Delta t)$ from $\mathbf{u}(t)$.

9.6.2 Simulating diffusion and reaction

To simulate the bleaching process – Eq. (2.63) – we use an operator-splitting method. This simply means that we simulate the diffusion and reaction processes separately for a finite time step Δt. First we use Eq. (9.19) to simulate the diffusion process:

$$\left(\mathbb{I} - \frac{D}{2\Delta t}\mathbb{L}\right)\tilde{\mathbf{u}}(t + \Delta t) = \left(\mathbb{I} + \frac{D}{2\Delta t}\mathbb{L}\right)\mathbf{u}(t),$$

where $\tilde{\mathbf{u}}(t + \Delta t)$ is an intermediate variable. Next we operate on $\tilde{\mathbf{u}}(t + \Delta t)$ to simulate the reaction processes. For the linear first-order bleaching process of Eq. (2.63), the reaction kinetics can be expressed exactly:

$$u_k(t + \Delta t) = \tilde{u}_k(t + \Delta t) \cdot \exp\left[-\alpha K(x(k), y(k))\Delta t\right],$$

which is the exact solution to

$$\frac{d\mathbf{u}}{dt} = -\alpha K(x, y)\mathbf{u}$$

at $t + \Delta t$ for $\mathbf{u}(t) = \tilde{\mathbf{u}}(t + \Delta t)$. In the above expression $x(k)$ and $y(k)$ are the x and y positions associated with the index k, and $K(x, y)$ is given by Eq. (2.64).

The following MATLAB code sets up the diffusion operator $D\mathbb{L}$ on a two-dimensional grid, simulates the FRAP processes without exchange using the numerical algorithm developed above, plots the concentration fields illustrated in Figure 2.16, and generates $F(t)$ from Eq. (2.65) for the 1 μm × 1 μm bleach window.

```
% Define rectangular 2D region in y-z plane:
S = 8.8623; % (microns) length/width of simulation domain
N = 80;      % number of grid steps
h = S/N;     % space step size
% simulation parameters
D = 0.075; % micron^2 /sec
Tbleach = 0.133; % seconds
wo = 0.2481; % microns
W = 1.0; % microns
alpha = 40; % 1/sec
X = 2; % (microns) width of sampled region for normalization
% Setting up 2D diffusion operator (with periodic BC's)
L = spalloc(N*N,N*N,5*N*N);
for i = 1:N
  ip =  1;
  im =  -1;
  if i == 1; im = N-1; end
  if i == N; ip = -(N-1); end
  for j = 1:N
    jp = N;
    jm = -N;
    if j == 1; jm = N*(N-1); end
    if j == N; jp = -N*(N-1); end
    k = i + N*(j-1);
    k_ip = k + ip;
    k_im = k + im;
    k_jp = k + jp;
    k_jm = k + jm;
    L(k,k) = -( 2/h^2 + 2/h^2 );
    L(k,k_ip) = + 1/h^2;
    L(k,k_im) = + 1/h^2;
    L(k,k_jp) = + 1/h^2;
    L(k,k_jm) = + 1/h^2;
  end
end
% Make diffusion operators for Eq. (9.xx)
dt = Tbleach/7; % seconds
Adiff = speye(N*N)-(dt*D).*L./2;
Bdiff = speye(N*N)+(dt*D).*L./2;
% Setting up bleaching and (2 X 2) window functions
xo = -S/2;
yo = -S/2;
x = xo + ((1:N)-1/2).*h;
y = yo + ((1:N)-1/2).*h;
for i = 1:N
  for j = 1:N
```

```
    I = i + N*(j-1);
    K1(I) = (1/4)*( ...
      erf(sqrt(2)*(W+2*y(j))/wo/2)*erf(sqrt(2)*(W+2*x(i))/wo/2) - ...
      erf(sqrt(2)*(W+2*y(j))/wo/2)*erf(sqrt(2)*(-W+2*x(i))/wo/2) - ...
      erf(sqrt(2)*(-W+2*y(j))/wo/2)*erf(sqrt(2)*(W+2*x(i))/wo/2) + ...
      erf(sqrt(2)*(-W+2*y(j))/wo/2)*erf(sqrt(2)*(-W+2*x(i))/wo/2) );
    K2(I) = (1/4)*( ...
      erf(sqrt(2)*(X+2*y(j))/wo/2)*erf(sqrt(2)*(X+2*x(i))/wo/2) - ...
      erf(sqrt(2)*(X+2*y(j))/wo/2)*erf(sqrt(2)*(-X+2*x(i))/wo/2) - ...
      erf(sqrt(2)*(-X+2*y(j))/wo/2)*erf(sqrt(2)*(X+2*x(i))/wo/2) + ...
      erf(sqrt(2)*(-X+2*y(j))/wo/2)*erf(sqrt(2)*(-X+2*x(i))/wo/2) );
  end
end
% Initialize and bleach for 0<t<Tbleach:
u = ones(N*N,1);
for i = 1:Tbleach/dt
  % Diffuse for 1 time step (by solving Eq. 9.xx)
  u = Adiff\(Bdiff*u);
  % Bleach for 1 time step
  u = u.*exp(-alpha*dt.*K1');
end
% Computing F(t) after bleaching
Froi = K1*u/W^2;
Fwc  = K2*u/X^2;
F(1) = Froi/Fwc;
t(1) = Tbleach;
% Plot concentration field
mesh(x,y,reshape(u,N,N)); fprintf('Strike any key.\n'); pause;
% Simulation of timecourse
for i = 1:(10-Tbleach)/dt
  % Diffuse for 1 time step (by solving Eq. 9.xx)
  u = Adiff\(Bdiff*u);
  % Compute F(t)
  Froi = K1*u/W^2;
  Fwc  = K2*u/X^2;
  F(i+1) = Froi/Fwc;
  t(i+1) = Tbleach + i*dt;
  % Plot concentration field
  if (i==46)||(i==151)
    mesh(x,y,reshape(u,N,N)); fprintf('Strike any key.\n'); pause;
  end
end
plot(t,F);
```

(The MATLAB commands `spalloc` and `speye` in the above code are used for generating and storing *sparse* matrices. The matrices `L`, `Adiff`, and `Bdiff` are called sparse because most of entries are zeros. It makes sense to store only the indices and values of the nonzero entries of these matrices. As described in the MATLAB documentation, the command `spalloc(M,N,NZMAX)` allocates an *M*-by-*N* sparse matrix with room to store NZMAX nonzeros. The command `speye(M)` creates an *M*-by-*M* sparse identity matrix, with 1s along the main diagonal.)

To simulate the exchange process of Eq. (2.62) we can use the same basic algorithm. However, the exact expression for the linear kinetics becomes slightly more complicated. In this case, we require the exact solution to

$$\frac{d\mathbf{u}}{dt} = -k_{off}\mathbf{u} + k_{on}\left(\frac{M}{V_c} - \frac{A_m}{V_c}\langle\mathbf{u}\rangle\right) \tag{9.20}$$

at $t + \Delta t$ for $\mathbf{u}(t) = \tilde{\mathbf{u}}(t + \Delta t)$, where $\langle\mathbf{u}\rangle$ is the average value on the grid. This equation has the exact solution

$$\mathbf{u}(t + \Delta t) = \mathbb{P}_1\left(\tilde{\mathbf{u}}(t + \Delta t) + \frac{1}{\lambda_1}\mathbf{b}\right)e^{\lambda_1\Delta t} - \frac{1}{\lambda_1}\mathbb{P}_1\mathbf{b}$$
$$+ \mathbb{P}_2\left(\tilde{\mathbf{u}}(t + \Delta t) + \frac{1}{\lambda_2}\mathbf{b}\right)e^{\lambda_2\Delta t} - \frac{1}{\lambda_2}\mathbb{P}_2\mathbf{b}, \tag{9.21}$$

where

$$\lambda_1 = -k_{on}\frac{A_m}{V_c} - k_{off},$$
$$\lambda_2 = -k_{off},$$
$$\mathbb{P}_1 = \frac{1}{N^2}\begin{bmatrix} 1 & 1 & \cdots & 1 \\ 1 & 1 & \cdots & 1 \\ \vdots & \vdots & \ddots & \\ 1 & 1 & \cdots & 1 \end{bmatrix},$$
$$\mathbb{P}_2 = \mathbb{I} - \mathbb{P}_1,$$

and

$$\mathbf{b} = \frac{k_{on}M}{V_c}\begin{bmatrix} 1 \\ 1 \\ \vdots \\ 1 \end{bmatrix}. \tag{9.22}$$

(See Exercise 2.8 of Chapter 2.)

9.7   Circuits of resistors, capacitors, and inductors

Mathematical models of circuits – including certain electrical and fluidic systems – are often conveniently cast in terms of networks of idealized component elements, including resistors, capacitors, and inductors. Indeed, one may purchase from an electronics supplier electrical circuit components that closely approximate the behavior of the idealized mathematical models used to represent their behavior. Such devices (either independent components soldered together on a breadboard or printed on a silicon wafer in an integrated circuit) are basic building blocks of electronic equipment. In Chapter 4 models of the circulatory system are constructed from fluid-mechanical elements that are analogous to electrical circuit elements. This section introduces and develops the basic mathematics of circuit analysis used in Chapter 4 and elsewhere.

9.7.1   Circuit components

Here we describe the basic resistor, capacitor, and inductor circuit elements, including the mathematical relationships implied by idealized devices of these sorts. We will illustrate analogous mathematical and diagrammatic concepts for electrical and fluidic systems. In electrical systems the currency that flows through a network is electrical charge, whereas it is fluid mass in a fluidic system. So electrical current and fluid flow are analogous variables in electrical and fluidic systems.

In an electrical system the thermodynamic potential driving electrical current is called *voltage*, which is measured in units of energy per unit charge, such as joules per coulomb of elementary charge (also called volts). The equivalent to voltage in a fluidic system is *pressure*. Positive charges tend to move down the gradient in voltage; mass tends to flow down the pressure gradient in a fluid. Most materials under most conditions have an intrinsic finite *resistance* to electrical current. This means that a finite voltage gradient in the material will bring about a finite current. Equivalently, fluid-mechanical systems offer intrinsic resistance to flow through viscous fluid interactions. We indicate a resistance element (called a *resistor*) with the zigzag line labeled R in the following diagram:

Here an electrical system (with current I and voltages V_1 and V_2) is indicated on the left and a fluidic system (with flow F and pressure P_1 and P_2) on the right. The relationship between the flow variables and the current variables in this diagram is, by definition, given by

$$V_1 - V_2 = IR$$
$$P_1 - P_2 = FR, \tag{9.23}$$

which is called *Ohm's law*.

Equation (9.23) predicts that an instantaneous change in the driving force (the voltage difference or pressure difference) across a resistor element results in an instantaneous (with no time delay) change in flow (current or mass flow).

Compliant or capacitive systems are systems for which flow is proportional to the rate of change of pressure or voltage. An example is an elastic balloon, where a positive flow of air in causes an increase in pressure, and the greater the volume of air in the balloon, the greater the pressure. In a linear regime, the pressure P and volume V are related by a proportionality constant C, the *compliance* of the balloon: $V = CP$. If F is the rate of mass flux into the system, then differentiating yields $dV/dt = F = CdP/dt$.

In a circuit a compliance element, called a *capacitor*, is indicated by the symbol marked "C" in the following diagram:

In the electrical and fluidic circuits the flows and driving pressure are related by

$$C\frac{d(V_1 - V_2)}{dt} = I$$
$$C\frac{d(P_1 - P_2)}{dt} = F. \tag{9.24}$$

Therefore flow through a capacitor is always zero in steady state. As applied in Section 4.2.1, the volume of a fluid capacitor is computed by the pressure drop multiplied by the capacitance. The higher the capacitance, the greater the volume at a given pressure drop. Electrical capacitors are devices that separate positive and negative electrical charge. The amount of charge separated is computed by the voltage difference across the capacitor multiplied by the capacitance. Thus the electrical capacitance is a measure of the amount of separated charge stored at a given voltage drop.

The final basic element that we need consider is the *inductor*, indicated by the coils labeled "L" in the diagram below.

The coil diagram derives from the electrical inductor, which is constructed from coiled wire. In fact, another name for an electrical inductor is a "coil". Electrical

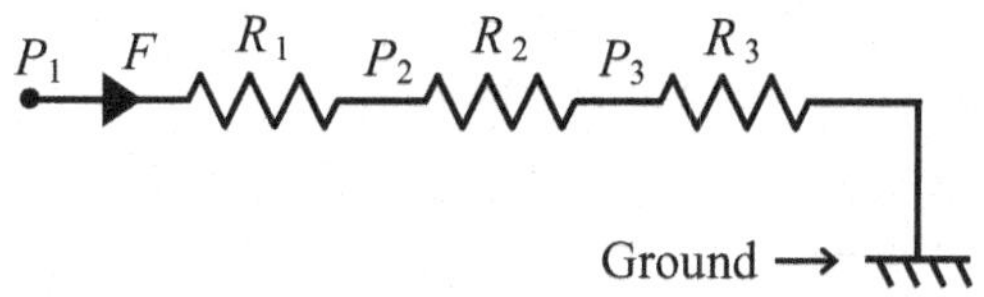

Figure 9.5

Simple network of three resistors connected in series.

current in an inductor sets up a magnetic field; a changing (time-varying) current sets up a changing magnetic field, which in turn sets up an electric field that tends to oppose the changing current. The greater the inductance L, the more the device resists changes in current. This phenomenon is captured in linear circuits by the equations

$$L\frac{dI}{dt} = (V_1 - V_2)$$

$$L\frac{dF}{dt} = (P_1 - P_2). \tag{9.25}$$

In fluidic circuits inductors capture inertial phenomena. Equation (9.25) indicates that flow changes in response to a pressure difference, but that a finite pressure differences does not result in an instantaneous pressure change in flow. In a steady state an ideal inductor cannot support a pressure drop with a finite flow, since the intrinsic resistance of an ideal inductor is zero. A rigid pipe, with finite resistance and inductance, may be modeled as a resistor and inductor in series. A distensible tube may be modeled using a resistor, an inductor, and a capacitor, as we do in Section 4.3.2.

9.7.2 Circuit analysis and simulation

Circuits are made by hooking up the elements of the sort illustrated above (and possibly other sorts) into networks. Connected circuits are analyzed by determining and solving and/or simulating the governing equations for the combined systems. Two general conservation relationships, called *Kirchhoff's current law* and *Kirchhoff's voltage law*, are applied extensively in circuit analysis.[3]

Kirchhoff's laws are conveniently illustrated using some simple examples. Consider first the resistor network illustrated in Figure 9.5, in which three resistors are connected together in series and the series of resistors is hooked up to a fixed

[3] The names of these laws (the voltage law and the current law) come from their application by Gustav Kirchhoff (1824–1917) to electrical circuit analysis. Here we see that they apply equally to fluidic circuits, although we use the electrical-circuit-centric names exclusively.

source at pressure P_1. (Here we have chosen to illustrate a fluidic circuit with pressures and flows as the network variables.)

The symbol labeled "ground" is commonly used to represent a common reference pressure (or voltage), here (as usual) taken to be the value 0. If P_1 is constant then there is no time dependence in the behavior of this circuit. Since there are no branching pathways, the flow through each resistor is the same and equal to F. By Ohm's law, we have equations for the pressure drops across the resistors:

$$P_3 = FR_3$$
$$P_2 - P_3 = FR_2$$
$$P_1 - P_2 = FR_1 \tag{9.26}$$

or, summing these equations,

$$P_1 = \sum_i FR_i. \tag{9.27}$$

This equation is a manifestation of Kirchhoff's voltage law, which says that the pressure (or voltage) drop along a series of circuit elements in a "closed loop" is zero. This means that if one starts at any arbitrary node in a network, and sums pressure differences as one travels around a loop, the total pressure difference for a closed loop is zero. To see this for the simple circuit of Figure 9.5, we can start at the ground and sum the pressure drop from ground to P_3, from P_3 to P_2, from P_2 to P_1, and from P_1 back to ground:

$$P_3 + (P_2 - P_3) + (P_1 - P_2) + (0 - P_1) = 0$$

or

$$FR_3 + FR_2 + FR_1 - P_1 = 0,$$

which is simply Eq. (9.27).

We can easily solve these equations for the flow F:

$$F = P_1/(R_1 + R_2 + R_3)$$

and using this expression, express the pressures

$$P_2 = P_1(R_2 + R_3)/(R_1 + R_2 + R_3)$$
$$P_3 = P_1 R_3/(R_1 + R_2 + R_3). \tag{9.28}$$

Here the application of Kirchhoff's voltage law was trivial, and it was more straightforward to simply write down Eq. (9.27). This is not always the case for more complex networks.

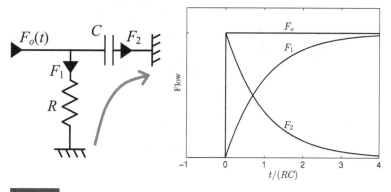

Figure 9.6

Simple network with one resistor and one capacitor.

The combined application of Kirchhoff's two laws is illustrated using the slightly more complex circuit of Figure 9.6. This basic resistor–capacitor (RC) circuit is used to represent compliant vessels in models of the circulatory system. The idea is that the vessels offer a finite resistance, R, and a finite compliance C. Total flow into a vessel pool is the sum of the flow out (F_1) plus the *transmural* flow (F_2), which is associated with swelling or shrinking of the vessels. Here we indicate a time-dependent input flow $F_o(t)$: thus F_1 and F_2 will in general be functions of time.

Kirchhoff's current law is a statement of conservation of mass or charge at a node in a network. Here, $F_o(t) = F_1(t) + F_2(t)$. We may apply the voltage law along the closed path from ground to ground indicated by the gray arrow in the figure:

$$R F_1(t) - V/C = 0, \qquad (9.29)$$

where V is the volume of the compliant vessel (the volume stored in the capacitor.) Differentiating and using the conservation statement of the current law yields a first-order differential equation for $F_1(t)$:

$$\frac{dF_1}{dt} + \frac{1}{RC} F_1 = \frac{1}{RC} F_o(t), \qquad (9.30)$$

which has the solution

$$F_1(t) = F_1(0)e^{-t/\tau} + e^{-t/\tau} \int_0^t e^{+s/\tau} F_o(s)ds, \qquad (9.31)$$

where $\tau = RC$.

If the system initially has no flow or volume ($F_1(0) = F_2(0) = 0$), and $F_o(t) = F_o$ (a constant) for $t > 0$, then we obtain the *step response function* for the

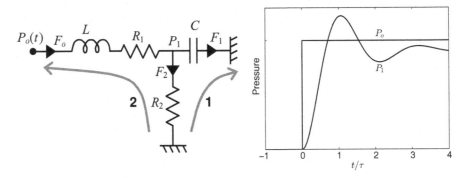

Simple network with one resistor, one capacitor, and one inductor. The solution indicated is for $R_1 = 0.5\,\text{mmHg s ml}^{-1}$, $R_2 = 5\,\text{mmHg s ml}^{-1}$, $C = 10\,\text{ml mmHg}^{-1}$, and $L = 50\,\text{mmHg s}^2\,\text{ml}^{-1}$.

system:

$$F_1(t) = F_o\left(1 - e^{-t/\tau}\right),\tag{9.32}$$

which is illustrated in Figure 9.6. For $t \ll RC$, $F_2 \approx F_o$, as all of the flow into the system goes into increasing the volume of the vessel. In the eventual steady state, $F_2 = 0$ and $F_1 = F_o$.

As a final example we consider the slightly more complicated network in Figure 9.7, which incorporates an inductor. Here the circuit is hooked up to a time-dependent pressure source $P_o(t)$ at the node indicated in the figure.

Applying Kirchhoff's voltage law along the path labeled "1" in the figure, we have

$$R_2 F_2 - V/C = 0$$
$$R_2 \frac{dF_2}{dt} - \frac{F_1}{C} = 0,\tag{9.33}$$

where V is the volume in the capacitor. Recalling that the pressure drop across an inductor with inductance L is $\Delta P = L\,dF/dt$, and applying Kirchhoff's voltage law along path "2", we have

$$R_2 F_2 + R_1 F_o + L\frac{dF_o}{dt} = P_o(t).\tag{9.34}$$

The current law requires

$$F_o = F_1 + F_2.\tag{9.35}$$

Combining Eqs (9.33), (9.34), and (9.35) yields a second-order differential equation for $F_2(t)$:

$$CR_2 L\frac{d^2 F_2}{dt^2} + (CR_1 R_2 + L)\frac{dF_2}{dt} + (R_1 + R_2)F_2 = P_o(t).\tag{9.36}$$

Noting that $P_1 = R_2 F_2$, we can express the above equation in terms of $P_1(t)$:

$$CL\frac{d^2 P_1}{dt^2} + (CR_1 + L/R_2)\frac{dP_1}{dt} + (1 + R_1/R_2) P_1 = P_o(t), \tag{9.37}$$

which is a linear constant-coefficient equation and thus has a closed-form analytical solution, given a prescribed forcing function $P_o(t)$.

Solution of this equation requires initial conditions for P_1 and dP_1/dt. If the initial volume in the capacitor is zero, then $P_1(0) = 0$. If the initial current throughout the circuit is zero, then from Eq. (9.33), $dP_1(0)/dt = R_2 dF_2(0)/dt = 0$. In the solution plotted in Figure 9.7 we also assume that $P_o(t) = P_o$ (a constant) for $t > 0$. Under these conditions,[4] Eq. (9.37) has solution

$$P_1(t) = \frac{P_o}{1 + R_1/R_2}\left[1 - e^{-t/\tau}\left(\cos(\omega t) + \frac{1}{\omega \tau}\sin(\omega t)\right)\right], \tag{9.38}$$

where

$$\tau = \frac{2LR_2 C}{L + R_1 R_2 C}$$

and

$$\omega = \frac{\sqrt{4CL(1 + R_1/R_2) - (CR_1 + L/R_2)^2}}{2CL},$$

which is plotted in Figure 9.7 for the parameter values given in the legend.

9.8 Rate laws and parameter values for glycogenolysis model

Glycogen phosphorylase

Two forms of glycogen phosphorylase are modeled: a phosphorylated form (glycogen phosphorylase a), which is the primarily active form of the enzyme; and an unphosphorylated form (glycogen phosphorylase b), which may be allosterically activated by AMP. The net J_{GP} is modeled by the following set of equations:

$$J_{GPa} = \left(\frac{1.404}{1 + 10^{5.94 - \text{pH}} + 10^{\text{pH} - 7.29}}\right)$$

$$\times \frac{V_{max,f}\left(\frac{[\text{GLY}][\text{Pi}]}{K_{GLYf} K_{Pi}}\right) - V_{max,r}\left(\frac{[\text{GLY}][\text{G1P}]}{K_{GLYb} K_{iG1P}}\right)}{1 + \frac{[\text{GLY}]}{K_{GLYf}} + \frac{[\text{Pi}]}{K_{iPi}} + \frac{[\text{GLY}]}{K_{iGLYb}} + \frac{[\text{G1P}]}{K_{iG1P}} + \frac{[\text{GLY}][\text{Pi}]}{K_{GLYf} K_{iPi}} + \frac{[\text{GLY}][\text{G1P}]}{K_{GLYb} K_{iG1P}}},$$

$$\tag{9.39}$$

---

[4] See Exercise 4.2 of Chapter 4.

where $V_{max,r}$ is expressed in terms of $V_{max,f}$ and $K_{eq,GP}$, the apparent equilibrium constant for the reaction

$$V_{max,r} = \frac{V_{max,f} K_{GLYb} K_{iG1P}}{K_{GLYf} K_{Pi} K_{eq,GP}};$$

$$J_{GPb} = \left(\frac{1.75}{1 + 10^{6.12-pH} + 10^{pH-7.03}} \right) \frac{\left(\frac{[AMP]}{K'_{AMP}} \right)^{n_H}}{1 + \left(\frac{[AMP]}{K'_{AMP}} \right)^{n_H}}$$

$$\times \frac{V_{max,f} \left(\frac{[GLY][Pi]}{K_{iGLYf} K_{Pi}} \right) - V_{max,r} \left(\frac{[GLY][G1P]}{K_{iGLYb} K_{G1P}} \right)}{1 + \frac{[GLY]}{K_{iGLYf}} + \frac{[Pi]}{K_{iPi}} + \frac{[GLY]}{K_{iGLYb}} + \frac{[G1P]}{K_{iG1P}} + \frac{[GLY][Pi]}{K_{iGLYf} K_{Pi}} + \frac{[GLY][G1P]}{K_{iGLYb} K_{G1P}}},$$

$$(9.40)$$

where $V_{max,r}$ is computed

$$V_{max,r} = \frac{V_{max,f} K_{iGLYb} K_{G1P}}{K_{iGLYf} K_{iPi} K_{eq,GP}}.$$

The net glycogen phosphorylase flux is the sum of contributions from the two forms of the enzyme:

$$J_{GP} = f_a J_{GPa} + (1 - f_a) J_{GPb}, \qquad (9.41)$$

where f_a is the fraction in the a form.

Phosphoglucose mutase

$$J_{PGLM} = \left(\frac{1.329}{1 + 10^{6.64-pH} + 10^{pH-8.36}} \right) \frac{V_{max,f} \left(\frac{[G1P]}{K_{G1P}} \right) - V_{max,r} \left(\frac{[G6P]}{K_{G6P}} \right)}{1 + \frac{[G1P]}{K_{G1P}} + \frac{[G6P]}{K_{G6P}}},$$

$$(9.42)$$

where $V_{max,r}$ is expressed in terms of $V_{max,f}$ and $K_{eq,PGLM}$, the apparent equilibrium constant for the reaction

$$V_{max,r} = \frac{V_{max,f} K_{G6P}}{K_{G1P} K_{eq,PGLM}}.$$

Phosphoglucose isomerase

$$J_{PGI} = \left(\frac{1}{1 + 10^{6.36-pH} + 10^{pH-9.91}} \right) \frac{V_{max,f} \left(\frac{[G6P]}{K_{G6P}} \right) - V_{max,r} \left(\frac{[F6P]}{K_{F6P}} \right)}{1 + \frac{[G6P]}{K_{G6P}} + \frac{[F6P]}{K_{F6P}}}, \quad (9.43)$$

where $V_{max,r}$ is expressed in terms of $V_{max,f}$ and $K_{eq,PGI}$, the apparent equilibrium constant for the reaction

$$V_{max,r} = \frac{V_{max,f} K_{G6P} K_{eq,PGI}}{K_{F6P}}.$$

Phosphofructose kinase

The quasi-steady flux expression for phosphofructose kinase accounts for allosteric regulation by ATP and AMP concentrations:

$$J_{PFK} = \left(\frac{1}{1 + (6.8/\text{pH})^{30}} \right)$$

$$\times \frac{V_{max,f} \left(\frac{[\text{ATP}][\text{F6P}]}{K_{ATP} K_{F6P}} \right) - V_{max,r} \left(\frac{[\text{ADP}][\text{FBP}]}{K_{ADP} K_{FBP}} \right)}{D} \cdot \frac{1 + \alpha L \left(\frac{D'}{D} \right)^3}{1 + L \left(\frac{D'}{D} \right)^4}, \quad (9.44)$$

where

$$D = \left(1 + \frac{[\text{F6P}]}{K_{F6P}} \right) \left(1 + \frac{[\text{ATP}]}{K_{ATP}} \right) + \frac{[\text{ADP}]}{K_{ADP}} + \frac{[\text{FBP}]}{K_{FBP}} \left(1 + \frac{[\text{ADP}]}{K_{ADP}} \right)$$

$$D' = \left(1 + \frac{[\text{F6P}]}{K'_{F6P}} \right) \left(1 + \frac{[\text{ATP}]}{K'_{ATP}} \right) + \frac{[\text{ADP}]}{K'_{ADP}} + \frac{[\text{FBP}]}{K'_{FBP}} \left(1 + \frac{[\text{ADP}]}{K'_{ADP}} \right)$$

$$\alpha = \frac{K_{F6P} K_{ATP}}{K'_{F6P} K'_{ATP}}$$

$$L = L_o \left[\left(\frac{1 + \frac{[\text{ATP}]}{K_{iATP}}}{1 + d \frac{[\text{ATP}]}{K_{iATP}}} \right) \left(\frac{1 + e \frac{[\text{AMP}]}{K_{aAMP}}}{1 + \frac{[\text{AMP}]}{K_{aAMP}}} \right) \right]^4$$

$$V_{max,r} = \frac{V_{max,f} K_{ADP} K_{FBP}}{K_{ATP} K_{F6P}}.$$

Aldolase

$$J_{ALD} = \left(\frac{1.013}{1 + 10^{5.32-\text{pH}} + 10^{\text{pH}-9.15}} \right) \frac{V_{max,f} \left(\frac{[\text{FBP}]}{K_{FBP}} \right) - V_{max,r} \left(\frac{[\text{DHAP}][\text{GAP}]}{K_{DHAP} K_{GAP}} \right)}{1 + \frac{[\text{FBP}]}{K_{FBP}} + \frac{[\text{DHAP}]}{K_{DHAP}} + \frac{[\text{GAP}]}{K_{GAP}}},$$

$$(9.45)$$

where $V_{max,r}$ is expressed in terms of $V_{max,f}$ and $K_{eq,ALD}$, the apparent equilibrium constant for the reaction

$$V_{max,r} = \frac{V_{max,f} K_{DHAP} K_{GAP}}{K_{FBP} K_{eq,ALD}}.$$

Trios phosphate isomerase

$$J_{TPI} = \frac{V_{max,f} \left(\frac{[\text{GAP}]}{K_{GAP}} \right) - V_{max,r} \left(\frac{[\text{DHAP}]}{K_{DHAP}} \right)}{1 + \frac{[\text{GAP}]}{K_{GAP}} + \frac{[\text{DHAP}]}{K_{DHAP}}}, \quad (9.46)$$

where $V_{max,r}$ is expressed in terms of $V_{max,f}$ and $K_{eq,TPI}$, the apparent equilibrium constant for the reaction

$$V_{max,r} = \frac{V_{max,f} K_{DHAP}}{K_{GAP} K_{eq,TPI}}.$$

Glycerol-3-phosphate dehydrogenase

$$J_{G3PDH} = \frac{V_{max,f} \left(\frac{[\text{GL3P}][\text{NAD}]}{K_{GL3P} K_{NAD}} \right) - V_{max,r} \left(\frac{[\text{DHAP}][\text{NADH}]}{K_{DHAP} K_{NADH}} \right)}{\left(1 + \frac{[\text{GL3P}]}{K_{GL3P}} + \frac{[\text{NADH}]}{K_{NADH}} \right) \left(1 + \frac{[\text{DHAP}]}{K_{DHAP}} + \frac{[\text{NAD}]}{K_{NAD}} \right)}, \tag{9.47}$$

where $V_{max,r}$ is expressed in terms of $V_{max,f}$ and $K_{eq,G3PDH}$, the apparent equilibrium constant for the reaction

$$V_{max,r} = \frac{V_{max,f} K_{GL3P} K_{NAD} K_{eq,G3PDH}}{K_{DHAP} K_{NADH}}.$$

Glyceraldehyde-3-phosphate dehydrogenase

$$J_{GAPDH} = \left(0.007 e^{0.8979 \cdot \text{pH}} \right) \frac{V_{max,f} \left(\frac{[\text{GAP}][\text{NAD}][\text{Pi}]}{K_{GAP} K_{NAD} K_{Pi}} \right) - V_{max,r} \left(\frac{[\text{BPG}][\text{NADH}]}{K_{BPG} K_{NADH}} \right)}{D}$$

$$\tag{9.48}$$

where

$$D = 1 + \frac{[\text{GAP}]}{K_{GAP}} + \frac{[\text{NAD}]}{K_{NAD}} + \frac{[\text{Pi}]}{K_{Pi}} + \frac{[\text{GAP}][\text{NAD}]}{K_{GAP} K_{NAD}} + \frac{[\text{GAP}][\text{NAD}][\text{Pi}]}{K_{GAP} K_{NAD} K_{Pi}}$$

$$+ \frac{[\text{BPG}]}{K_{BPG}} + \frac{[\text{NADH}]}{K_{NADH}} + \frac{[\text{BPG}][\text{NADH}]}{K_{BPG} K_{NADH}}$$

$$V_{max,r} = \frac{V_{max,f} K_{BPG} K_{NADH}}{K_{GAP} K_{NAD} K_{Pi} K_{eq,G3PDH}}.$$

Phosphoglycerate kinase

$$J_{PGK} = \frac{V_{max,f} \left(\frac{[\text{BPG}][\text{ADP}]}{K_{BPG} K_{ADP}} \right) - V_{max,r} \left(\frac{[\text{PG3}][\text{ATP}]}{K_{PG3} K_{ATP}} \right)}{1 + \frac{[\text{BPG}]}{K_{BPG}} + \frac{[\text{ADP}]}{K_{ADP}} + \frac{[\text{BPG}][\text{ADP}]}{K_{BPG} K_{ADP}} + \frac{[\text{PG3}]}{K_{PG3}} + \frac{[\text{ATP}]}{K_{ATP}} + \frac{[\text{PG3}][\text{ATP}]}{K_{PG3} K_{ATP}}}, \tag{9.49}$$

where $V_{max,f}$ is expressed in terms of $V_{max,r}$ and $K_{eq,PGK}$, the apparent equilibrium constant for the reaction

$$V_{max,f} = \frac{V_{max,r} K_{BPG} K_{ADP} K_{eq,PGK}}{K_{PG3} K_{ATP}}.$$

Phosphoglycerol mutase

$$J_{PGM} = \left(\frac{0.989}{1 + 10^{5.62 - \text{pH}} + 10^{\text{pH} - 8.74}} \right) \frac{V_{max,f} \left(\frac{[\text{PG3}]}{K_{PG3}} \right) - V_{max,r} \left(\frac{[\text{PG2}]}{K_{PG2}} \right)}{1 + \frac{[\text{PG3}]}{K_{PG3}} + \frac{[\text{PG2}]}{K_{PG2}}}, \tag{9.50}$$

where $V_{max,r}$ is expressed in terms of $V_{max,f}$ and $K_{eq,PGM}$, the apparent equilibrium constant for the reaction

$$V_{max,r} = \frac{V_{max,f} K_{PG2}}{K_{PG3} K_{eq,PGM}}.$$

Enolase

$$J_{EN} = \frac{V_{max,f} \left(\frac{[PG2]}{K_{PG2}}\right) - V_{max,r} \left(\frac{[PEP]}{K_{PEP}}\right)}{1 + \frac{[PG2]}{K_{PG2}} + \frac{[PEP]}{K_{PEP}}}, \tag{9.51}$$

where $V_{max,r}$ is expressed in terms of $V_{max,f}$ and $K_{eq,EN}$, the apparent equilibrium constant for the reaction

$$V_{max,r} = \frac{V_{max,f} K_{PEP}}{K_{PG2} K_{eq,EN}}.$$

Pyruvate kinase

$$J_{PYK} = \left(\frac{1.05}{1 + 10^{5.58-pH} + 10^{pH-8.79}}\right)$$

$$\times \frac{V_{max,f} \left(\frac{[PEP][ADP]}{K_{PEP} K_{ADP}}\right) - V_{max,r} \left(\frac{[PYR][ATP]}{K_{PYR} K_{ATP}}\right)}{1 + \frac{[PEP]}{K_{PEP}} + \frac{[ADP]}{K_{ADP}} + \frac{[PEP][ADP]}{K_{PEP} K_{ADP}} + \frac{[PYR]}{K_{PYR}} + \frac{[ATP]}{K_{ATP}} + \frac{[PYR][ATP]}{K_{PYR} K_{ATP}}},$$

$$\tag{9.52}$$

where $V_{max,r}$ is expressed in terms of $V_{max,f}$ and $K_{eq,PYK}$, the apparent equilibrium constant for the reaction

$$V_{max,r} = \frac{V_{max,f} K_{ATP} K_{PYR}}{K_{PEP} K_{ADP} K_{eq,PYK}}.$$

Lactate dehydrogenase

$$J_{LDH} = (1.6069 - 0.1134 \cdot pH)$$

$$\times \frac{V_{max,f} \left(\frac{[PYR][NADH]}{K_{PYR} K_{NADH}}\right) - V_{max,r} \left(\frac{[LAC][NAD]}{K_{LAC} K_{NAD}}\right)}{1 + \frac{[PYR]}{K_{PYR}} + \frac{[NADH]}{K_{NADH}} + \frac{[PYR][NADH]}{K_{PYR} K_{NADH}} + \frac{[LAC]}{K_{LAC}} + \frac{[NAD]}{K_{NAD}} + \frac{[LAC][NAD]}{K_{LAC} K_{NAD}}},$$

$$\tag{9.53}$$

where $V_{max,r}$ is expressed in terms of $V_{max,f}$ and $K_{eq,LDH}$, the apparent equilibrium constant for the reaction

$$V_{max,r} = \frac{V_{max,f} K_{LAC} K_{NAD}}{K_{PYR} K_{NADH} K_{eq,LDH}}.$$

Creatine kinase

$$J_{CK} = \frac{V_{max,r}\left(\frac{[\text{Cr}][\text{ATP}]}{K_{Cr}K_{iATP}}\right) - V_{max,f}\left(\frac{[\text{PCr}][\text{ADP}]}{K_{PCr}K_{iADP}}\right)}{1 + \frac{[\text{PCr}]}{K_{iPCr}} + \frac{[\text{ADP}]}{K_{iADP}} + \frac{[\text{PCr}][\text{ADP}]}{K_{PCr}K_{iADP}} + \frac{[\text{ATP}]}{K_{iATP}} + \frac{[\text{Cr}][\text{ATP}]}{K_{Cr}K_{iATP}}}, \tag{9.54}$$

where $V_{max,f}$ is expressed in terms of $V_{max,r}$ and $K_{eq,CK}$, the apparent equilibrium constant for the reaction

$$V_{max,r} = \frac{V_{max,f}K_{iATP}K_{Cr}}{K_{PCr}K_{iADP}K_{eq,CK}}.$$

Adenylate kinase

$$J_{AK} = \frac{V_{max,f}\left(\frac{[\text{ATP}][\text{AMP}]}{K_{ATP}K_{AMP}}\right) - V_{max,r}\left(\frac{[\text{ADP}]^2}{K_{ADP}^2}\right)}{1 + \frac{[\text{ATP}]}{K_{ATP}} + \frac{[\text{AMP}]}{K_{AMP}} + \frac{[\text{ATP}][\text{AMP}]}{K_{ATP}K_{AMP}} + \frac{2[\text{ADP}]}{K_{ADP}} + \frac{[\text{ADP}]^2}{K_{ADP}^2}}, \tag{9.55}$$

where $V_{max,r}$ is expressed in terms of $V_{max,f}$ and $K_{eq,AK}$, the apparent equilibrium constant for the reaction

$$V_{max,r} = \frac{V_{max,f}K_{ADP}^2}{K_{ATP}K_{AMP}K_{eq,AK}}.$$

ATP hydrolysis

ATP hydrolysis is assumed to proceed by simple mass-action proportionality to ATP concentration:

$$J_{ATPase} = k[\text{ATP}], \tag{9.56}$$

where k is a constant. Since ATP concentration in the experiments of Vinnakota *et al.* [65] remains essentially constant, the ATP hydrolysis rate remains constant.

Oxidative phosphorylation

The rate of oxidative synthesis of ATP is modeled by a phenomenological dependence on cellular ADP concentration:

$$J_{OxPhos} = V_{max,f} \frac{([\text{ADP}]/K_{ADP})^{n_H}}{1 + ([\text{ADP}]/K_{ADP})^{n_H}}, \tag{9.57}$$

where n_H is an effective Hill coefficient.

Carbon dioxide generation

The rate of CO_2 generation from oxidative phosphorylation depends on the relative proportion of fat to carbohydrate utilized by mitochondria. In the model J_{CO2} is computed

$$J_{CO2} = J_{OxPhos} RQ/PO_2 ratio, \tag{9.58}$$

where RQ and PO_2ratio are fixed parameters defined as follows. The respiratory quotient (RQ) is the ratio of carbon dioxide generated to oxygen consumed. Its value ranges from approximately 0.7 to 1, depending on the relative proportion of fatty acids versus carbohydrates consumed for oxidative phosphorylation. The parameter PO_2ratio is the ration of ATP synthesized to oxygen molecule consumed by oxidative phosphorylation.

Uncatalyzed hydration of carbon dioxide
This reaction is assumed to proceed via mass action:

$$J_{CO2hyd} = k_f[CO_2] - k_b[H^+][HCO_3^-]. \tag{9.59}$$

Carbonic anhydrase
Similar to the uncatalyzed reaction, this reaction proceeds via mass action:

$$J_{CA} = X_{CA}\left(k_f[CO_2] - k_b[H^+][HCO_3^-]\right), \tag{9.60}$$

where X_{CA} is the activity of CA measured relative to that of the uncatalyzed reaction.

Monocarboxylate transporter

$$J_{MCT}$$
$$= V_{max}\left(\frac{\frac{[LAC][H^+]}{K_{LAC}K_H} - \frac{[LAC]_{cham}[H^+]_{cham}}{K_{LAC}K_H}}{2 + \frac{[LAC]}{K_{LAC}} + \frac{[H^+]}{K_H} + \frac{[LAC][H^+]}{K_{LAC}K_H} + \frac{[LAC]_{cham}}{K_{LAC}} + \frac{[H^+]_{cham}}{K_H} + \frac{[LAC]_{cham}[H^+]_{cham}}{K_{LAC}K_H}}\right). \tag{9.61}$$

The parameter values used in the model are listed in Table 9.1.

The hydrogen and metal cation dissociation constants used in the model are listed in Table 9.2.

The terrific complexity of this model and the depth of prior knowledge that went into its construction should be appreciated. Compare the simple kinetic model of Section 1.2.3, captured by Eqs (1.13)–(1.16), with the large number of equations and biophysical parameters used to capture the kinetics of this system. While both Sections 1.2.3 and 7.2 apply differential-equation-based models that arise from mass conservation and chemical kinetics to analyze the behavior of biochemical systems, these examples represent extremes in terms of the level of detail at which the chemistry is represented. The model in Section 1.2.3 lumps chemical species into pooled reactants, and therefore cannot capture the detailed pH dynamics captured by the glycogenolysis model of Section 7.2. Similar to the model fermentation model of Section 2.3.1, the aquarium model of

Table 9.1: List of all parameter values for cellular glycolysis model of Vinnakota *et al.* [65].

Enzymes	Parameter	Value	Units
Glycogen phosphorylase a	$V_{max,f}$	0.0127	M min^{-1}
	f_a	0.0001	Unitless
	K_{GLYf}	1.7	mM
	K_{iGLY}	2	mM
	K_{Pi}	4	mM
	K_{iPi}	4.7	mM
	K_{GLYb}	0.15	mM
	K_{G1P}	2.7	mM
	K_{iG1p}	10.1	mM
	$K_{eq,GP}$	0.2932	Unitless
Glycogen phosphorylase b	$V_{max,f}$	0.0127	M min^{-1}
	f_a	0.0001	Unitless
	K_{iGLYf}	1.7	mM
	K_{iGLY}	2	mM
	K_{Pi}	4	mM
	K_{iPi}	4.7	mM
	K_{GLYb}	0.15	mM
	K_{G1P}	2.7	mM
	K_{iG1p}	10.1	mM
	K_{AMP}	9.7×10^{-5}	mM
	n_H	1.75	Unitless
Phosphoglucose mutase	$V_{max,f}$	0.0177	M min^{-1}
	K_{G1P}	0.0063	mM
	K_{G6P}	0.030	mM
	$K_{eq,PGM}$	14.54	Unitless
Phosphoglucose isomerase	$V_{max,r}$	1.3155	M min^{-1}
	K_{G6P}	0.48	mM
	K_{F6P}	0.119	mM
	$K_{eq,PGI}$	0.2733	Unitless
Phosphofructose kinase	$V_{max,f}$	0.0127	M min^{-1}
	K_{F6P}	0.18	mM
	K'_{F6P}	20	mM
	K_{FBP}	4.02	mM
	K'_{FBP}	4.02	mM
	K_{ATP}	0.08	mM
	K'_{ATP}	0.25	mM
	K_{ADP}	2.7	mM
	K'_{ADP}	2.7	mM
	K_{iATP}	0.87	mM

Table 9.1: (*cont.*)

Enzymes	Parameter	Value	Units
	$K_{eq,PFK}$	7.81×10^3	Unitless
	L_o	13	Unitless
	d	0.01	Unitless
	e	0.01	Unitless
Aldolase	$V_{max,f}$	0.0412	M min^{-1}
	K_{FBP}	0.05	mM
	K_{DHAP}	2.0	mM
	K_{GAP}	1.0	mM
	$K_{eq,ALD}$	2.287×10^{-4}	M
Triose phosphate isomerase	$V_{max,f}$	0.3480	M min^{-1}
	K_{GAP}	0.05	mM
	K_{DHAP}	2.0	mM
	$K_{eq,ALD}$	19.87	Unitless
Glycerol-3-phosphate dehydrogenase	$V_{max,r}$	0.631	M min^{-1}
	K_{G3P}	0.18	mM
	K_{NAD}	0.012	mM
	K_{DHAP}	0.22	mM
	K_{NADH}	0.008	mM
	$K_{eq,G3PDH}$	1.282×10^{-4}	Unitless
Glyceraldehyde-3-phosphate dehydrogenase	$V_{max,f}$	0.1865	M min^{-1}
	K_{GAP}	0.0025	mM
	K_{NAD}	0.09	mM
	K_{Pi}	0.29	mM
	K_{PG3}	0.0008	mM
	K_{NADH}	0.0033	mM
	$K_{eq,GAPDH}$	0.7164	M^{-1}
Phosphoglycerate kinase	$V_{max,r}$	0.0212	M min^{-1}
	K_{BPG}	2.0	mM
	K_{ADP}	0.008	mM
	K_{PG3}	1.2	mM
	K_{ATP}	0.35	mM
	$K_{eq,PGK}$	27.40	Unitless
Phosphoglycerate mutase	$V_{max,f}$	0.2046	M min^{-1}
	K_{PG3}	0.2	mM
	K_{PG2}	0.014	mM
	$K_{eq,PGM}$	0.3086	Unitless
Enolase	$V_{max,f}$	1.943	M min^{-1}
	K_{PG2}	0.1	mM
	K_{PEP}	0.37	mM
	$K_{eq,EN}$	4.4111	Unitless

(cont.)

Table 9.1: *(cont.)*

Enzymes	Parameter	Value	Units
Pyruvate kinase	$V_{max,f}$	0.0996	M min^{-1}
	K_{PEP}	0.08	mM
	K_{ADP}	0.30	mM
	K_{PYR}	7.05	mM
	K_{ATP}	1.13	mM
	$K_{eq,PYK}$	9.195×10^4	Unitless
Lactate dehydrogenase	$V_{max,f}$	0.2687	M min^{-1}
	K_{PYR}	0.335	mM
	K_{NADH}	0.002	mM
	K_{LAC}	17.0	mM
	K_{NAD}	0.849	mM
	$K_{eq,LDH}$	1.357×10^4	Unitless
Creatine kinase	$V_{max,f}$	11.137	M min^{-1}
	K_{iATP}	3.5	mM
	K_{iADP}	0.135	mM
	K_{Cr}	3.80	mM
	K_{PCr}	1.11	mM
	K_{iPCr}	3.90	mM
	$K_{eq,CK}$	149.65	Unitless
Adenylate kinase	$V_{max,f}$	2.563	M min^{-1}
	K_{AMP}	0.32	mM
	K_{ATP}	0.27	mM
	K_{ADP}	0.35	mM
	$K_{eq,AK}$	2.66	Unitless
Carbonic anhydrase and uncatalyzed CO_2 hydration	X_{CA}	740	Unitless
	k_f	2.814	min^{-1}
	k_b	4.04×10^6	min^{-1}
Transporters and transport processes:			
Monocarboxylate transporter	V_{max}	0.00469	M min^{-1}
	K_{LAC}	25.5	mM
	K_H	$10^{-8.89}$	M
CO_2 transport	PS	2.3084	min^{-1}
Other processes:			
Oxidative phosphorylation	V_{max}	0.003747	M min^{-1}
	n_H	2.429	Unitles
	K_{ADP}	48.34	μM
CO_2 generation	RQ	1.0	Unitless
	$PO_2 ratio$	4.2	Unitless

Table 9.1: *(cont.)*

Enzymes	Parameter	Value	Units
Cell volume	V_{cell}	3.685×10^{-5}	Liter
Chamber volume	V_{cham}	9.711×10^{-5}	Liter
Chamber flow	F	4.0×10^{-3}	$l\,min^{-1}$
MOPS buffer	K_{MOPS}	$10^{-7.184}$	M
	A_{MOPS}	0	mM
Input concentrations	$[H^+]_{input}$	$10^{-7.44}$	M
	$[CO_2{}^+]_{input}$	1.564	mM
	$[HCO_3{}^+]_{input}$	30.0	mM

Table 9.2: Cation dissociation constants used for the glycogenolysis model of Vinnakota *et al.* [65]. Values in Vinnakota *et al.* are computed based on ionic strength $I = 0.17$ M and $T = 25\,^\circ$C. The base-10 logarithms of the first proton, magnesium, and potassium dissociation constants are reported in the columns pKa, pKm, and pKk.

Reactant	Abbreviation	Reference species	pKa	pKm	pKk
Phosphate	Pi	HPO_4^{2-}	6.672	1.493	0.50
ATP	ATP	ATP^{4-}	6.323	3.876	1.013
ADP	ADP	ADP^{3-}	6.262	3.014	1.0
AMP	AMP	AMP^{2-}	6.212	1.736	—
Phosphocreatine	PCr	PCr^{2-}	4.422	1.443	0.310
Creatine	Cr	Cr^0	2.30	—	—
Glucose-1-phosphate	G1P	$G1P^{2-}$	6.012	2.323	—
Glucose-6-phosphate	G6P	$G6P^{2-}$	6.032	—	—
Fructose-6-phosphate	F6P	$F6P^{2-}$	5.812	—	—
Fructose-1,6-bisphosphate	FBP	FBP^{4-}	6.40	2.70	—
Glyceraldehyde-3-phosphate	GAP	GAP^{2-}	6.142	1.630	—
Dihydroxyacetone phosphate	DHAP	$DHAP^{2-}$	5.90	1.570	—
1,3-Bisphosphoglycerate	BPG	BPG^{4-}	7.50	—	—
3-Phosphoglycerate	PG3	$PG3^{3-}$	6.210	—	—
2-Phosphoglycerate	PG2	$PG3^{3-}$	7.0	2.450	1.180
Phosphoenolpyruvate	PEP	PEP^{3-}	6.350	2.260	1.080
Pyruvate	PYR	PYR^-	2.490	—	—
Lactate	LAC	LAC^-	3.631	0.980	—
NADH	NADH	$NADH^{2-}$	—	—	—
NAD	NAD	NAD^-	—	—	—

Section 1.2.3 treats the kinetics phenomenologically, and does not attempt to capture the mechanisms underlying the individual reaction processes.

These simple models have the practical advantage that they provide useful tools to analyze the data available. They represent minimally complex, or something close to minimally complex, representations of the underlying kinetics. The more detailed class of model, which is more costly to develop and analyze, does provide broader utility. By formally representing the biochemical kinetics, models of this sort can effectively integrate prior information to make predictions regarding the integrated operation of the system. The goals of this sort of model, to "synthesize all appropriate available data and mechanisms at a consistent level of detail" and to "be used as the foundation for systematically integrating and expanding our ability to reliably simulate cellular biochemical function" [5], are beyond the humble goals of simpler minimal models.

References

[1] R. A. Alberty. Standard transformed formation properties of carbon dioxide in aqueous solutions at specified pH. *J. Phys. Chem.*, 99: 11028–11034, 1995.

[2] R. A. Alberty. Thermodynamic properties of weak acids involved in enzyme-catalyzed reactions. *J. Phys. Chem. B*, 110: 5012–5016, 2006.

[3] D. Axelrod, D. E. Koppel, J. Schlessinger, E. Elson, and W. W. Webb. Mobility measurement by analysis of fluorescence photobleaching recovery. *Biophys. J.*, 16: 1055–1069, 1976.

[4] J. N. Bazil, F. Qi, and D. A. Beard. A parallel algorithm for reverse engineering of biological networks. *Integr. Biol. (Camb.)*, 3: 1215–1223, 2011. Personal communication.

[5] D. A. Beard. Simulation of cellular biochemical system kinetics. *Wiley Interdiscip. Rev. Syst. Biol. Med.*, 3: 136–146, 2010.

[6] D. A. Beard and E. O. Feigl. Understanding Guyton's venous return curves. *Am. J. Physiol. Heart Circ. Physiol.*, 301: H626–H633, 2011.

[7] D. A. Beard and H. Qian. Relationship between thermodynamic driving force and one-way fluxes in reversible processes. *PLoS One*, 2: e144, 2007.

[8] D. A. Beard and H. Qian. *Chemical Biophysics: Quantitative Analysis of Cellular Systems*. Cambridge University Press, Cambridge, UK, 2008.

[9] M. Bébarová, T. O'Hara, J. L. M. C. Geelen, R. J. Jongbloed, C. Timmermans, Y. H. Arens, L.-M. Rodriguez, Y. Rudy, and P. G. A. Volders. Subepicardial phase 0 block and discontinuous transmural conduction underlie right precordial ST-segment elevation by a SCN5A loss-of-function mutation. *Am. J. Physiol.*, 295: H48–H58, 2008.

[10] R. B. Bird, W. E. Stewart, and E. N. Lightfoot. *Transport Phenomena*. John Wiley & Sons, Inc., New York, NY, second edition, 2001.

[11] G. L. Brengelmann. A critical analysis of the view that right atrial pressure determines venous return. *J. Appl. Physiol.*, 94: 849–859, 2003.

[12] R. P. Brown, M. D. Delp, S. L. Lindstedt, L. R. Rhomberg, and R. P. Beliles. Physiological parameter values for physiologically based pharmacokinetic models. *Toxicol. Ind. Health.*, 13: 407–484, 1997.

[13] S. M. Bugenhagen, A. W. Cowley, and D. A. Beard. Identifying physiological origins of baroreflex dysfunction in salt-sensitive hypertension in the Dahl SS rat. *Physiol. Genomics*, 42: 23–41, 2010.

[14] A. C. Burton. *Physiology and Biophysics of the Circulation*. Year Book Medical Publishers, Inc., Chicago, IL, 1965.

[15] E. C. W. Clarke and D. N. Glew. Evaluation of Debye-Hückel limiting slopes for water between 0 and 150°C. *J. Chem. Soc., Faraday Trans. 1*, 76: 1911–1916, 1980.

[16] A. Cornish-Bowden. *Fundamentals of Enzyme Kinetics*. Portland Press, London, UK, third edition, 2004.

[17] B. J. Diamond. Temperature and pH dependence of the cyclization of creatine: A study via mass spectroscopy. Master's thesis, Marshall University, West Virginia, February 2005.

[18] L. M. Ellwein. *Cardiovascular and Respiratory Regulation, Modeling and Parameter Estimation*. PhD thesis, North Carolina State University, Raleigh, NC, USA, May 2008.

[19] G. Gao and F. C. P. Law. Physiologically based pharmacokinetics of matrine in the rat after oral administration of pure chemical and ACAPHA. *Drug. Metab. Disp.*, 37: 884–891, 2009.

[20] A. S. Greene and A. A. Shoukas. Changes in canine cardiac function and venous return curves by the carotid baroreflex. *Am. J. Physiol.*, 251: H288–H296, 1986.

[21] D. E. Gregg, E. M. Khouri, and C. R. Rayford. Systemic and coronary energetics in the resting unanesthetized dog. *Circ. Res.*, 16: 102–113, 1965.

[22] A. C. Guyton. Determination of cardiac output by equating venous return curves with cardiac response curves. *Physiol. Rev.*, 35: 123–129, 1955.

[23] A. C. Guyton. *Circulatory Physiology: Cardiac Output and Its Regulation*. W. B. Saunders Company, Philadelphia, PA, first edition, 1963.

[24] A. C. Guyton, B. Abernathy, J. B. Langston, B. N. Kaufmann, and H. M. Fairchild. Relative importance of venous and arterial resistances in controlling venous return and cardiac output. *Am. J. Physiol.*, 196: 1008–1014, 1959.

[25] A. C. Guyton and J. E. Hall. *Textbook of Medical Physiology*. W. B. Saunders Company, Philadelphia, PA, twelth edition, 2010.

[26] A. C. Guyton, A. W. Lindsey, and B. N. Kaufmann. Effect of mean circulatory filling pressure and other peripheral circulatory factors on cardiac output. *Am. J. Physiol.*, 180: 463–468, 1955.

[27] B. Hille. *Ion Channels of Excitable Membranes*. Sinauer Associates, Inc., Sunderland, MA, third edition, 2001.

[28] A. L. Hodgkin and A. F. Huxley. A quantitative description of membrane current and its application to conduction and excitation in nerve. *J. Physiol.*, 117: 500–544, 1952.

[29] A. L. Hodgkin and A. F. Huxley. Currents carried by sodium and potassium ions through the membrane of the giant axon of *Loligo*. *J. Physiol.*, 116: 449–472, 1952.

[30] A. L. Hodgkin and A. F. Huxley. The dual effect of membrane potential on sodium conductance in the giant axon of *Loligo*. *J. Physiol.*, 116: 497–506, 1952.

[31] N. Iranfar, D. Fuller, R. Sasik, T. Hwa, M. Laub, and W. F. Loomis. Expression patterns of cell-type-specific genes in *Dictyostelium*. *Mol. Biol. Cell*, 12: 2590–2600, 2001.

[32] A. M. Katz. Ernest Henry Starling, his predecessors, and the "Law of the Heart". *Circulation*, 106: 2986–2992, 2002.

[33] L. A. Katz, J. A. Swain, M. A. Portman, and R. S. Balaban. Relation between phosphate metabolites and oxygen consumption of heart in vivo. *Am. J. Physiol.*, 256: H265–H274, 1989.

[34] O. Kedem and A. Katchalsky. Permeability of composite membranes. Part 1: Electric current, volume flow and flow of solute through membranes. *Trans. Faraday Soc.*, 59: 1918–1930, 1963.

[35] J. Keener and J. Sneyd. *Mathematical Physiology: I: Cellular Physiology*. Springer, New York, NY, second edition, 2009.

[36] J. W. R. Lawson and R. L. Veech. Effects of pH and free Mg^{2+} on the K_{eq} of the creatine kinase reaction and other phosphate hydrolyses and phosphate transfer reacions. *J. Biol. Chem.*, 254: 6528–6537, 1979.

[37] R. J. LeVeque. *Finite Difference Methods for Ordinary and Partial Differential Equations*. SIAM, Philadelphia, PA, 2007.

[38] M. N. Levy. The cardiac and vascular factors that determine systemic blood flow. *Circ. Res.*, 44: 739–747, 1979.

[39] X. Li, R. K. Dash, R. K. Pradhan, F. Qi, M. Thompson, K. C. Vinnakota, F. Wu, F. Yang, and D. A. Beard. A database of thermodynamic quantities for the reactions of glycolysis and the tricarboxylic acid cycle. *J. Phys. Chem. B*, 114: 16068–16082, 2010.

[40] D. A. McQuarrie and J. D. Simon. *Physical Chemistry: A Molecular Approach*. University Science Books, Sausalito, CA, 1997.

[41] M. Mescam, K. C. Vinnakota, and D. A. Beard. Identification of the catalytic mechanism and estimation of kinetic parameters for fumarase. *J. Biol. Chem.*, 286: 21100–21109, 2011.

[42] L. Michaelis and M. Menten. Die Kinetik der Invertinwirkung. *Biochem. Z.*, 49: 333–369, 1913.

[43] D. Noble. From the Hodgkin-Huxley axon to the Virtual Heart. *J. Physiol.*, 580: 15–22, 2007.

[44] J. B. Olansen, J. W. Clark, D. Khoury, F. Ghorbel, and A. Bidani. A closed-loop model of the canine cardiovascular system that includes ventricular interaction. *Comput. Biomed. Res.*, 33: 260–295, 2000.

[45] M. S. Olufsen, H. T. Tran, J. T. Ottesen, Research Experiences for Undergraduates Program, L. A. Lipsitz, and V. Novak. Modeling baroreflex regulation of heart rate during orthostatic stress. *Am. J. Physiol.*, 291: R1355–R1368, 2006.

[46] L. Onsager. Reciprocal relations in irreversible processes. I. *Phys. Rev.*, 37: 405–426, 1931.

[47] L. Onsager. Reciprocal relations in irreversible processes. II. *Phys. Rev.*, 38: 2265–2279, 1931.

[48] S. W. Patterson and E. H. Starling. On the mechanical factors which determine the output of the ventricles. *J. Physiol*, 48: 357–379, 1914.

[49] B. Peng, J. Andrews, I. Nestorov, B. Brennan, P. Nicklin, and M. Rowland. Tissue distribution and physiologically based pharmacokinetics of antisense phosphorothioate oligonucleotide ISIS 1082 in rat. *Antisense Nucleic Acid Drug Dev.*, 11: 15–27, 2001.

[50] B. Petschacher and B. Nidetzky. Altering the coenzyme preference of xylose reductase to favor utilization of NADH enhances ethanol yield from xylose in a metabolically engineered strain of *Saccharomyces cerevisiae*. *Microbial Cell Factories*, 7: 9, 2008.

[51] H. Qian. Cellular biology in terms of stochastic nonlinear biochemical dynamics: emergent properties, isogenetic variations and chemical system inheritability. *J. Stat. Phys.*, 141: 990–1013, 2010.

[52] R. F. Rushmer. *Cardiovascular Dynamics*. W. B. Saunders Company, Philadelphia, PA, fourth edition, 1976.

[53] A. H. J. Salmon, C. R. Neal, D. O. Bates, and S. J. Harper. Vascular endothelial growth factor increases the ultrafiltration coefficient in isolated intact Wistar rat glomeruli. *J. Phyiol.*, 570: 141–156, 2006.

[54] V. J. Savin, R. Sharma, H. B. Lovell, and D. J. Welling. Measurement of albumin reflection coefficient with isolated rat glomeruli. *J. Am. Soc. Nephrol.*, 3: 1260–1269, 1992.

[55] O. Shimomura, F. H. Johnson, and Y. Saiga. Extraction, purification and properties of aequorin, a bioluminescent protein from the luminous hydromedusan, *Aequorea*. *J. Cell Physiol.*, 59: 223–239, 1962.

[56] B. S. Shin, C. H. Kim, Y. S. Jun, D. H. Kim, B. M. Lee, C. H. Yoon, E. H. Park, K. C. Lee, S.-Y. Han, K. L. Park, H. S. Kim, and S. D. Yoo. Physiologically based pharmacokinetics of bisphenol A. *J. Toxicol. Environ. Health A*, 67: 1971–1985, 2004.

[57] A. Silverberg. The mechanics and thermodynamics of separation flow through porous, molecularly disperse, solid media: The Poiseuille Lecture 1981. *Biorheology*, 19: 111–127, 1982.

[58] W. E. Teague Jr and G. P. Dobson. Effect of temperature on the creatine kinase equilibrium. *J. Biol. Chem.*, 267: 14084–14093, 1992.

[59] V. J. Savin and D. A. Terreros. Filtration in single isolated mammalian glomeruli. *Kidney Int.*, 20: 188–197, 1981.

[60] M. Thompson and D. A. Beard. Development of appropriate equations for physiologically based pharmacokinetic modeling of permeability-limited and flow-limited transport. *J. Pharmacokinet. Pharmacodyn.*, 38: 405–421, 2011.

[61] G. J. Tortora and S. R. Grabowski. *Principles of Anatomy and Physiology*. HarperCollins, New York, NY, seventh edition, 1993.

[62] G. A. Truskey, F. Yuan, and D. F. Katz. *Transport Phenomena in Biological Systems*. Pearson Prentice Hall Bioengineering. Pearson Prentice Hall, 2004.

[63] K. Uemura, T. Kawada, A. Kamiya, T. Aiba, I. Hidaka, K. Sunagawa, and M. Sugimachi. Prediction of circulatory equilibrium in response to changes in stressed blood volume. *Am. J. Physiol.*, 289: H301–H307, 2005.

[64] K. C. Vinnakota, D. A. Mitchell, T. Wakatsuki, R. J. Deschenes, and D. A. Beard. Analysis of the diffusion of Ras2 in *Saccharomyces cerevisiae* using fluorescence recovery after photobleaching. *Phys. Biol.*, 7: 026011, 2010.

[65] K. C. Vinnakota, J. Rusk, L. Palmer, E. Shankland, and M. J. Kushmerick. Common phenotype of resting mouse extensor digitorum longus and soleus muscles: equal ATPase and glycolytic fux during transient anoxia. *J. Physiol.*, 588: 1961–1983, 2010.

[66] F. Wu, E. Y. Zhang, J. Zhang, R. J. Bache, and D. A. Beard. Phosphate metabolite concentrations and atp hydrolysis potential in normal and ischaemic hearts. *J. Physiol.*, 586: 4193–4208, 2008.

[67] X. Wu, F. Yamashita, M. Hashida, X. Chen, and Z. Hu. Determination of matrine in rat plasma by high-performance liquid chromatography and its application to pharmacokinetic studies. *Talanta*, 59: 965–971, 2003.

Index